Physical Design for
3D Integrated Circuits

Devices, Circuits, and Systems

Series Editor
Krzysztof Iniewski
ET CMOS Inc.
Vancouver, British Columbia, Canada

PUBLISHED TITLES:

FORTHCOMING TITLES:

FORTHCOMING TITLES:

Radio Frequency Integrated Circuit Design
Sebastian Magierowski

Silicon on Insulator System Design
Bastien Giraud

Semiconductor Devices in Harsh Conditions
Kirsten Weide-Zaage and Malgorzata Chrzanowska-Jeske

Smart eHealth and eCare Technologies Handbook
Sari Merilampi, Lars T. Berger, and Andrew Sirkka

Structural Health Monitoring of Composite Structures Using Fiber Optic Methods
Ginu Rajan and Gangadhara Prusty

Tunable RF Components and Circuits: Applications in Mobile Handsets
Jeffrey L. Hilbert

Wireless Medical Systems and Algorithms: Design and Applications
Pietro Salvo and Miguel Hernandez-Silveira

Physical Design for
3D Integrated Circuits

EDITED BY
AIDA TODRI-SANIAL
CNRS-LIRMM, France

CHUAN SENG TAN
Nanyang Technological University, Singapore

KRZYSZTOF INIEWSKI MANAGING EDITOR
Emerging Technologies CMOS Inc.
Vancouver, British Columbia, Canada

CRC Press
Taylor & Francis Group
Boca Raton   London   New York

CRC Press is an imprint of the
Taylor & Francis Group, an **informa** business

CRC Press
Taylor & Francis Group
6000 Broken Sound Parkway NW, Suite 300
Boca Raton, FL 33487-2742

First issued in paperback 2020

ISBN-13: 978-1-4987-1036-7 (hbk)
ISBN-13: 978-0-367-77887-3 (pbk)

Library of Congress Cataloging-in-Publication Data

Physical design for 3D integrated circuits / edited by Aida Todri-Sanial and Chuan Seng Tan.
    pages cm
  Includes bibliographical references and index.
  ISBN 978-1-4987-1036-7 (hardcover : alk. paper)
  1. Three-dimensional integrated circuits--Design and construction. I. Todri-Sanial, Aida, editor. II. Tan, Chuan Seng, editor. III. Title: Physical design fort three-D integrated circuits.

TK7874.893.P48 2016
621.39'5--dc23

2015035061

Visit the Taylor & Francis Web site at
http://www.taylorandfrancis.com

and the CRC Press Web site at
http://www.crcpress.com

Contents

SECTION I 3D Integration Overview

SECTION II Physical Design Methods for 3D Integration

SECTION III   Reliability Concerns for 3D Integration

SECTION IV   CAD Design Tools and Future Directions for 3D Physical Design

Contents

SECTION I   3D Integration Overview

SECTION II   Physical Design Methods for 3D Integration

SECTION III   Reliability Concerns for 3D Integration

SECTION IV   CAD Design Tools and Future Directions for 3D Physical Design

Preface

Three-dimensional (3D) stacking and 2.5D interposer side-by-side integration are very attractive contenders as we head toward an incommensurate return of interconnects and packaging technology. As demand for on-chip functionalities and requirements for low power operation continue to increase as a result of the emergence of mobile, wearable, and Internet of Things products, 3D/2.5D integration has been identified as inevitable in moving forward. The advent of 3D/2.5D integration is a direct result of active research in academia, research laboratories, and industry over the past decade. Today, 3D/2.5D integration takes many forms, depending on the applications. At the time of this writing, there are already commercial products driven by the needs for form factor and density.

As a direct result of many years of active research, there is substantial documentation on 3D/2.5D technology. A book dedicated to the physical design for 3D integrated circuits (ICs), however, is lacking. The idea for a book on physical design for 3D ICs dates back more than a year ago. While the initial idea was to write a book, we soon realized that such an endeavor would be extremely challenging given the various expertises in this field. We revisited the plan and decided to edit a book instead, with contributions from experts in academia, research laboratories, and industry. After careful planning, we identified and invited contributions from an impressive lineup of highly qualified researchers. The task took a full year from planning, writing, editing, and printing.

This book aims to unveil how to effectively and optimally design such 3D circuits. It also presents the design tools for 3D circuits, while exploiting the benefits of 3D technology. Initially, an overview of physical design challenges with respect to conventional 2D circuits is provided, and then each chapter is dedicated to provide an in-depth look into each physical design topic. *Physical Design for 3D Integrated Circuits* is the first book to analyze the design tools for 3D ICs covering all design aspects and explaining the challenges and solutions unique for 3D circuits. This book is particularly beneficial to researchers and engineers who are already working or are beginning to work on 3D technology.

This book would not have been possible without a team of highly qualified and dedicated people. We are particularly grateful to Kris Iniewski for initiating this undertaking and for his encouragement. Nora Konopka and Jessica Vakili worked alongside us and provided us with the necessary editorial support. Aida Todri-Sanial is grateful for the continued support for her work on 3D integration from the French National Research Agency (ANR) and strong collaborations with CEA-LETI, France. Chuan Seng Tan is grateful for the continued support for his work on 3D integration in Singapore from the Ministry of Defense (MINDEF), the Ministry of Education (MOE), the Agency for Science, Technology and Research (A*STAR), and the National Research Foundation (NRF). This book would not have been possible without this extended research support. Last but not least, we are extremely thankful to the authors who accepted our invitation and contributed chapters to this book.

We hope that the readers will find this book useful in their pursuit of 3D/2.5D technology. Please do not hesitate to contact us if you have any comments or suggestions.

Aida Todri-Sanial dedicates this book to her family. Chuan Seng Tan dedicates this book to his wife and sons.

Aida Todri-Sanial
Montpellier, France

Chuan Seng Tan
Singapore, Singapore

Acknowledgments

We would like to thank all the contributing authors. Without their dedication, this book would not have been possible. We would also like to thank Nora Konopka of Taylor & Francis Group for taking charge of this book. Finally, the resulting book has been possible by the hard work of Deepa Kalaichelvan. Thank you very much!

Aida Todri-Sanial
Chuan Seng Tan

Editors

Aida Todri-Sanial earned a PhD in electrical and computer engineering at the University of California, Santa Barbara, in 2009. She is currently a research scientist (Chargée de Recherché) at CNRS (French National Center for Scientific Research) and a member of the Microelectronics Department at Laboratoire d'Informatique, de Robotique, de Microelectronique de Montpellier (LIRMM), where she is the group leader of Integration and Design of Energy-Aware Circuits and Systems. Prior to joining CNRS, she was an R&D engineer at Fermilab. She also has several visiting research positions at Cadence Design Systems, Mentor Graphics, IBM TJ Watson Research Center, and STMicroelectronics. Dr. Todri-Sanial has received several awards, including the CNRS Scientific Excellence Award (PES) 2012, the John Bardeen Fellow in Engineering at Fermilab 2009, and Best Teaching Assistant at the University of California, Santa Barbara. She has published more than 100 papers in very-large-scale integration design area. She serves on the technical program committees for the IEEE (Emerging Technologies Track), IEEE ISQED, IEEE NEWCAS, IEEE ISVLSI, IEEE GLSVLSI, and as an expert reviewer for IEEE/ACM DAC. She also serves as a technical reviewer for *IEEE Transactions on VLSI*, Computers, CAD, CAS-I, CAS-II, TNS, and IET. She is an associate editor for the *IEEE Transactions on Very Large Scale Integration* journal. She is also engaged with European agencies, such as the European Platform of Women Scientist (EPWS) and the European Association for Women in Science, Engineering and Technology (WiTEC). She is a member of the IEEE and the ACM.

Chuan Seng Tan earned a BE in electrical engineering at the University of Malaya, Malaysia, in 1999. Subsequently, he earned an ME in advanced materials at the National University of Singapore under the Singapore-MIT Alliance (SMA) program in 2001. He then joined the Institute of Microelectronics, Singapore, as a research engineer where he worked on process integration of strained-Si/relaxed-SiGe heterostructure devices. In the fall of 2001 he began his doctoral work at the Massachusetts Institute of Technology, Cambridge, Massachusetts, and earned a PhD in electrical engineering in 2006. He was the recipient of the Applied Materials Graduate Fellowship for 2003–2005. In 2003, he interned at Intel Corporation, Oregon. He joined Nanyang Technological University, Singapore, in 2006 as a Lee Kuan Yew postdoctoral fellow, and since July 2008 he has held the inaugural Nanyang assistant professorship. In February 2014 he was promoted to associate professor (with tenure). His research interests are semiconductor process technology and device physics. He is working on the process technology of three-dimensional integrated circuits (3D ICs). He has edited two books, *Wafer Level 3-D ICs Process Technology* and *3D Integration of VLSI Systems*. He has numerous publications on 3D technology. He is a committee member of the International Conference on Wafer Bonding, IEEE-3DIC, IEEE-EPTC, IEEE-ECTC, ECS-Wafer Bonding, SSDM, and ISTDM. He is an associate editor for Elsevier's *Microelectronics Journal* (*MEJ*). He is a member of the IEEE.

Krzysztof Iniewski is managing R&D at Redlen Technologies Inc., a start-up company in Vancouver, Canada. Redlen's revolutionary production process for advanced semiconductor materials enables a new generation of more accurate, all-digital, radiation-based imaging solutions. He is also a president of CMOS Emerging Technologies Research Inc. (www.cmosetr.com), an organization of high-tech events covering communications, microsystems, optoelectronics, and sensors. He also has held numerous faculty and management positions at the University of Toronto, the University of Alberta, Simon Frazer University, and PMC-Sierra Inc. He has published more than 100 research papers in international journals and conferences. He holds 18 international patents granted in the United States, Canada, France, Germany, and Japan. He is a frequent invited speaker and has consulted for multiple organizations internationally. He has written and edited several books for CRC Press, Cambridge University Press, IEEE Press, Wiley, McGraw-Hill, Artech House, and Springer. He seeks to contribute to healthy living and sustainability through innovative engineering solutions. He can be reached at kris.iniewski@gmail.com.

Contributors

Ra'ed Al-Dujaily
School of Electrical and Electronic
Engineering
Newcastle University
Newcastle upon Tyne, United Kingdom

Shih-Chieh Chang
Department of Computer Science
National Tsing Hua University
Hsinchu, Taiwan

Yu-Guang Chen
Department of Computer Science
National Tsing Hua University
Hsinchu, Taiwan

Ayse K. Coskun
Department of Electrical and Computer
Engineering
Boston University
Boston, Massachusetts

Nizar Dahir
Department of Electronics
University of York
Heslington, York

Robert Fischbach
Design Automation Division
Fraunhofer Institute for Integrated Circuits
Dresden, Germany

Brad Gaynor
Draper Laboratory
Cambridge, Massachusetts

Soha Hassoun
Department of Computer Science
Tufts University
Medford, Massachusetts

Andy Heinig
Design Automation Division
Fraunhofer Institute for Integrated Circuits
Dresden, Germany

Fulya Kaplan
Department of Electrical and Computer
Engineering
Boston University
Boston, Massachusetts

Nauman Khan
Intel Corporation
Hillsboro, Oregon

Taewhan Kim
School of Electrical and Computer Engineering
Seoul National University
Seoul, South Korea

Tony Tae-Hyoung Kim
School of Electrical and Electronic
Engineering
Nanyang Technological University
Singapore, Singapore

Johann Knechtel
Institute of Electromechanical and Electronic
Design
Dresden University of Technology
Dresden, Germany

Sumeet S. Kumar
Faculty of Electrical Engineering, Mathematics
and Computer Science
Delft University of Technology
Delft, The Netherlands

Jens Lienig
Institute of Electromechanical and Electronic
Design
Dresden University of Technology
Dresden, Germany

Sung Kyu Lim
School of Electrical and Computer Engineering
Georgia Institute of Technology
Atlanta, Georgia

Aung Myat Thu Linn
School of Electrical and Electronic
  Engineering
Nanyang Technological University
Singapore, Singapore

Tiantao Lu
Department of Electrical and Computer
  Engineering
University of Maryland
College Park, Maryland

Terrence Mak
Faculty of Physical Sciences and Engineering
University of Southampton
Southampton, United Kingdom

Heechun Park
School of Electrical and Computer Engineering
Seoul National University
Seoul, South Korea

Emre Salman
Department of Electrical and Computer
  Engineering
Stony Brook University
Stony Brook, New York

Sachin S. Sapatnekar
Department of Electrical and Computer
  Engineering
University of Minnesota
Minneapolis, Minnesota

Yiyu Shi
Missouri University of Science and Technology
Rolla, Missouri

Ankur Srivastava
Department of Electrical and Computer
  Engineering
University of Maryland
College Park, Maryland

Cliff C.N. Sze
IBM Research
Austin, Texas

Chuan Seng Tan
School of Electrical and Electronic
  Engineering
Nanyang Technological University
Singapore, Singapore

Aida Todri-Sanial
Laboratoire d'Informatique, de Robotique et de
  Microélectronique de Montpellier
Centre National de la Recherche Scientifique
University of Montpellier
Montpellier, France

Rene van Leuken
Faculty of Electrical Engineering, Mathematics
  and Computer Science
Delft University of Technology
Delft, The Netherlands

Alex Yakolev
School of Electrical and Electronic
  Engineering
Newcastle University
Newcastle upon Tyne, United Kingdom

Tiansheng Zhang
Department of Electrical and Computer
  Engineering
Boston University
Boston, Massachusetts

Pingqiang Zhou
School of Information Science and Technology
ShanghaiTech University
Pudong, Shanghai, People's Republic of China

Amir Zjajo
Faculty of Electrical Engineering, Mathematics
  and Computer Science
Delft University of Technology
Delft, The Netherlands

Section I

3D Integration Overview

1 2.5D/3D ICs
Drivers, Technology, Applications, and Outlook

Chuan Seng Tan

CONTENTS

ABSTRACT

2.5D interposer side-by-side assembly and three-dimensional (3D) stacking of functional integrated circuits (ICs) are identified as inevitable solutions for future system miniaturization and functional diversification. 2.5D/3D integration offers a long list of benefits in terms of system form factor, density scaling and multiplication, reduced interconnection latency and power consumption, band-width enhancement, and heterogeneous integration of disparate technologies. In 2.5D integration, ICs are placed side-by-side in close proximity on a suitable substrate such as silicon interposer. On the other hand in 3D implementation, thinned IC layers are seamlessly bonded with a reliable bonding medium and vertically interconnected with electrical through-strata-via (TSV). This chapter discusses the drivers and new integration capabilities brought about by 2.5D/3D technology, enabling technology platforms, and potential applications made possible by 2.5D/3D technology. A future outlook is provided at the end.

1.1 BACKGROUND AND INTRODUCTION

Beginning with the invention of integrated circuits (ICs) in 1959, higher computing power was achieved primarily through density scaling and commensurate performance enhancement of transistors as a result of continuously scaling down the device dimensions in a harmonious manner. This has resulted in a steady doubling of device density from one technology node to another [1]. This observation was famously known as "Moore's Law," first coined by Carver Mead of Caltech. The scaling law was based on a set of rules proposed by Robert Dennard of IBM [2]. Improvement in transistor switching speed and density are two of the most direct contributors to the historical performance growth in ICs (particularly in silicon-based digital CMOS). This scaling approach has been so effective in many aspects (performance and cost) that ICs have essentially remained a planar platform throughout this period of rigorous scaling. In the recent years, pitch scaling is augmented with a number of performance boosters such as strain engineering, high-κ/metal gate, and nonplanar 3D transistors. At the time of this writing, the industry is already manufacturing the 16/14 nm node devices. There is a consensus that geometrical scaling cannot be sustained indefinitely as the manufacturing cost will be prohibitively high. The industry is actively exploring several promising options. The focus of attention is on "system scaling" and 2.5D/3D integration has been favorably singled out. Several "low hanging" products that leverage on slim form factor and high density afforded by 3D integration, such as CMOS image sensor and memory stack, are already in the market. It is anticipated that the next phase of development will deliver high-bandwidth memory/logic stack and heterogeneous systems to meet the insatiable demands.

1.2 DRIVERS

This section examines the role of 3D integration in ensuring that system-level performance growth enjoyed by the semiconductor industry can continue in the future to support emerging applications. Scaling alone has met with diminishing return due to fundamental and economics barriers (noncommensurate scaling). 3D integration explores the third dimension of IC and offers new dimension for performance growth. The concept of 3D integration should not be confused with nonplanar 3D transistor such as finFET. 3D integration also enables integration of disparate chips in a more compact form factor, and it is touted by many as an attractive method for system miniaturization and functional diversification commonly known as heterogeneous integration.

1.2.1 SUSTAINABLE SYSTEM PERFORMANCE GROWTH

Beginning with the invention of the first IC by Kilby and Noyce, the world has witnessed sustainable performance growth in IC. The trend is best exemplified by the exponential improvement in computing power (measured in million instructions per second, MISP) in Intel's microprocessors over the past 40 years as shown in Figure 1.1 [3].

This continuous growth is a result of the ability to scale silicon transistor to smaller dimension in every new technology nodes. The growth has continued, instead of hitting a plateau, in more recent nodes, thanks to the addition of performance boosters (e.g., strained-Si, high-κ and metal gate, finFET)

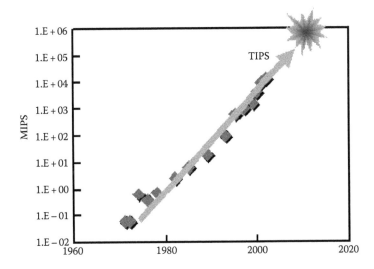

FIGURE 1.1 Evolution of computing performance. (From Intel, www.intel.com, accessed on December 2014.)

on top of conventional geometrical scaling. Scaling doubles the number of transistors on IC in every generation and allows us to integrate more functions on IC and to increase its computing power. We are now in the giga-scale integration era featured by billions of transistors, GHz operating frequency, etc. Going forward to tera-scale integration, however, there are a number of imminent show-stoppers (described in the next section) that pose serious threat to continuous performance enhancement in IC, and a new paradigm shift in IC technology and architecture is needed to sustain the historical growth. It is widely recognized that the growth can be sustained if one utilizes the vertical (i.e., the third) dimension of IC to build a 3D IC, a departure from today's planar IC as illustrated in Figure 1.2.

Three-dimensional integrated circuits (3D ICs) refer to a stack consisting of multiple ultrathin layers of IC that are vertically bonded and interconnected with through-silicon-via (TSV). It is also possible to stack multiple thin dies and connect them using conventional wire-bonding as commonly found in flash memory stack. This method is better known as 3D packaging and it will not be the main discussion point in this text. In 3D implementation, each block can be fabricated and optimized using their respective technologies and assembled to form a vertical stack. In today's technology,

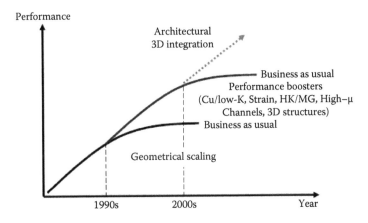

FIGURE 1.2 Historical IC performance growth can be sustained with a new paradigm shift to 3D integration.

thin IC layer in the range of 20–50 µm and TSV diameter of 2–5 µm are often reported. The scaling of these dimensions in the future nodes can be readily found in various technology road maps.

1.2.2 SHOW-STOPPERS AND 3D INTEGRATION AS A REMEDY

1.2.2.1 Transistor Scaling Barriers

There are at least two barriers that will slow down or impede further geometrical scaling. The first one relates to the fundamental properties of transistor in extremely scaled devices. Experimental and modeling data suggest that performance improvement in devices is no longer commensurate with ideal scaling in the past due to high leakage and parasitic, hence they consume more power. This is shown in Figure 1.3 by Khakifirooz and Antoniadis of MIT [4]. The intrinsic delay of n-MOS is shown to increase beyond 45 nm despite continuous down scaling of the transistor pitch.

Another issue related to scaled devices is variability [5,6]. Variability in transistor performance and leakage is a critical challenge to the continued scaling and effective utilization of CMOS technologies with nanometer-scale feature sizes. Some of the factors contributing to the variability increase are fundamental to the planar CMOS transistor architecture. Random dopant fluctuations (RDFs) and line-edge roughness (LER) are two examples of such intrinsic sources of variation. Other reasons for the variability increase include advanced resolution-enhancement techniques (RETs) used to print patterns with feature sizes smaller than the wavelength of lithography. Transistor variation affects many aspects of IC manufacturing and design. Increased transistor variability (e.g., in leakage current and threshold voltage) can have negative impact on product performance and yield. Variability worsens as we continue to scale in future technology nodes and it is a severe challenge.

The second barrier concerns the economic aspect of scaling. The development and manufacturing cost has increased multiple folds from one node to another making scaling a less favorable option in future nodes of IC. In addition, delay in lithography capability seriously slows down geometrical scaling. The capital investment in a new technology node is sky-rocketing and that limits the number of users. 3D integration on the other hand achieves device density multiplication by

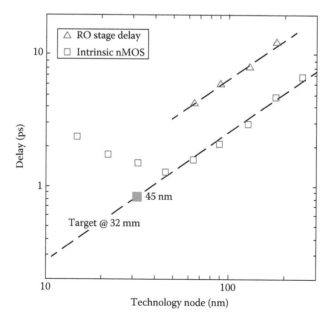

FIGURE 1.3 Intrinsic delay in n-MOS transistor is projected to increase in future nodes despite continuous down scaling of the device pitch. (From Khakifirooz, A., Transport enhancement techniques for nanoscale MOSFETs, PhD thesis, Cambridge, MA, MIT, 2008, http://dspace.mit.edu/handle/1721.1/42907.)

stacking IC layers in the third dimension without aggressive scaling. 3D integration can be manufactured using existing nodes. Therefore, it can be a viable and immediate remedy as conventional scaling becomes less cost effective.

1.2.2.2 On-Chip Interconnect

While dimensional scaling has consistently improved device performance in terms of gate switching delay, it has a reverse effect on semi-global and global interconnect latency [7]. The global interconnect RC delay has increasingly become the circuit performance limiting factor especially in the deep submicron regime. Even though Cu/low-κ multilevel interconnect structures improve interconnect RC delay, they are not a long-term solution since the diffusion barrier required in Cu metallization has a finite thickness that is not readily scaled. The effective resistance of interconnect is larger than it would be in bulk copper, and the difference increases with reduced interconnect width. Surface electron scattering further increases the Cu line resistance, and hence the RC delay suffers [8]. When the chip size continues to increase to accommodate for more functionalities, the total interconnects' length increases at the same time. This causes a tremendous amount of power to be dissipated unnecessarily in interconnects and repeaters are used to minimize delay and latency. On-chip signals also require more clock cycles to travel across the entire chip as a result of increasing chip size and operating frequency.

Rapid rise in interconnects delay and power consumption due to smaller wire cross-section, tighter wire pitch, and longer lines that transverse across larger chips is severely limiting IC performance enhancement in current and future nodes. 3D IC with multiple active Si layers stacked vertically is a promising method to overcome this scaling barrier as it replaces long inter-block global wires with much shorter vertical interlayer interconnects as shown in Figure 1.4.

1.2.2.3 Off-Chip Interconnect

Figure 1.5a depicts the memory hierarchy in today's computer system in which the processor core is connected to the memory (DRAM) via power-hungry and slower off-chip buses on the board

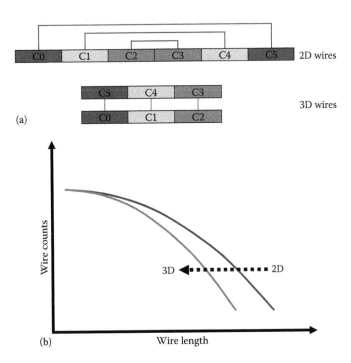

FIGURE 1.4 (a) Long global wires on IC can be shortened by chip partitioning and stacking. (b) 3D integration reduces the number of long wires on IC.

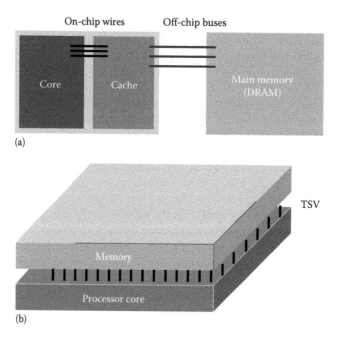

FIGURE 1.5 (a) Memory hierarchy in today computer system. (b) Direct placement of memory on processor improves the data bandwidth.

level. Data transmission on these buses experiences severe delay and consumes significant amount of power. The number of available bus channels is also limited by the amount of external pin count available on the packaged chips. As a consequence, the data bandwidth suffers. As the computing power in processor increases in each generation, the limited bandwidth between processor core and memory places a severe limitation on the overall system performance [9]. The problem is even more pressing in multi-core architecture as every core will demand for data supply. To close this gap, the most direct way is to shorten the connections and to increase the number of data channels. By placing memory directly on processor, the close proximity shortens the connections and the density of connections can be increased by using more advanced CMOS processes (as opposed to packaging/assembly processes) to achieve fine-pitch TSV. This massively parallel interconnection is shown in Figure 1.5b. Table 1.1 is a comparison between 2D and 3D implementations in terms of connection density and power consumption. Clearly, 3D can provide bandwidth enhancement (100× increment at the same frequency) at lower power consumption (10× reduction). Effectively, this translates into 1000× improvement in bandwidth/power efficiency, an extremely encouraging and impressive number.

TABLE 1.1
Comparison of 2D and 3D Implementations

	2D	3D	Remark
Connections density	<1e3 per cm²	~1e5 per cm²	100× increment
Power consumption[a] per pin/via	30–40 mW	~25 μW	
Total power consumption (per cm²)	30–40 W	2.5 W	10× reduction

[a] Data from Tezzaron, http://www.tezzaron.com.

TABLE 1.2
3D Integration Provides the Best Features as Shown in the Following

	Monolithic	2.5D/3D Integration	Package/Hybrid
Cost	–	+	+
Performance	+	+	–
Power	+	+	–
Functionality	+	+	+
Time to market	–	+	+
Modularity	– Sequential	+ Parallel	+ Parallel
Hermetic seal	*Ex situ*	*In situ*	*Ex situ*

Note: Since hermetic seal is formed during stacking, the MEMS sensor does not need to be encapsulated separately.

1.2.3 HETEROGENEITY

Increasingly, there is a need to co-integrate several functional blocks to form a functional system in a very slim form factor to meet emerging applications in the Wearables and Internet of Things (IoT) era. One example is to integrate CMOS electronics and MEMS sensor. There exist a myriad of methods for CMOS and sensor integration to realize a smart microsystem such as single chip monolithic approach or multi-package board approach. While monolithic integration offers the best in terms of performance, power, and functionality, this method is highly complex and expensive with long time-to-market. On the other end, CMOS and sensor chips can be individually packaged or co-packaged. While this method is low cost, one compromises the system performance and needs to put up with high power consumption due mostly to the extra parasitic loads presented by the external wire bond interconnect. An emerging method for CMOS-sensor integration that can potentially reap the best merits of both monolithic and package implementations is 2.5D silicon interposer or 3D stacking. The merits are summarized in Table 1.2.

1.3 OPTIONS OF 3D IC

1.3.1 CLASSIFICATION

There are a number of technology options to assemble ICs in a vertical stack. It is possible to stack ICs in a vertical fashion at various stages of processing: (1) post-singulation 3D packaging (e.g., chip-to-chip) and (2) pre-singulation wafer-level 3D integration (e.g., chip-to-wafer, wafer-to-wafer, and monolithic approaches). Active layers can be vertically interconnected using physical contact such as bond wire or interlayer vertical via (including TSV). It is also possible to establish chip-to-chip connection via noncontact (or wireless) links such as capacitive and inductive couplings [11]. Capacitive coupling utilizes a pair of electrodes that are formed using conventional IC fabrication. The inductive-coupling input/output (I/O) is formed by placing two planar coils (planar inductors) above each other and is also made using conventional IC fabrication. The advantages of these approaches are fewer processing steps, hence lower cost, no requirement for ESD protection, low power, and smaller area I/O cell. Since there is substantial overlap between various options and lack of standardization in terms of definition, classification of 3D IC technology is often not straight forward. This section makes an attempt to classify 3D IC based on the processing stage when stacking takes place.

1.3.2   Monolithic Approaches

Monolithic, or sequential, 3D IC is described as a build-up process to fabricate vertical layers of transistors on the same starting substrate using nanoscale interlayer vias as vertical interconnects, rather than stacking substrates and interconnecting them with TSV. In these approaches, devices in each active layer are processed sequentially starting from the bottom-most layer. Devices are built on a substrate wafer by mainstream process technology. After proper isolation, a second device layer is formed and devices are processed by conventional means on the second layer. This sequence of isolation, layer formation, and device processing can be repeated to build a multilayer structure.

The key technology in this approach is forming a high-quality active layer isolated from the bottom substrate. This bottom-up approach has the advantage that precision alignment between layers can be accomplished. However, it suffers from a number of drawbacks. The crystallinity of upper layers is usually low and imperfect. As a result, high-performance devices cannot be built in the upper layers. Thermal cycling during upper layer crystallization and device processing can degrade underlying devices and therefore a tight thermal budget must be imposed. Due to the sequential nature of this method, manufacturing throughput is low. A simpler front-end-of-line (FEOL) process flow is feasible if polycrystalline silicon can be used for active devices; however, a major difficulty is to obtain high-quality electrical devices and interconnects. While obtaining single-crystal device layers in a generic IC technology remains in the research stage, polycrystalline devices suitable for nonvolatile memory (NVM) have not only been demonstrated but have been commercialized (e.g., by SanDisk). A key advantage of FEOL-based 3D integration is that IC BEOL and packaging technologies are unchanged; all the innovation occurs in 3D stacking of active layers. A number of FEOL techniques include: laser-beam recrystallization [12,13], seeding-assisted recrystallization [14,15], selective epitaxy and overgrowth [16], and grapho-exitaxy [17].

CEA-Leti has recently unveiled the test results on multilayer transistors stacking for true 3D monolithic integration using a "CoolCube" technology. This technology uses a cold layer build-up process that eliminates the need for high temperatures that could degrade performance of transistors or metal interconnects between the layers [18]. The technology no longer relies on high aspect ratio TSVs. The key difference with the conventional TSVs, where two or more processed dies are assembled one on top of another, is the transfer and molecular bonding of a thin Si wafer film. Since the transferred film is so thin and optically transparent (~1 μm as compared to tens of μm in TSV), the layer of transistors that are processed on top can be aligned to the bottom transistors with lithographic precision. Hence the stacked layers can be connected at the transistor scale rather than going through the entire die. Using solid phase epitaxial regrowth (SPER), dopant activation can be accomplished at temperatures between 450°C and 600°C, about half the typical thermal budget for dopant activation. Researches from Stanford University have investigated monolithic 3D ICs using carbon nanotube (CNT) structures [19]. In this method, wafer-scale low temperature CNT transfer processes can be done at 120°C, much cooler than the CoolCube method. The process is Si CMOS compatible, in that the Si FETS can be used to build only the bottom layer because of the high temperatures required for subsequent Si layers. In this design, the next layers are fabricated with CNT layers. The density achieved by monolithic 3D nanoscale interlayer via is 1000× that of TSV. Additionally, bonding is not required.

Perhaps one of the most impressive demonstrations to date is that of Samsung's V-NAND technology featuring 32 active cell layers at ISSCC'2015 [20]. Readers are encouraged to refer to numerous articles on the monolithic approach at http://www.monolithic3d.com/blog.

1.3.3   Assembly Approaches

This is a parallel integration scheme in which fully processed or partially processed ICs are assembled in a vertical fashion in the case of 3D stacking or placed side-by-side in close proximity in the case of 2.5D silicon interposer. Stacking can be achieved with one of these methods: (1) chip-to-chip, (2) chip-to-wafer, and (3) wafer-to-wafer. Vertical connection in chip-to-chip stacking can be achieved using wire bond or TSV. The bulk of the discussion in this text is on 3D stacking using TSV.

Wafer-level 3D integration, such as chip-to-wafer and wafer-to-wafer stacking, use TSV as the vertical interconnect. This integration approach often involves a sequence of wafer thinning and handling, alignment, TSV formation, and bonding. The key differentiators are

- *Bonding medium*: Metal-to-metal, dielectric-to-dielectric (oxide, adhesive, etc.), or hybrid bonding
- *TSV formation*: Via first, via middle, or via last
- *Stacking orientation*: Face-to-face or back-to-face stacking
- *Singulation level*: Chip-to-chip, chip-to-wafer, or wafer-to-wafer

The types of wafer bonding potentially suitable for wafer-level 3D integration are depicted in Figure 1.6. Dielectric-to-dielectric bonding is most commonly accomplished using silicon oxide or BCB polymer as the bonding medium. These types of bonding provide primary function as a mechanical bond and the inter-wafer via is formed after wafer-to-wafer alignment and bonding (Figure 1.6a). When metallic copper-to-copper bonding is used (Figure 1.6b), the inter-wafer via is completed during the bonding process; note that appropriate interconnect processing within each wafer is required to enable 3D interconnectivity. Besides providing electrical connections between IC layers, dummy pads can also be inserted at the bonding interface at the same time to enhance

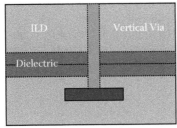

Dielectric = SiO$_2$, BCB
- Dielectric–dielectric bonding
- Mechanical bond
(a) • Via after bonding

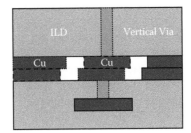

- Cu–Cu bonding
- Mechanical and electrical bonds
(b) • Via during bonding

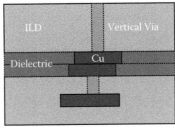

Dielectric = SiO$_2$, BCB
- Hybrid bonding
- Mechanical and electrical bonds
(c) • Via during bonding

FIGURE 1.6 Wafer-bonding techniques for wafer-level 3D integration: (a) dielectric-to-dielectric; (b) metal-to-metal; and (c) dielectric/metal hybrid.

the overall mechanical bond strength. This bonding scheme inherently leaves behind isolation gap between Cu pads and this could be a source of concern for moisture corrosion and compromise the structural integrity especially when IC layers above the substrate is thinned down further. Figure 1.6c shows a bonding scheme utilizing a hybrid medium of dielectric and Cu. This scheme in principle provides a seamless bonding interface consisting of dielectric bond (primarily a mechanical bond) and Cu bond (primarily an electrical bond). However, very stringent requirements with regards to surface planarity (dielectric and Cu) and Cu contamination control in the dielectric layer due to misalignment are needed.

The selection of the optimum technology platform is depending on applications. Cu-to-Cu bonding has significant advantages for highest interwafer interconnectivity. As a result, this approach is desirable for microprocessors and digitally based system-on-a-chip (SoC) technologies. Polymer-to-polymer bonding is attractive when heterogeneous integration of diverse technologies is the driver and the interwafer interconnect density is more relaxed; benzocyclobute (BCB) is the polymer most widely investigated. Taking advantage of the viscosity of the polymer, this method is more forgiving in terms of surface planarity and particle contamination. Oxide-to-oxide bonding of fully processed IC wafers requires atomic-scale smoothness of the oxide surface. In addition, wafer distortions introduced by FEOL and BEOL processing introduces sufficient wafer bowing and warping that prevents sufficient contact area to achieve the required bonding strength. While oxide-to-oxide bonding after FEOL and local interconnect processing has been shown to be promising (particularly with SOI wafers that allows for extreme thinning down to the buried oxide layer), the increased wafer distortion and oxide roughness after multilevel interconnect processing require extra attention during processing. A discussion on SOI-based 3D integration can be found in [21].

TSV can be formed at various stages during the 3D IC process as shown in Figure 1.7. When TSV is formed before any CMOS processes, the process sequence is known as "via first." As the TSV is formed before any FEOL processes, one can consider doped poly-silicon or tungsten (W) as the TSV filler material. Since Cu is unable to withstand high processing temperature during the FEOL steps, it is not a desirable choice of filler. In the case of silicon interposer with no FEOL devices, Cu-TSV can be used and by definition this is a "via first" process. It is also possible to form the TSV when the front-end processes are completed. In this "via middle" process, back-end processes will continue after the TSV process is completed. Typically, the connection between TSV and BEOL metal layer can be done at the bottom most or middle metal layers. This "via-middle" process can be part of a foundry process. When TSV is formed after the CMOS processes are completed, it is known as "via last" process. TSV can be formed from the front side or the back side of the wafer. The schemes mentioned earlier have different requirements in terms of process parameters and materials selection. The choice depends on final application requirements and infrastructures in the supply chain.

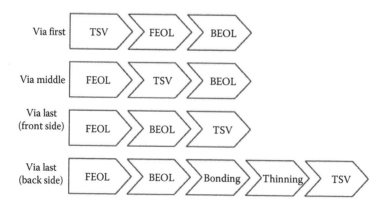

FIGURE 1.7   TSV can be formed at various stages of IC processing.

Another key differentiator in 3D IC integration is related to the stacking orientation. One option is to perform face-to-face (F2F) alignment and bonding with all required I/Os brought to the thinned backside of the top wafer (which becomes the face of the two-wafer stack). Another approach is to temporarily bond the top wafer to a handling wafer, after which the device wafer is thinned from the back side and permanently bonded to the full-thickness bottom wafer; after this permanent bonding the handling wafer is removed. This is also called a back-to-face (B2F) stacking. These two stacking orientations are shown in Figure 1.8.

F2F stacking allows a high-density layer-to-layer interconnection, which is limited by the alignment accuracy. Handle wafer is not required in F2F stacking and this imposes more stringent requirement on the mechanical strength of the bonding interface in order to sustain shear force during wafer thinning, which is often achieved by mechanical grinding or polishing. Since one of the IC layers is facing down in the final assembly, F2F stacking also complicates the layout design as opposed to more conventional layout design whereby IC layers are facing up. Another potential disadvantage of F2F stacking relates to the thickening of the effective ILD layer at the bonding interface, which presents higher barrier for effective heat dissipation. However, since the active layer is flipped and buffered by the thin Si layer, variability and reliability issues related

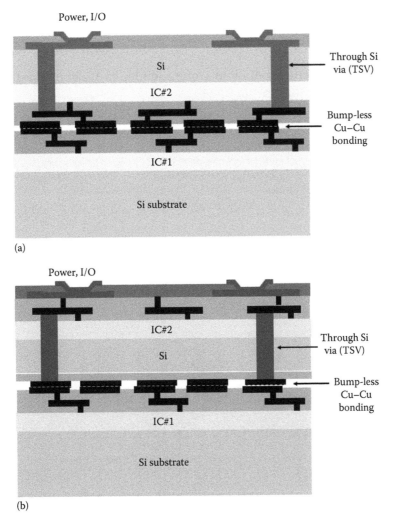

FIGURE 1.8 (a) Face-to-face or face-down stacking; (b) back-to-face or face-up stacking.

to die/substrate thermo-mechanical stress is reduced. B2F stacking requires the use of a temporary handle, and the layer-to-layer interconnection density is limited by the TSV pitch. Since the device layer is bonded to a temporary handle, the final permanent bond does not need to withstand damages resulting from wafer thinning. It requires the use of a temporary bonding medium that can provide sufficient strength during wafer handling and can be readily released after successful device layer permanent transfer on the substrate. The design methodology and thermal barrier is better compared to that of F2F stacking. The active device layer is more susceptible to variability and reliability issues due to thermo-mechanical interaction between the die and the substrate at the micro-bumps.

In wafer-level 3D integration, permanent bonding can be done either in chip-to-wafer (C2W) or wafer-to-wafer (W2W) stacking. A comparison of these two methods is summarized in Table 1.3. As shown in Figure 1.9, the option of C2W or W2W depends on two key requirements on chip size and alignment accuracy. When high-precision alignment is desired in order to achieve high-density layer-to-layer interconnections, W2W is a preferred choice to maintain acceptable throughput by performing a wafer level alignment. W2W is also preferred when chip size gets smaller.

TABLE 1.3

Comparison between Wafer-to-Wafer and Chip-to-Wafer Stacking

	Wafer-to-Wafer	Chip-to-Wafer
Wafer/die size	Wafer/die of common size in order to avoid silicon area wastage	Dissimilar wafer/die size is acceptable
Throughput	Wafer scale	Die scale
Yield	Lower than lowest yield wafer, therefore high-yield wafer must be used	Known good die can be used if pre-stacking testing is available
Alignment accuracy	<2 μm global alignment	~10 μm for >1000 dph <2 μm for <100 dph

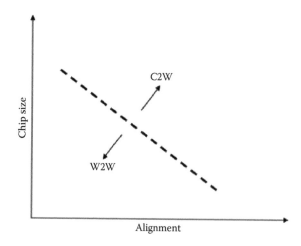

FIGURE 1.9   Choice between C2W and W2W depends on the chip size and the required alignment accuracy.

1.3.4 3D INTERCONNECT TECHNOLOGY DEFINITIONS BY ITRS

Since 3D technology is actively pursued by almost all players (such as IDM, fab-less, IC foundry, semiconductor assembly and test, printed circuit board, and assembly) in the electronic manufacturing supply chain, a broad variety of technology option is being proposed. As a result, the traditional interfaces between all these players are blurring. In order to come to a clear vision on road maps for 3D technologies, it is important to come to a clear definition of what is understood as "3D interconnect" technology. The International Technology Roadmap for Semiconductor (ITRS), in the 2009 report on Interconnect, has proposed a classification of the wide variety of 3D technologies (International Technology Roadmap for Semiconductor, Interconnect, 2009, http://www.itrs.net/) that captures the functional requirements of 3D technology at the different hierarchical levels of the system and correspond to the supply chain manufacturing capabilities. The following is a summary of 3D definitions and naming conventions proposed by ITRS.

> *3D-Interconnect Technology*: Technology which allows for the vertical stacking of layers of "basic electronic components" that are connected using a 2D-interconnect fabric are listed as following. "Basic electronic components" are elementary circuit devices such as transistors, diodes, resistors, capacitors, and inductors. A special case of 3D-interconnect technology is the Si interposer structures that may only contain interconnect layers, although in many cases other basic electronic components (in particular decoupling capacitors) may be embedded.
>
> *3D Bonding*: Operation that joins two die or wafer surfaces together.
>
> *3D Stacking*: Operation that also realizes electrical interconnects between the two device levels.
>
> *3D-Packaging (3D-P)*: 3D integration using "traditional" packaging technologies, such as wire-bonding, package-on-package stacking, or embedding in printed circuit boards.
>
> *3D-Wafer-Level-Packaging (3D-WLP)*: 3D integration using wafer-level-packaging technologies, performed after wafer fabrication, such as flip-chip redistribution, redistribution interconnect, fan-in chip-size packaging, and fan-out reconstructed wafer chip-scale packaging.
>
> *3D-System-on-Chip (3D-SOC)*: Circuit designed as a system-on-chip, SOC, but realized using multiple stacked die. 3D-interconnects directly connect circuit tiles in different die levels. These interconnects are at the level of global on-chip interconnects. This allows for extensive use/reuse of IP-blocks.
>
> *3D-Stacked-Integrated-Circuit (3D-SIC)*: 3D approach using direct interconnects between circuit blocks in different layers of the 3D die stack. Interconnects are on the global or intermediate on-chip interconnect levels. The 3D stack is characterized by a sequence of alternating front-end (devices) and back-end (interconnect) layers.
>
> *3D-Integrated-Circuit (3D-IC)*: 3D approach using direct stacking of active devices. Interconnects are on the local on-chip interconnect levels. The 3D stack is characterized by a stack of front-end devices, combined with a common back-end interconnect stack.

Table 1.4 presents a structured definition of 3D interconnect technologies based on the interconnect hierarchy. This structure also refers to the industrial semiconductor supply chain and allows definition of meaningful road maps and targets for each layer of the interconnect hierarchy.

1.4 TECHNOLOGY PLATFORMS AND STRATEGIES

A number of new enabling technologies must be developed and introduced into the existing fabrication process flow to make 3D integration a reality. Depending on the level of granularity, new capabilities include permanent wafer bonding, temporary bonding, and de-bonding,

TABLE 1.4

3D Interconnect Technologies Based on Interconnect Hierarchy

Level	Name	Supply Chain	Key Features
Package	3D-Packaging (3D-P)	OSAT Assembly PCB	• Traditional packaging of interconnect technologies, for example, wire-bonded die stacks, package-on-package stacks • Also includes die in PCB integration • No through-Si-vias (TSVs)
Bond-pad	3D-Wafer-Level Package (3D-WLP)	Wafer-level Packaging	• WLP infrastructure, such as redistribution layer (RDL) and bumping • 3D interconnects are processed after the IC fabrication, "post-IC-passivation" (via last process). Connections on bond-pad level • TSV density requirements follow bond-pad density road maps
Global	3D-Stacked Integrated Circuit/3D-System-on-Chip (3D-SIC/3D-SoC)	Wafer Fab	• Stacking of large circuit blocks (tiles, IP-blocks, memory-banks), similar to an SoC approach but having circuits physically on different layers • Un-buffered I/O drivers (Low C, little or no ESD protection on TSVs) • TSV density requirement significantly higher than 3D-WLP: Pitch requirement down to 4–16 μm
Intermediate	3D-SIC	Wafer Fab	• Stacking of smaller circuit blocks, parts of IP-blocks stacked in vertical dimensions • Mainly wafer-to-wafer stacking • TSV density requirements very high: Pitch requirement down to 1–4 μm
Local	3D-Integrated Circuit (3D-IC)	Wafer Fab	• Stacking of transistor layers • Common BEOL interconnect stack on multiple layers of FEOL • Requires 3D connections at the density level of local interconnects

Source: ITRS, http://www.itrs.net/, 2009.

through-silicon/strata-via (TSV), wafer thinning and handling, and precision alignment. There are a number of references on technology platforms available in the literature and the references therein [22–24]. This section describes a 3D integration process based on work on Cu thermo-compression bonding originally proposed by researchers at MIT. In this scheme, two FEOL active device wafers are stacked in a back-to-face fashion and bonded by means of low temperature Cu-to-Cu thermo-compression bonding. Interlayer vertical vias electrically interconnect the device layers. Low-temperature wafer-bonding is necessary since the pre-bonding device layers already have Al or Cu metal interconnect lines. Process sequence in this proposed 3D integration scheme is illustrated in Figure 1.10a through f. Using this process flow as a baseline, key process steps that have to be developed for 3D IC are discussed. In addition, this section will primarily discuss low temperature Cu–Cu permanent bonding, which is the author's core research expertise.

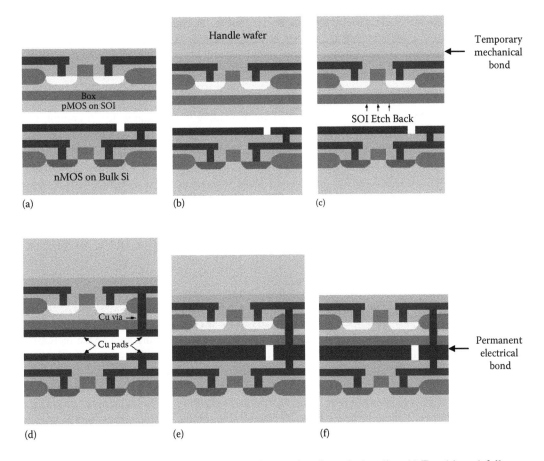

FIGURE 1.10 Process flow for 3D integration scheme using Cu wafer bonding. (a) Two (almost) fully processed wafers; (b) SOI wafer is attached to a handle wafer; (c) SOI wafer backside thinning; (d) Cu vias and pads are created; (e) Cu-to-Cu wafer bonding; and (f) Handle wafer release.

Even though a back-to-face stacking orientation is used for the purpose of illustration, most of the new process capabilities, except temporary bonding and de-bonding, can be applied for a face-to-face stacking process.

1.4.1 HANDLE WAFER ATTACHMENT

Figure 1.10a through f depicts a possible process sequence to fabricate a 3D CMOS inverter. The bottom device layer is an n-MOS device fabricated on bulk Si while the top device layer is a p-MOS device fabricated on SOI wafer independently prior to stacking. Note that SOI wafer is proposed for ease of subsequent wafer thinning in this proposed flow and bulk Si wafer is widely used in the industry. To start with, the front side of the top layer is attached to a handle wafer as shown in Figure 1.10b to provide mechanical support for ease of wafer handling. Therefore, the bonding has to be strong enough to hold the SOI wafer during subsequent processes. Note that this bonding is a temporary one, as the handle wafer will be released from the final 3D stack. This dictates the ease of handle wafer release at the end. Low-temperature oxide wafer bonding [25] is used in this original proposal. Note that the recent advances in temporary bonding using adhesive and additional discussion can be found in [26]. These adhesives include two main types: thermo-plastic or thermo-set. It is important to note that in the case of face-to-face bonding, a handle wafer is not required.

The favorite choice of handle wafer is silicon as it is thermo-mechanically matched to the device wafer and for ease of manufacturing in a silicon-based manufacturing environment. There has been attempt to use glass handle due to its transparency and lower cost. However, it is not widely used.

1.4.2 WAFER THIN BACK

In Figure 1.10c, the SOI substrate is thinned back after bonding to a handle wafer. A combination of mechanical grinding, plasma dry etch, and chemical wet etch can be used for this thinning step. In order to achieve good etch stop behavior, it is typical to etch the final 50–100 µm of Si using wet chemical etch. The buried oxide (BOX) serves as the etch stop layer as there is an excellent selectivity between Si and oxide in wet etchant. The handle wafer has to be protected against chemical attack by SiO_2 coating. When a bulk Si wafer is used, the grinding is controlled using precision thickness measurement as there is no BOX etch stop layer. Usually a coarse grinding step is used to remove the bulk of the wafer and fine grinding is used at the end to remove the surface damages and to release the stress. The major consideration during this step includes warpage and total thickness variation.

1.4.3 VERTICAL VIA AND BUMP FORMATION

Backside interlayer vertical vias and Cu pads are created on the thinned SOI wafer. There are two sets of Cu pads. The first set is the via landing pads to form electrical connection between both device layers and the second set is the dummy pads to increase the bonding area hence increasing the bonding strength. This is schematically shown in Figure 1.10d. In the current "via-middle" TSV process widely pursued by the industry, high aspect ratio TSV are formed on the top donor wafer during step 10a. During wafer thinning step in Figure 10c, the TSV must be delicately exposed and stringent control is required to prevent Cu diffusion. Depending on the interconnect routing, a redistribution layer (RDL) might be formed and bumping is completed to connect the top layer to the bottom substrate wafer. A short description of TSV formation is appended as follows.

1.4.3.1 Through-Silicon-Via

Figure 1.11 is a generic process flow of TSV fabrication flow using Cu as the filler metal. It begins with high aspect ratio deep etching of Si. Dielectric liner layer is then deposited on the via sidewall followed by barrier and Cu seed layers deposition. Liner layer, which is made of dielectric layer such as silicon dioxide, provides electrical isolation between Cu core and Si substrate. The liner thickness must be chosen appropriately to control leakage current and capacitance between Cu core and Si substrate. Cu super conformal filling is then achieved with electroplating process. Super conformal filling is required to prevent void formation in the Cu-TSV. Finally, Cu over-burden is removed by chemical mechanical polishing. More information on TSV fabrication can be found in literature such as [22–24].

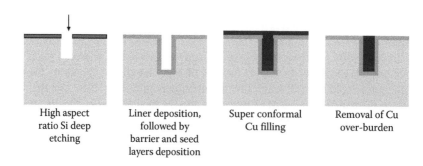

| High aspect ratio Si deep etching | Liner deposition, followed by barrier and seed layers deposition | Super conformal Cu filling | Removal of Cu over-burden |

FIGURE 1.11 Generic process flow of Cu-filled TSV fabrication.

1.4.4 Alignment

Precision alignment is another key technology for 3D stacking. Usually, alignment can be accomplished with different degree of accuracy using: (1) stored image to live image, or (2) stored image to stored image. Sources of alignment error include Δx, Δy, and $\Delta\theta$ (rotational). These errors are often caused by inherent inaccuracy and additional error during handling or processing. Depending on the density of 3D interconnects, alignment accuracy <1 μm is often required with some applications call for <0.5 μm accuracy.

1.4.5 Bonding

Figure 1.10e shows that the top device layer is aligned to the bottom device layer, presumably with Cu pads already created on it, and bonded at low temperature with a constant down force in an inert ambient. A final post-bonding annealing step allows inter-diffusion at the Cu–Cu interface and promotes grain growth. While Cu–Cu bonding offers the most desirable physical properties, reliability, and scalability, the bonding requirements are typically demanding and low throughput is often cited. Cu/Sn micro-bump and Cu pillar are two promising options pursued in the industry.

Using metal as bonding interface between active layers is an attractive choice because metal is a good heat conductor and this will help circumvent the heat dissipation problem encountered in a 3D IC. At the same time, a metal interface allows additional wiring and routing. Cu is a metal of choice because it is a mainstream CMOS material, and it has good electrical ($\rho_{Cu} = 1.7$ mΩ cm vs $\rho_{Al} = 2.65$ mΩ cm) and thermal ($K_{Cu} = 400$ W/m/K vs. $K_{Al} = 235$ W/m/K) conductivities and longer electro-migration lifetime. Another advantage offered by metal bonding interface is that the metal layer can act as a ground shield if properly grounded, hence providing better noise isolation between device layers on the stack.

1.4.5.1 Cu–Cu Permanent Bonding

3D integration of ICs by means of bump-less Cu–Cu direct bonding is an attractive choice as one accomplishes both electrical and mechanical bonds simultaneously. Cu–Cu direct bonding is desired compared to solder-based connections because: (1) Cu–Cu bond is more scalable and ultra-fine pitch can be achieved; (2) Cu has better electrical and thermal conductivities; and (3) Cu has much better electro-migration resistance and can withstand higher current density in future nodes. Direct Cu–Cu bonding has been demonstrated using thermo-compression bonding (also known as diffusion bonding). As the name implies, thermo-compression bonding involves simultaneous mechanical pressing (~200 kPa) and heating of the wafers (~300°C–400°C). Two wafers can be held together when the Cu thin films bond together to form a uniform bonded layer. In order for this technique to be applicable to wafers that carry device and interconnect layers, an upper bound of temperature step is set at 400°C to prevent undesired damages particularly to the interconnects. The main objective of this Cu thermo-compression bonding study is to explore its suitability for utilization as a permanent bond that holds active device layers together in a multilayer ICs stack. Cu is a metal of choice for 3D ICs application because it is a mainstream CMOS material, and it has better electrical and thermal conductivities compared to Al-based interconnect. Most importantly, Cu bonds to itself under conditions compatible with CMOS backend processes as initially demonstrated by Fan et al. [27].

Since Cu is an electrically conductive medium, isolation is needed. One solution is to form damascene Cu lines and to perform hybrid bonding of Cu and dielectric. A few examples are

1. Jourdain et al. [28] at IMEC have successfully demonstrated the 3D stacking of an extremely thinned IC chip onto a Cu/oxide landing substrate using simultaneous Cu–Cu thermo-compression and compliant glue-layer (BCB) bonding. The goal of this intermediate BCB glue layer between the two dies is to reinforce the mechanical and thermal

stability of the bonded stack and to enable separation of die pick-and-place operations from a collective bonding step.

2. Gutmann et al. [29] at RPI have demonstrated another scheme of hybrid bonding using face-to-face bonding of Cu/BCB redistribution layers. The first step is to prepare the single-level damascene-patterned structures (Cu and BCB) by CMP in the two Si wafers to be bonded. The second step is to align the two wafers and bond the two aligned wafers.

3. Researchers at Ziptronix have developed a Cu/oxide hybrid bonding technology known as Direct Bond Interconnect (DBI™) [30]. Vertical interconnections in direct oxide bond DBI are achieved by preparing a heterogeneous surface of nonconductive oxide and conductive Cu. The surfaces are aligned and placed together to effect a bond. The high-bond energies possible with the direct oxide bond between the heterogeneous surfaces result in vertical DBI electrical interconnections.

1.4.5.2 Low-Temperature Cu Bonding

Thermo-compression bonding of Cu layers is typically performed at a temperature of 300°C or higher. There is a strong motivation to move the bonding temperature to even lower range primarily from the point of view of thermal stress induced due to CTE mismatch of dissimilar materials in a multilayer stack and temperature swing. A number of approaches have been explored:

1. *Surface-activated bonding* [31]: In this method, a low energy Ar ion beam is used to activate the Cu surface prior to bonding. Contacting two surface-activated wafers enables successful Cu–Cu direct bonding. The bonding process is carried out under an ultrahigh vacuum (UHV) condition. No thermal annealing is required to increase the bonding strength. Tensile test results show that a high bonding strength equivalent to bulk material is achieved at room temperature. In [32], adhesion of Cu–Cu bonded at room temperature in UHV condition was measured to be about ~3 J/m^2 using AFM tip pull-off method.

2. *Cu nanorod* [33]: Recent investigation on surface melting characteristics of copper-nanorod arrays shows that the threshold of the morphological changes of the nanorod arrays occurs at a temperature significantly below the copper bulk melting point. With this unique property of the copper nanorod arrays, wafer bonding using copper nanorod arrays as a bonding intermediate layer is investigated at low temperatures (400°C and lower). Silicon wafers, each with a copper nanorod array layer, are bonded at 200°C–400°C. The FIB/SEM results show that the copper nanorod arrays fuse together accompanied by a grain growth at a bonding temperature of as low as 200°C.

3. *Solid–liquid inter-diffusion bonding (SLID)* [34]: This method involves the use of a second solder metal with low-melting temperature such as Tin (Sn) in between two sheets of Cu with high-melting temperature. Typically a short reflow step is followed by a longer curing step. The required temperature is often slightly higher than Sn melting temperature (232°C). The advantages of SLID is that the inter-metallic phase is stable up to 600°C and the requirement of contact force is not critical.

4. In the DBI technology described in [30], a moderate post-oxide bonding anneal may be used to effect the desired bonding between Cu. Due to the difference in coefficient of expansion between the oxide and Cu and the constraint of the Cu by the oxide, Cu compresses each other during heating and a metallic bond can be formed.

5. Direct Cu–Cu bonding at atmospheric pressure is investigated by researchers at LETI [35]. By means of CMP, the roughness and hydrophily (measured by contact angle) of Cu film are improved from 15 to 0.4 nm and from 50° to 12°. Blanket wafers were successfully bonded at room temperature with an impressive bond strength of 2.8 J/m^2. With a post-bonding annealing at 100°C for 30 min, the bonding strength was improved to 3.2 J/m^2.

6. A novel Cu–Cu bonding process has been developed and characterized to create all-copper chip-to-substrate I/O connections [36]. Electroless copper plating followed by low-temperature annealing in a nitrogen environment was used to create an all-copper bond between copper pillars. The bond strength for the all-copper structure exceeded 165 MPa after annealing at 180°C. While this technique is demonstrated as a packaging solution, it is an attractive low temperature process for Cu–Cu bonding.

7. In the author's research group, a method of Cu surface passivation using self-assembled monolayer (SAM) of alkane-thiol has been developed. This method has been shown to be effective to protect the Cu surface from particle contamination and to retard surface oxidation. The SAM layer can be thermally desorbed *in situ* in the bonding chamber rather effectively, hence providing clean Cu surface for successful low-temperature bonding. Cu wafers bonded at 250°C present significant reduction in micro-void and substantial Cu grain growth at the bonding interface [37–40].

1.4.6 HANDLE WAFER RELEASE

In the proposed flow presented in Figure 1.10, the top donor wafer is bonded to the silicon handle wafer using oxide fusion bonding. Handle wafer can be removed from the final 3D stack by a combination of mechanical grinding and wet etch. Alternatively, a less abusive method such as hydrogen-induced wafer splitting can be used. Hydrogen is implanted into the handle wafer prior to bonding. The handle wafer can be released by annealing at a temperature higher than temperature during Cu thermo-compression bonding to form the permanent bond. When adhesive is used as the temporary bonding medium, handle wafer release can be achieved by thermal release followed by mechanical lift-off, laser ablation, or chemical release. The clear advantage of these release methods is that the handle wafer can be recovered and reused.

1.4.7 2.5D SILICON INTERPOSER

Another approach which is widely considered is the 2.5D silicon interposer (Figure 1.12). In this approach, processed dies (without TSV) are placed side-by-side in close proximity. Interconnections between dies are done using fine-pitch horizontal interconnects on the silicon substrate afforded by advanced silicon processing. TSV are fabricated in the silicon interposer to connect the dies to the conventional package for power, ground, and I/O. Besides silicon, glass is also being considered as a strong contender due to its low cost and insulating property. Collaboration between TSMC and Xilinx has demonstrated a 28 nm FPGA chip that can be partitioned into four smaller dies and assembled on silicon interposer. The interconnects on the silicon interposer, including TSV, are fabricated in a 65 nm technology node.

1.4.8 DEMONSTRATOR

Heterogeneous integration of MEMS and CMOS is critical in future development of multisensor data fusion in a low-cost chip size system for Wearable and IoT applications. MEMS/CMOS integration was primarily done using monolithic and hybrid/package approaches until recently. In this demonstration [41], 3D CMOS-on-MEMS stacking without TSV using direct (i.e., solder-less) metal bonding, as shown in Figure 1.13, is discussed. This MEMS/CMOS integration leads to a simultaneous formation of electrical, mechanical, and hermetic bonds, eliminates chip-to-chip wire-bonding, and hence presents competitive advantages over hybrid or monolithic solutions. This stacking method makes use of an active capping layer (which is the CMOS layer), hence it eliminates the need for an *ex situ* hermetic seal.

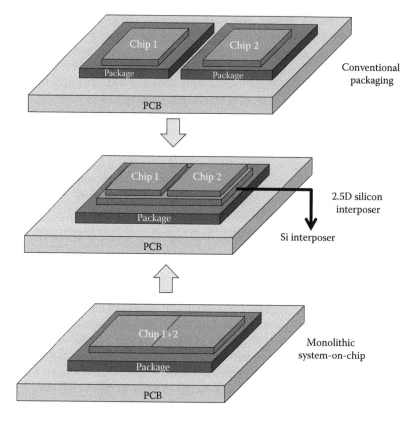

FIGURE 1.12 2.5D silicon interposer offers the benefits of packaging and monolithic approaches. It provides flexibility in manufacturing especially for disparate dies as compared to 3D stacking.

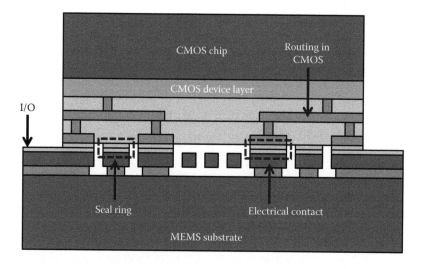

FIGURE 1.13 Heterogeneous 3D CMOS-on-MEMS stacking with face-to-face direct metal bonding (no solder) that realizes electrical, mechanical, and hermetic bonds simultaneously. Besides acting as the processing unit, the CMOS chip also acts as an active cap for the MEMS chip. There is no wire-bonding from chip to chip, hence this will improve delay and power consumption. Since I/O count is low, TSV is not used. Electrical feed-through is accomplished by peripheral pads.

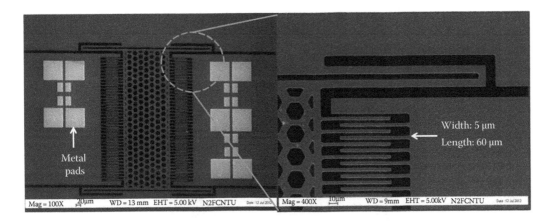

FIGURE 1.14 MEMS capacitive accelerator is fabricated on SOI wafer using DRIE and release. Single layer of patterned metal consists of electrical contact pad and hermetic seal is used. Electrical feed-throughs are routed through the on-chip interconnect in the CMOS chip.

MEMS: SOI-based (resistivity ~0.002 Ω/cm) capacitive MEMS accelerometer (sensitivity ~4.88 fF/g) is designed and fabricated by DRIE as shown in Figure 1.14. A metal layer is patterned on the MEMS prior to etching for electrical contact and hermetic seal. The resonant frequency is ~136 kHz.

Readout circuit: A CMOS readout circuit (2 mm × 2 mm) is designed and implemented using 0.35 μm (2P4M) MPW process. The readout circuit comprises of a low-noise gain stage, a fully differential synchronous demodulator, and an off-chip low-pass filter. The block diagram is presented in Figure 1.15. The low-noise, band-pass gain stage is realized using two single-ended output amplifiers, based on folded-cascode architecture. Tunable feedback capacitance allows variable gain so that the same readout circuit can be used with accelerometers of different sensitivities. Synchronous demodulation is achieved using a four-switch, full-wave rectifier. Figure 1.16 shows the die micrograph.

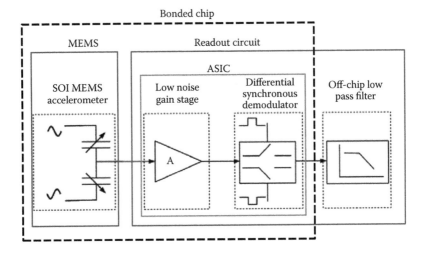

FIGURE 1.15 System block diagram: The MEMS and the ASIC have been seamlessly integrated to form a single bonded chip. There is no wire-bonding from chip to chip. Chip-to-chip electrical connections are done using direct metal bonding without solder.

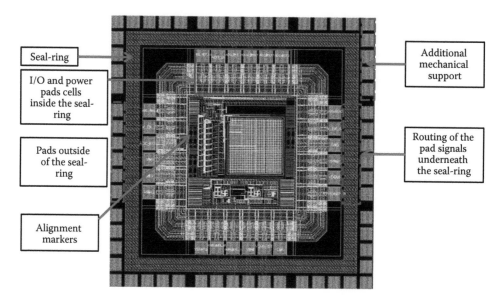

FIGURE 1.16 Die micrograph of the readout circuit fabricated through MPW (0.35 μm, 2P4M process). Key components are highlighted and specific ones are sealing ring, mechanical support, and alignment mark to assist CMOS–MEMS bonding.

Stacking process: In order to ensure proper operation, the delicate micro-structures (MEMS) should be protected from the ambient. In this approach, a hermetic seal ring is formed simultaneously during stacking of CMOS on MEMS and hence eliminating the need for post-processing hermetic encapsulation. Effectively, the CMOS layer acts as an "active cap." In addition, I/Os to the MEMS chip are routed through the CMOS metal layers to simplify the MEMS process (Figure 1.1). Since the I/O count is low, TSV is not used and electrical feed-through is achieved by peripheral pads. As no solder is applied, the top passivation layer of the CMOS chip is partially recessed to expose the CMOS metal layer for ease of direct bonding with the MEMS metal layer. The metal surfaces are carefully treated and bonded (thermo-compression at 300°C, 50N, 10 min). The bonded samples were packaged inside 44-pin J-leaded ceramic package for testing as shown in Figure 1.17.

Validation of the bonded CMOS/MEMS chip: The frequency range over which the readout circuit works is decided by the corner frequencies of the low-noise, band-pass gain stage. The measurement results for the bonded MEMS-CMOS chip are shown in Figure 1.18, when the chip was excited by 1 Vpp, 50 kHz differential sinusoids. The variation in the peak-to-peak amplitude of the gain-stage output with respect to an excitation carrier is observed as the chip is flipped between −1 *g*/+1 *g* orientations. The minimum amplitude corresponds to the chip at 0 *g* orientation. The amplitude of the gain-stage output grows in-phase and anti-phase with respect to carrier in +*g* and −*g* flip directions, respectively, which suggests that the bonded chip is working as desired. The maximum amplitudes (pp) observed in the two flip directions are roughly equal, thereby implying an approximately symmetrical behavior of the accelerometer.

Bonding reliability: The metal bonding quality is investigated using four-point bending test and helium leak test. The mechanical strength and hermeticity are studied based on the MIL-STD-883E, 1014.9 standard. The reliability of the hermetic seal is studied using thermal cycling (−40°C to 125°C), humidity (IPC/JEDEC J-STD-020), and corrosion tests. No serious degradation is found and the hermeticity is maintained at below 5×10^{-8} atm cm^3/s.

FIGURE 1.17   The CMOS chip is bonded face-to-face on the MEMS chip. The bonded chip is then wire bonded to the package for electrical testing. The vertically stacked CMOS and MEMS chip has a thickness of 1155 μm.

1.5   APPLICATIONS, STATUS, AND OUTLOOK

Applications enabled by 3D fall into a few categories depending on the use of TSV and bonding. At the time of this writing, there are already commercial products made with TSV, such as CMOS image sensor and memory stack, and many others are still under research and development. This section samples a number of closely watched applications enabled by 3D technology including image sensor, high-density memory stack, memory/logic integration, and more futuristic ones like 3D heterogeneous systems. It ends with a positive note on the current status and future outlook of 3D integration.

The main drivers for 3D applications include: (1) form factor, such as replacing wire bond with TSV in CMOS image sensor, (2) high density, such as stand-alone memory stack, (3) performance, such as bandwidth enhancement in memory on logic, and (4) heterogeneous integration of disparate chips. Regardless of the main driver, the feasibility and key consideration of any 3D application for consumer products has always been low cost manufacturing.

Broadly, applications enabled by 3D technology can be classified into three categories as shown in Figure 1.19. The first group of products only utilizes TSV such as CMOS image sensor (at the time of writing, there are commercial products from companies such as ST Microelectronics, Toshiba, OKI, etc.), backside ground (e.g., SiGe power amplifier by IBM), and silicon interposer (focus on the interposer before functional dies placement). In this class of devices, chip-to-chip bonding is not required. In another group, 3D devices are implemented by bonding chip on chip in a face-to-face fashion (see demonstration Section 1.4). I/O is formed using conventional wire bond or flip chip at the nonbonding periphery area. One such example is the Sony Play Station featuring memory on logic. The real 3D devices that make use of both TSV and bonding include stand-alone high density memory stack, memory on logic, logic on logic, and heterogeneous systems. At the time of this writing, there has been announcement and engineering products on a multilayer DRAM stack using 3D stacking technology. There are a number of emerging drivers and trends that are closely related to 2.5D/3D technology. Two of them are highlighted and discussed.

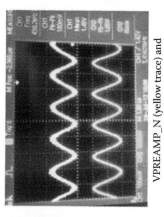

VPREAMP_N (yellow trace) and
carrier (blue trace) at −g cos(45°)

VPREAMP_N (yellow trace) and
carrier (blue trace) at −1 g

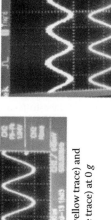

VPREAMP_N (yellow trace) and
carrier (blue trace) at 0 g

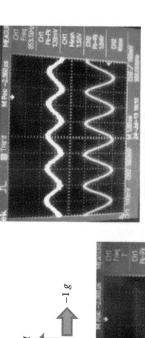

VPREAMP_N (yellow trace) and
carrier (blue trace) at +g cos(45°)

VPREAMP_N (yellow trace) and
carrier (blue trace) at +1 g

0 g   −1 g   +1 g

FIGURE 1.18 Gain-stage output (yellow, upper trace) when excitation carriers are at 1 Vpp (blue, lower trace) under: 0 g, +g cos(45°), +1 g, −g cos(45°) and −1 g orientations.

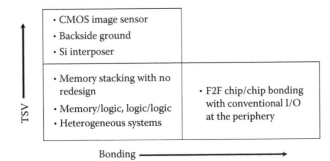

FIGURE 1.19 Applications enabled by 3D technology.

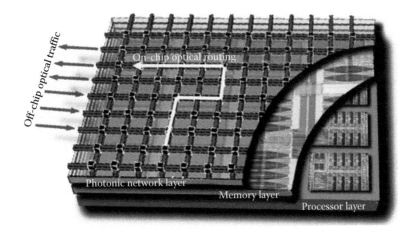

FIGURE 1.20 Conceptual diagram of a 3D integrated photonic interconnect network. (From IBM, http://www.research.ibm.com/, accessed on December 2014.)

1.5.1 Silicon Photonics

The integration of 3D photonic interconnects with multi-core processors is expected to dramatically reduce the power of interconnects particularly in the regime where the interconnect distance is long, and the data rates are high. A conceptual 3D optical interconnect scheme is shown in Figure 1.20 from IBM. Here a complete network-on-a-chip is shown, which utilizes a bottom multi-core processor layer, an intermediate memory layer, and an optical interconnect layer on top. In such a system, local interconnects could be provided by standard metal wiring, whereas global connections would be made using a photonic interconnect fabric consisting of silicon waveguides. The photonic network provides the additional advantage that off-chip I/O is achieved at the same bandwidth with little additional power. Apart from this 3D stacking approach, one could also think about leveraging on the benefits afforded by 2.5D silicon interposer to co-integrated photonics and electronics.

1.5.2 ITRS 2.0

ITRS 2.0 [42] was first introduced in April of 2014 and full documentation is expected in the end of 2015. Many elements of semiconductor manufacturing have been mapped into seven focus topics: System integration, Outside System Connectivity, Heterogeneous Integration, Heterogeneous Components, Beyond CMOS, More Moore, and Factory Integration. As made clear in the seven focus areas, 2.5D/3D will play a bigger role in ITRS 2.0 as these technologies are critical to high-density heterogeneous integration.

For more than 40 years, performance growth in IC is realized primarily by geometrical scaling. In more recent nodes, performance boosters are used to sustain this historical growth. Moving forward, 3D integration is an inevitable path. There has been significant investment in 3D technology by various sectors and the development has been both rewarding and encouraging. While 3D technology is not without its challenges, it is likely that the industry will continue to come out with solutions in thermal management, EDA tools, testing, and standardization. Wide adoption of 2.5D/3D technology is expected to happen when a competitive cost can be achieved.

ACKNOWLEDGMENT

The author (CST) is supported by funding from the Nanyang Technological University through an award of Nanyang Assistant Professorship, Defense Science and Technology Agency (DSTA, Singapore), Agency for Science, Technology and Research (ASTAR), Semiconductor Research Corporation (SRC, USA) through a subcontract from the Interconnect and Packaging Center at the Georgia Institute of Technology, and Defense Advanced Research Projects Agency (DARPA, USA).

REFERENCES

1. G. Moore, Cramming more components onto integrated circuits, *Electron. Mag.*, 38(8), 1965.
2. R.H. Dennard, F.H. Gaensslen, H.N. Yu, V.L. Rideout, E. Bassous, and A.R. LeBlanc, Design of ion-implanted MOSFETs with very small physical dimensions, *IEEE J. Solid-State Circuits*, 9, 256–268, 1974.
3. Intel Corporation. http://www.intel.com.
4. A. Khakifirooz, Transport enhancement techniques for nanoscale MOSFETs, PhD thesis, MIT, Cambridge, MA, http://dspace.mit.edu/handle/1721.1/42907, 2008.
5. S. Saxena et al., Variation in transistor performance and leakage in nanometer-scale technologies, *IEEE Trans. Electron Devices*, 55(1), 131, 2008.
6. S. Nassif et al., High performance CMOS variability in the 65 nm regime and beyond. In: *IEDM*, Washington, DC, 2007, p. 569.
7. D. Sylvester and C. Hu, Analytical modeling and characterization of deep-submicrometer interconnect, *Proc. IEEE*, 89(5), 634, 2001.
8. P. Kapur, J.P. McVittie, and K.C. Saraswat, Realistic copper interconnect performance with technological constraints. In: *Proceedings of the IEEE Interconnect Technology Conference*, San Francisco, CA, 2001, p. 233.
9. P.G. Emma, Is 3D chip technology the next growth engine for performance improvement?, *IBM J. Res. Dev.*, 52(6), 541, 2008.
10. Tezzaron. http://www.tezzaron.com.
11. N. Miura et al., Capacitive and inductive-coupling I/Os for 3D chips. In: *Integrated Interconnect Technologies for 3D Nanoelectronic Systems*, M.S. Bakir and J.D. Meindl (Eds.), Artech House, Norwood, MA, 2009, p. 449.
12. S. Kawamura, N. Sasaki, T. Iwai, M. Nakano, and M. Takagi, Three-dimensional CMOS ICs fabricated by using beal recrystallization, *IEEE Electron Device Lett.*, 4(10), 366, 1983.
13. T. Kunio, K. Oyama, Y. Hayashi, and M. Morimoto, Three dimensional ICs, having four stacked active device layers. In: *IEDM Technical Digest*, Washington, DC, 1989, p. 837.
14. V. Subramanian, M. Toita, N.R. Ibrahim, S.J. Souri, and K.C. Saraswat, Low-leakage germanium-seeded laterally-crystallized single-grain 100-nm TFTs for vertical integration applications, *IEEE Electron Device Lett.*, 20(7), 341, 1999.
15. V.W.C. Chan, P.C.H. Chan, and M. Chan, Three-dimensional CMOS SOI integrated circuit using high-temperature metal-induced lateral crystallization, *IEEE Trans. Electron Devices*, 48(7), 1394, 2001.
16. S. Pae, T. Su, J.P. Denton, and G.W. Neudeck, Multiple layers of silicon-on-insulator islands fabrication by selective epitaxial growth, *IEEE Electron Device Lett.*, 20(5), 194, 1999.

17. B. Rajendran, R.S. Shenoy, D.J. Witte, N.S. Chokshi, R.L. DeLeon, G.S. Tompa, and R.F.W. Pease, Low temperature budget processing for sequential 3-D IC fabrication, *IEEE Trans. Electron Devices*, 54(4), 707, 2007.

18. P. Batude et al., Advances in 3D CMOS sequential integration. In: *IEDM*, Washington, DC, 2009, pp. 345–348.

19. M. Shulaker, T. Wu, A. Pal, K. Saraswat, H.-S.P. Wong, and S. Mitra, Monolithic 3D integration of logic and memory: Carbon nanotube FETs, resistive RAM, and silicon FETs. In: *IEEE International Electron Devices Meeting (IEDM)*, San Francisco, CA, December 15–17, 2014.

20. J.-W. Im et al., A 128 Gb 3b/cell V-NAND flash memory with 1Gb/s I/O rate. In: *ISSCC*, San Francisco, CA, 2015.

21. C.S. Tan, SOI-based 3D integration. In: *Silicon on Insulator (SOI) Technology*, O. Kononchuk and B.-Y. Nguyen (Eds.), Woodhead Publishing, Cambridge, UK, 2014. ISBN10: 0 85709 526 9, ISBN13: 978 0 85709 526 8. (DOI: 10.1533/9780857099259.2.358)

22. C.S. Tan, R.J. Gutmann, and R. Reif, *Wafer Level 3-D ICs Process Technology*, Springer, New York, 2008. ISBN: 978-0-387-76532-7.

23. C.S. Tan, K.-N. Chen, and S.J. Koester (Eds.), *3D Integration for VLSI Systems*, Pan Stanford, Singapore, 2011. ISBN: 978-981-4303-81-1.

24. P. Garrou, C. Bower, and P. Ramm, *Handbook of 3D Integrations: Technology and Applications of 3D Integrated Circuits*, Wiley-VCH, Weinheim, Germany, 2008. ISBN: 978-3-527-32034-9.

25. C.S. Tan, A. Fan, K.N. Chen, and R. Reif, Low-temperature thermal oxide to plasma-enhanced chemical vapor deposition oxide wafer bonding for thin-film transfer application, *Appl. Phys. Lett.*, 82(16), 2649–2651, 2003.

26. M. Privett, 3D technology platform: Temporary bonding and release. In: *3D Integration for VLSI System*, C.S. Tan (Ed.), Pan Stanford, Singapore, 2012, pp. 121–138, ISBN: 978-981-4303-81-1.

27. A. Fan, A. Rahman, and R. Reif, Copper wafer bonding, *Electrochem. Solid-State Lett.*, 2(10), 534–536, 1999.

28. A. Jourdain, S. Stoukatch, P. De Moor, W. Ruythooren, S. Pargfrieder, B. Swinnen, and E. Beyne, Simultaneous Cu–Cu and compliant dielectric bonding for 3D stacking of ICs. In: *Proceedings of IEEE International Interconnect Technology Conference*, Burlingame, CA, 2007, pp. 207–209.

29. R.J. Gutmann, J.J. McMahon, and J.-Q. Lu, Damascene-patterned metal-adhesive (Cu-BCB) redistribution layers, *Mater. Res. Soc. Symp. Proc.*, 970, 205–214, 2007.

30. P. Enquist, High density bond interconnect (DBI) technology for three dimensional integrated circuit applications, *Mater. Res. Soc. Symp. Proc.*, 970, 13–24, 2007.

31. T.H. Kim, M.M.R. Howlader, T. Itoh, and T. Suga, Room temperature Cu–Cu direct bonding using surface activated bonding method, *J. Vac. Sci. Technol. A: Vac. Surf. Films*, 21(2), 449–453, 2003.

32. R. Tadepalli and C.V. Thompson, Formation of Cu–Cu interfaces with ideal adhesive strengths via room temperature pressure bonding in ultrahigh vacuum, *Appl. Phys. Lett.*, 90, 151919, 2007.

33. P.-I. Wang, T. Karabacak, J. Yu, H.-F. Li, G.G. Pethuraja, S.H. Lee, M.Z. Liu, and T.-M. Lu, Low temperature copper-nanorod bonding for 3D integration, *Mater. Res. Soc. Symp. Proc.*, 970, 225–230, 2007.

34. P. Benkart, A. Kaiser, A. Munding, M. Bschorr, H.-J. Pfleiderer, E. Kohn, A. Heittmann, and U. Ramacher, 3D chip stack technology using through-chip interconnects, *IEEE Design Test Comput.*, 22(6), 512–518, 2005.

35. P. Gueguen, L. Di Cioccio, M. Rivoire, D. Scevola, M. Zussy, A.M. Charvet, L. Bally, and L. Clavelier, Copper direct bonding for 3D integration. In: *IEEE International Interconnect Technology Conference*, Burlingame, CA, 2008, pp. 61–63.

36. T. Osborn, A. He, H. Lightsey, and P. Kohl, All-copper chip-to-substrate interconnects. In: *Proceedings of IEEE Electronic Components and Technology Conference*, Lake Buena Vista, FL, 2008, pp. 67–74.

37. D.F. Lim, S.G. Singh, X.F. Ang, J. Wei, C.M. Ng, and C.S. Tan, Achieving low temperature Cu to Cu diffusion bonding with self assembly monolayer (SAM) passivation. In: *IEEE International Conference on 3D System Integration*, San Francisco, CA, 2009, art. no. 5306545.

38. D.F. Lim, S.G. Singh, X.F. Ang, J. Wei, C.M. Ng, and C.S. Tan, Application of self assembly monolayer (SAM) in Cu–Cu bonding enhancement at low temperature for 3-D integration. In: D.C. Edelstein and S.E. Schulz (Eds.), *Advanced Metallization Conference*, Baltimore, MD, October 13–15, 2009, pp. 259–266, Materials Research Society, Warrendale, PA.

39. C.S. Tan, D.F. Lim, S.G. Singh, S.K. Goulet, and M. Bergkvist, Cu–Cu diffusion bonding enhancement at low temperature by surface passivation using self-assembled monolayer of alkane-thiol, *Appl. Phys. Lett.*, 95(19), 192108, 2009.

40. D.F. Lim, J. Wei, C.M. Ng, and C.S. Tan, Low temperature bump-less Cu–Cu bonding enhancement with self assembled monolayer (SAM) passivation for 3-D integration. In: *IEEE Electronic Components and Technology Conference (ECTC)*, Las Vegas, NV, June 1–4, 2010, pp. 1364–1369.
41. S.L. Chua, A. Razzaq, K.H. Wee, K.H. Li, H. Yu, and C.S. Tan, 3D CMOS-MEMS stacking with TSV-less and face-to-face direct metal bonding. In: *Symposium on VLSI Technology (VLSI)*, Honolulu, HI, June 9–12, 2014.
42. P. Singer, Reframing the Roadmap: ITRS 2.0, February 2015, http://electroiq.com/petes-posts/2015/02/02/reframing-the-roadmap-itrs-2-0/, accessed on March 2014.

2 Overview of Physical Design Issues for 3D-Integrated Circuits

Aida Todri-Sanial

CONTENTS

ABSTRACT

Three-dimensional integration presents a broad spectrum of challenges to designers of high-performance and energy-efficient circuits. Seeking to improve both the energy efficiency and throughput density, designers have turned to physical design and optimization methods targeting devices—interconnects to system-level power savings. This chapter provides an overview of the different physical design challenges and solutions that span various levels from devices and circuits to architectures.

2.1   INTRODUCTION

Three-dimensional integrated circuits (3D ICs) present a novel design paradigm and opportunity to address the current challenges the semiconductor industry is facing with aggressive Moore's law scaling. 3D ICs can provide potential benefits such as reduced power consumption, improvement in delay, and higher integration density. Through-Silicon-Vias (TSVs) are the key enablers for 3D stacking by allowing direct and less resistive signal paths between vertically stacked circuit layers. TSVs can help reduce the wire lengths from 2D ICs and also allow replacement of chip-to-chip interconnections by intra-chip connections. Such advancements have led to design and implementation of heterogeneous systems on the same platform, that is, Flash, DRAM, SRAM-placed atop logic devices, and microprocessor cores.

However, TSV parasitics contribute to the power, signal, and thermal integrity of the full systems. Moreover, densely packaged vertical circuit layers introduce significant thermal and power integrity challenges compared to 2D systems.

Fundamentally, 3D ICs change how circuits are designed, analyzed, and verified. Physical design takes a new dimension as new constraints and cost functions become more important and conventional extensions of 2D design approaches are simply not sufficient to solving these problems.

In this chapter, we provide an overview of the various design challenges that 3D integration introduces and the present state-of-art solutions for addressing these challenges. The following sections present the current landscape for each 3D physical design challenge.

2.2   OVERVIEW OF 3D PHYSICAL DESIGN CHALLENGES

As modern processor chips will demand for even larger cache sizes and bandwidth to get the maximum system performance, industry is looking into 3D integration technology as a viable option for increasing cache capacity and bandwidth requirements. 3D integration overcomes many limitations from the traditional 2D IC design. It offers tighter integration of heterogeneous technologies and significant density benefit with an increase in chip interconnects resulting in short wires that are also low power as well.

There are also many design issues that 3D integration arises that impact the overall physical design methodology. 3D integration poses fundamental differences on computer-aided design approaches specifically on clock and power distribution, architectural planning and partitioning, TSV sizing and insertion, placement and routing, high-level design system, and several others, which we will discuss on the following subsections and more in-depth in the rest of this book.

2.3   CHALLENGES AND SOLUTIONS FOR 3D POWER DELIVERY NETWORKS

Design of 3D power delivery networks is one of the main challenges in 3D integration. Power delivery in 3D systems draws much larger current from package and power/ground networks than in conventional 2D systems due to multiple circuit layers. The large current demand leads to significant voltage droop (accumulated *IR* drop, *L di/dt*, and *RC dv/dt* effects) due to the parasitics of power and ground networks and TSVs. Furthermore, the surge of current can lead to considerable *di/dt* effects due to on-chip and package inductances.

Due to the increased power density and greater thermal resistance to heat sink, thermal integrity is a crucial challenge for reliable 3D integration. High temperatures can degrade the reliability and performance of interconnects and devices. Power/ground network resistivity is a function of temperature, thus at nodes with high temperature, voltage droop values become even worse. Furthermore, the large amount of current on power and ground networks flowing for significant amount of time can ultimately elevate the temperature and cause Joule heating phenomena and electromigration. Thus, voltage droop and temperature are interdependent and should be considered simultaneously for reliable design of 3D power delivery networks. In 3D systems, voltage droop and temperature increase in opposite direction. Voltage droop tends to increase for tiers further away from package and close to heat sink while temperature increases for tiers further away from heat sink and near to package. Additionally, each vertical tier can be of different technology and power demand, which would require a customized power delivery network. Thus, there are several challenges and constraints to be taken into account for designing reliable 3D power delivery networks.

There are several research papers that have addressed 3D power delivery design problem, and they can be categorized into two groups. First group investigates TSV topology, power, and thermal integrity analysis of 3D PDNs [14,15,33,37,43–45], whereas in the second group optimization methods are proposed for reliable 3D PDNs [5,19,35,38]. More in-depth overview of 3D power delivery design is given in Chapters 6 and 8.

2.4   CHALLENGES AND SOLUTIONS FOR 3D CLOCK DELIVERY NETWORKS

Clock distribution is another challenging problem for 3D ICs. Clock signals are distributed over the entire 3D stack to feed all the sequential logic. Clock skew, also defined as the maximum difference between clock source and sink arrival times, is required to be within 5% of the clock period for high performance systems. Hence, clock skew control is one of the main challenges for reliable 3D clock network design and it can also cause setup/hold time violations. Moreover, the clock signal is distributed not only on each circuit layer but also on the vertical direction through the TSVs from one tier to another. Overall, 3D clock network needs to drive a large capacitive load at high switching frequency. This also leads to a large portion of power consumption dissipated on clock networks. In some designs, the power consumption of clock networks can reach up to 50% of total chip power. Thus, skew control and low power are the primary constraints that need to be considered for designing 3D clock networks.

TSVs not only provide short connections between tiers, they also contribute with their parasitics, and their placement and sizing play a key role on overall 3D clock network performance and overall power consumption. Additionally, the different types of TSVs such as via-first, middle, or last can indicate the number and sizes of TSVs, which will further impact the TSV parasitics and overall clock skew and power consumption of 3D clock networks. There are several works that addressed the challenges of 3D clock network design [24,25,30,34,47–50]. More in-depth overview of 3D clock networks is given in Chapter 7.

2.5 CHALLENGES AND SOLUTIONS FOR 3D PLACEMENT AND ROUTING

Placement and routing are one of the most imminent problems to solve with migrating from 2D to 3D circuits. As already mentioned, one concern for 3D ICs is that power density is much higher than in 2D circuits. Thus, high levels of temperature build-up and thermal gradients become a first-order objective to be considered during placement and routing.

During placement, logic cells are arranged in row-based topology to satisfy the layout constraints and achieve a reasonable temperature distribution while also satisfying traditional topological placement constraints. The challenge also arises from the TSV placement and circuit area space that they occupy. The placement tool for 3D integration become more complex trying to satisfy the thermal and power density issues, layout, and 3D topological constraints, while also ensuring that the algorithms are scalable and runtime efficient to handle large size problems.

Placement algorithms will try to equalize the thermal profile however to a certain degree, otherwise it will create unbalance on other criteria, that is, wire lengths and signal integrity. Moreover, creating a uniform thermal profile is unrealistic to the varying switching activity of the circuits and lack of heat dissipation paths for middle tiers. One solution to this issue is by inserting thermal vias to alleviate some of the heat build-up and improve heat dissipation. Thermal vias are inter-tier connections but with no electrical connectivity. Their role is to remove heat from hot spots to the heat sink. Several works have investigated [3,4,9,10,20,26,36] optimal insertion of thermal vias and cell placement. Their insertion can be addressed at different levels of the design to ensure overall thermal integrity of the 3D system.

Once placement is performed, the circuit needs to undergo routing to obtain a complete circuit layout. Similar to 2D routing algorithms, 3D routing algorithms need to address the same constraints such as avoiding blockages due to areas occupied by thermal vias, taking into account the thermal impact on wire delays, overall wire lengths, timing, congestion, and routing completion. Several works have looked into 3D routing problem [6,8,12,31,32,46]. Additionally, Chapter 5 provides more insight to the 3D placement and routing methodology.

2.6 CHALLENGES AND SOLUTIONS FOR 3D FLOORPLANNING

One of the first steps for determining which circuit blocks should be placed close together and allocating the space is floorplanning. In 3D ICs, finding a suitable floorplan becomes more complex due to TSVs and nonuniform thermal gradients between metal, dielectric, and circuit layers. Thus, consideration of TSV impact on overall power and thermal integrity of the system should be considered from an early phase of the design cycle in order to avoid re-design and costly re-spins. The impact of

TSVs on interconnect delay depends on its dimensions, technology and filling material, their placement, and number of neighboring TSVs.

Several approaches have been published in the literature [1,2,11,13,16,29,40–42,51] where the TSV impacts are considered for both power and thermal management while having a minimal impact on net delay and wire lengths. During the floorplanning process, it is important to consider TSV dimensions and their placement in order to perform multivariable optimization of the circuit for better performance and thermal dissipation.

2.7 CHALLENGES AND SOLUTIONS ON TSV SIZING AND PLACEMENT

As aforementioned, TSV plays an important role on the overall 3D system performance, power, and thermal integrity. Their sizing, placement, and distribution are key knobs for being able to control and ease power consumption and excessive thermal dissipation. Typical diameters of via-first TSVs range from 1 to 5 μm and via-last TSVs range from 5 to 20 μm. Also, TSVs are fabricated in bulk silicon, which consumes silicon area for logic gates. Additionally, there is a keep-out-zone area between TSVs to avoid any coupling or thermal stress on devices, which further limits the circuit area space. Because of these constraints, inserting a large number of TSVs into 3D ICs can lead to important area overhead. Additionally, connecting TSVs to the intermediate and local level interconnects might cause routing congestions. Therefore, their sizing and placement should be considered carefully during the early design stages.

Several papers have considered the impact of TSVs during the placement and routing design flow [7,17,18,21–23,27,28,39]. TSV assignment algorithms have been developed and compare the impact of TSVs on wire lengths, circuit area, power consumption, and thermal dissipation.

2.8 SUMMARY AND CONCLUSIONS

In comparison to 2D ICs, 3D ICs impose fundamentally different physical design methodologies to cope with the complexity introduced by the vertical dimension. As summarized in this chapter (and in the next chapters), several challenges are unique to 3D integration such as heat build-up on middle layers, TSV thermal stress, and keep-out-zone requirement around TSVs.

In the recent years, academia and industry have devoted a lot of efforts into understanding the implications and complexity introduced by TSVs. There are novel challenges in all aspects of physical design such as floorplanning, placement and routing, power and clock delivery networks, TSV sizing and distribution, etc. While, 3D integration is becoming a more viable technology for addressing the power wall from aggressive scaling, several physical design solutions already exist to allow exploration of optimal system design for performance, power, and thermal integrity. The rest of this book is dedicated to address each individual's physical design challenge and provide an in-depth view on the dedicated solutions.

REFERENCES

1. M.A. Ahmed and M. Chrzanowska-Jeske. Delay and power optimization with TSV-aware 3D floorplanning. In *15th International Symposium on Quality Electronic Design (ISQED), 2014,* Hangzhou, China, pp. 189–196, March 2014.
2. M.A. Ahmed, S. Mohapatra, and M. Chrzanowska-Jeske. 3D floorplanning with nets-to-TSVs assignment. In *21st IEEE International Conference on Electronics, Circuits and Systems (ICECS), 2014,* Marseille, France, pp. 578–581, December 2014.
3. K. Athikulwongse, M. Pathak, and S.K. Lim. Exploiting die-to-die thermal coupling in 3D IC placement. In *49th ACM/EDAC/IEEE Design Automation Conference (DAC), 2012,* San Francisco, CA, pp. 741–746, June 2012.
4. P. Budhathoki, A. Henschel, and I.M. Elfadel. Thermal-driven 3D floor-planning using localized TSV placement. In *IEEE International Conference on IC Design Technology (ICI-CDT), 2014,* Austin, TX, pp. 1–4, May 2014.

5. H.-T. Chen, H.-L. Lin, Z.-C. Wang, and T.T. Hwang. A new architecture for power network in 3D IC. In *Design, Automation Test in Europe Conference Exhibition (DATE), 2011*, Grenoble, France, pp. 1–6, March 2011.
6. K.-C. Chen, S.-Y. Lin, H.-S. Hung, and A.A. Wu. Topology-aware adaptive routing for nonstationary irregular mesh in throttled 3D NoC systems. *IEEE Transactions on Parallel and Distributed Systems*, 24(10):2109–2120, October 2013.
7. Y. Chen, E. Kursun, D. Motschman, C. Johnson, and Y. Xie. Through silicon via aware design planning for thermally efficient 3-D integrated circuits. *IEEE Transactions on Computer-Aided Design of Integrated Circuits and Systems*, Tampa, FL, 32(9):1335–1346, September 2013.
8. L. Cheng, W.N.N. Hung, G. Yang, and X. Song. Congestion estimation for 3D routing. In *Proceedings of the IEEE Computer Society Annual Symposium on VLSI, 2004*, Yokohama, Japan, pp. 239–240, February 2004.
9. J. Cong and G. Luo. A multilevel analytical placement for 3D ICS. In *Asia and South Pacific Design Automation Conference, 2009. ASP-DAC 2009*, Yokohama, Japan, pp. 361–366, January 2009.
10. J. Cong, G. Luo, J. Wei, and Y. Zhang. Thermal-aware 3D IC placement via transformation. In *Asia and South Pacific Design Automation Conference, 2007. ASP-DAC '07*, San Jose, CA, pp. 780–785, January 2007.
11. J. Cong, J. Wei, and Y. Zhang. A thermal-driven floorplanning algorithm for 3D ICS. In *IEEE/ACM International Conference on Computer Aided Design, 2004. ICCAD-2004*, Shanghai, China, pp. 306–313, November 2004.
12. J. Cong and Y. Zhang. Thermal-driven multilevel routing for 3D ICS. In *Asia and South Pacific Design Automation Conference, 2005. Proceedings of the ASP-DAC 2005*, San Jose, CA, Vol. 1, pp. 121–126, January 2005.
13. X. He, S. Dong, Y. Ma, and X. Hong. Simultaneous buffer and interlayer via planning for 3D floorplanning. In *Quality of Electronic Design, 2009. ISQED 2009. Quality Electronic Design*, Grenoble, France, pp. 740–745, March 2009.
14. M.B. Healy and S.K. Lim. A novel TSV topology for many-tier 3D power-delivery networks. In *Design, Automation Test in Europe Conference Exhibition (DATE), 2011*, pp. 1–4, March 2011.
15. M.B. Healy and S.K. Lim. Distributed TSV topology for 3-D power-supply networks. *IEEE Transactions on Very Large Scale Integration (VLSI) Systems*, Taipei, Taiwan, 20(11):2066–2079, November 2012.
16. H. Hong, J. Lim, and S. Kang. Process variation-aware floorplanning for 3D many-core processors. In *Electrical Design of Advanced Packaging and Systems Symposium (EDAPS), 2012 IEEE*, pp. 193–196, December 2012.
17. M.-K. Hsu, V. Balabanov, and Y.-W. Chang. TSV-aware analytical placement for 3-D IC designs based on a novel weighted-average wirelength model. *IEEE Transactions on Computer-Aided Design of Integrated Circuits and Systems*, San Diego, CA, 32(4):497–509, April 2013.
18. M.-K. Hsu, Y.-W. Chang, and V. Balabanov. TSV-aware analytical placement for 3D IC designs. In *48th ACM/EDAC/IEEE Design Automation Conference (DAC), 2011*, San Jose, CA, pp. 664–669, June 2011.
19. G. Huang, M. Bakir, A. Naeemi, H. Chen, and J.D. Meindl. Power delivery for 3D chip stacks: Physical modeling and design implication. In *Electrical Performance of Electronic Packaging, 2007 IEEE*, Izmir, Turkey, pp. 205–208, October 2007.
20. R. Jagtap, S.S. Kumar, and R. van Leuken. A methodology for early exploration of TSV placement topologies in 3D stacked ICS. In *15th Euromicro Conference on Digital System Design (DSD), 2012*, San Jose, CA, pp. 382–388, September 2012.
21. D.H. Kim, K. Athikulwongse, and S.K. Lim. A study of through-silicon-via impact on the 3D stacked IC layout. In *IEEE/ACM International Conference on Computer-Aided Design—Digest of Technical Papers, 2009. ICCAD 2009*, pp. 674–680, November 2009.
22. D.H. Kim, K. Athikulwongse, and S.K. Lim. Study of through-silicon-via impact on the 3-D stacked IC layout. *IEEE Transactions on Very Large Scale Integration (VLSI) Systems*, Jeju, Korea, 21(5):862–874, May 2013.
23. D.H. Kim, R.O. Topaloglu, and S.K. Lim. TSV density-driven global placement for 3D stacked ICS. In *SoC Design Conference (ISOCC), 2011 International*, Taipei, Taiwan, pp. 135–138, November 2011.
24. T.-Y. Kim and T. Kim. Clock tree embedding for 3D ICS. In *15th Asia and South Pacific Design Automation Conference (ASP-DAC), 2010*, Seoul, Korea, pp. 486–491, January 2010.
25. T.-Y. Kim and T. Kim. Clock network design techniques for 3D ICS. In *IEEE 54th International Midwest Symposium on Circuits and Systems (MWSCAS), 2011*, New Orleans, LO, pp. 1–4, August 2011.

26. J. Li and H. Miyashita. Efficient thermal via planning for placement of 3D integrated circuits. In *IEEE International Symposium on Circuits and Systems, 2007. ISCAS 2007*, Taipei, Taiwan, pp. 145–148, May 2007.

27. C.-T. Lin, D.-M. Kwai, Y.-F. Chou, T.-S. Chen, and W.-C. Wu. Cad reference flow for 3D via-last integrated circuits. In *15th Asia and South Pacific Design Automation Conference (ASP-DAC), 2010*, pp. 187–192, January 2010.

28. W.-K. Mak and C. Chu. Rethinking the wirelength benefit of 3-D integration. *IEEE Transactions on Very Large Scale Integration (VLSI) Systems*, 20(12):2346–2351, December 2012.

29. R.K. Nain and M. Chrzanowska-Jeske. Placement-aware 3D floorplanning. In *IEEE International Symposium on Circuits and Systems, 2009. ISCAS 2009*, Taipei, Taiwan, pp. 1727–1730, May 2009.

30. M.M. Navidi and G.-S. Byun. Comparative analysis of clock distribution networks for TSV-based 3D IC designs. In *15th International Symposium on Quality Electronic Design (ISQED), 2014*, Hangzhou, China, pp. 184–188, March 2014.

31. A.-M. Rahmani, K.R. Vaddina, P. Liljeberg, J. Plosila, and H. Tenhunen. Power and thermal analysis of stacked mesh 3D NoC using adaptive xyz routing algorithm. In *15th Euromicro Conference on Digital System Design (DSD), 2012*, Izmir, Turkey, pp. 208–215, September 2012.

32. D. Roy, P. Ghosal, and S.P. Mohanty. Fuzzroute: A method for thermally efficient congestion free global routing in 3D ICS. In *IEEE Computer Society Annual Symposium on VLSI (ISVLSI), 2014*, Tampa, FL, pp. 71–76, July 2014.

33. I. Savidis, S. Kose, and E.G. Friedman. Power noise in TSV-based 3-D integrated circuits. *IEEE Journal of Solid-State Circuits*, 48(2):587–597, February 2013.

34. Y. Shang, C. Zhang, H. Yu, C.S. Tan, X. Zhao, and S.K. Lim. Thermal-reliable 3D clock-tree synthesis considering nonlinear electrical–thermal-coupled TSV model. In *18th Asia and South Pacific Design Automation Conference (ASP-DAC), 2013*, Yokohama, Japan, pp. 693–698, January 2013.

35. P. Singh, R. Sankar, X. Hu, W. Xie, A. Sarkar, and T. Thomas. Power delivery network design and optimization for 3D stacked die designs. In *IEEE International 3D Systems Integration Conference (3DIC), 2010*, Munich, Germany, pp. 1–6, November 2010.

36. T. Thorolfsson, G. Luo, J. Cong, and P.D. Franzon. Logic-on-logic 3D integration and placement. In *IEEE International 3D Systems Integration Conference (3DIC), 2010*, Paris, France, pp. 1–4, November 2010.

37. A. Todri-Sanial. Frequency domain power and thermal integrity analysis of 3D power delivery networks. In *17th IEEE Workshop on Signal and Power Integrity (SPI), 2013*, pp. 1–4, May 2013.

38. A. Todri-Sanial, S. Kundu, P. Girard, A. Bosio, L. Dilillo, and A. Virazel. Globally constrained locally optimized 3-D power delivery networks. *IEEE Transactions on Very Large Scale Integration (VLSI) Systems*, 22(10):2131–2144, October 2014.

39. M.-C. Tsai and T.T. Hwang. A study on the trade-off among wirelength, number of TSV and placement with different size of TSV. In *2011 International Symposium on VLSI Design, Automation and Test (VLSI-DAT)*, Taipei, Taiwan, pp. 1–4, April 2011.

40. C.-C. Wen, Y.-J. Chen, and S.-J. Ruan. Cluster-based thermal-aware 3D-floorplanning technique with post-floorplan TTSV insertion at via-channels. In *Fifth Asia Symposium on Quality Electronic Design (ASQED), 2013*, Penang, Malaysia, pp. 200–207, August 2013.

41. E. Wong and S.K. Lim. 3D floorplanning with thermal vias. In *Proceedings of the Design, Automation and Test in Europe, 2006. DATE '06*, Dresden, Germany, Vol. 1, pp. 1–6, March 2006.

42. E. Wong, J. Minz, and S.K. Lim. Power supply noise-aware 3D floorplanning for system-on-package. In *IEEE 14th Topical Meeting on Electrical Performance of Electronic Packaging, 2005*, San Jose, CA, pp. 259–262, October 2005.

43. J. Xie, D. Chung, M. Swaminathan, M. Mcallister, A. Deutsch, L. Jiang, and B.J. Rubin. Electrical–thermal co-analysis for power delivery networks in 3D system integration. In *IEEE International Conference on 3D System Integration, 2009. 3DIC 2009*, San Francisco, CA, pp. 1–4, September 2009.

44. J. Xie and M. Swaminathan. DC IR drop solver for large scale 3D power delivery networks. In *IEEE 19th Conference on Electrical Performance of Electronic Packaging and Systems (EPEPS), 2010*, San Jose, CA, pp. 217–220, October 2010.

45. J. Xie and M. Swaminathan. Simulation of power delivery networks with joule heating effects for 3D integration. In *Third Electronic System-Integration Technology Conference (ESTC), 2010*, Helsinki, Finland, pp. 1–6, September 2010.

46. T. Zhang, Y. Zhan, and S.S. Sapatnekar. Temperature-aware routing in 3D ICS. In *Asia and South Pacific Conference on Design Automation, 2006*, Yokohama, Japan, 6pp., January 2006.

47. X. Zhao and S.K. Lim. Power and slew-aware clock network design for through-silicon-via (TSV) based 3D ICS. In *15th Asia and South Pacific Design Automation Conference (ASP-DAC), 2010*, Taipei, Taiwan, pp. 175–180, January 2010.
48. X. Zhao and S.K. Lim. Through-silicon-via-induced obstacle-aware clock tree synthesis for 3D ICS. In *17th Asia and South Pacific Design Automation Conference (ASP-DAC), 2012*, Sydney, Australia, pp. 347–352, January 2012.
49. X. Zhao, J. Minz, and S.K. Lim. Low-power and reliable clock network design for through-silicon via (TSV) based 3D ICS. *IEEE Transactions on Components, Packaging and Manufacturing Technology*, 1(2):247–259, February 2011.
50. X. Zhao, S. Mukhopadhyay, and S.K. Lim. Variation-tolerant and low-power clock network design for 3D ICS. In *IEEE 61st Electronic Components and Technology Conference (ECTC), 2011*, Lake Buena Vista, FL, pp. 2007–2014, May 2011.
51. P. Zhou, Y. Ma, Z. Li, R.P. Dick, L. Shang, H. Zhou, X. Hong, and Q. Zhou. 3D-staf: Scalable temperature and leakage aware floorplanning for three-dimensional integrated circuits. In *IEEE/ACM International Conference on Computer-Aided Design, 2007. ICCAD 2007*, San Jose, CA, pp. 590–597, November 2007.

3 Detailed Electrical and Reliability Study of Tapered TSVs

Tiantao Lu and Ankur Srivastava

CONTENTS

ABSTRACT

This chapter investigates through-silicon via (TSV)'s tapering effect, an inevitable by-product of the nonideal dry reactive ion etching–based manufacturing, and its impact on TSV's electrical performance and reliability properties. For TSV's electrical performance, we show that TSV delay (estimated by the Elmore delay model) is not bidirectionally symmetric for a tapered TSV. Then we study several reliability properties of tapered TSV such as its current density and thermal mechanical stress distribution, TSV's self-heating effect, and its electromigration trend. We show that the current density and thermal mechanical stress distributions inside the TSV are more nonuniform in realistic tapered TSVs than ideal cylindrical TSVs, both of which lead to faster material fatigue and severe electromigration. Based on current, stress, and thermal simulation, we build a multiphysics simulation framework to quantify TSV's electromigration trend, which is capable of estimating TSV's electromigration lifetime. From this electromigration simulation framework, we can easily recognize the most vulnerable positions in a 3D structure, which would be valuable during TSV manufacturing and design optimization.

3.1 INTRODUCTION

Three-dimensional integrated circuits (3D-ICs) stack multiple planar wafers together and expand the design space to the vertical dimension. The through-silicon-via (TSV), the vertical connection between neighboring wafers, is the key enabling technology for 3D-ICs. TSVs provide a fast and power-efficient path, and replace the long and slow off-chip wires so that the signals can travel in the vertical direction freely.

One important aspect about the TSVs is the TSV tapering effect. TSV tapering effect refers to the empirical observation that the diameter of the TSV shrinks as it penetrates deeper into the silicon substrate, making the TSV resemble a cone rather than an ideal cylinder [2,5,8,14]. The tapering effect is inevitable during the deep reactive-ion etching (DRIE) process due to its intrinsic isotropic etching process, and the TSV tapering, even with a very small angle, significantly changes TSV's geometrical structure.

According to some previous experimental data [2,5,8,14], the TSV taper angle θ could vary from 85° to 90°. A mere 3° reduction of θ (see Figure 3.1f) from the ideal value of 90° causes significant variation between the top and bottom surface areas. For a TSV with top diameter 4 μm and a height of 30 μm, an 87° tapering will result in the bottom diameter to be only 0.856 μm. This causes the bottom surface area to be 4.6% of the top surface area. Hence even small deviations from the ideal cylindrical structure cause significant variation between the top and bottom area of cross-sections.

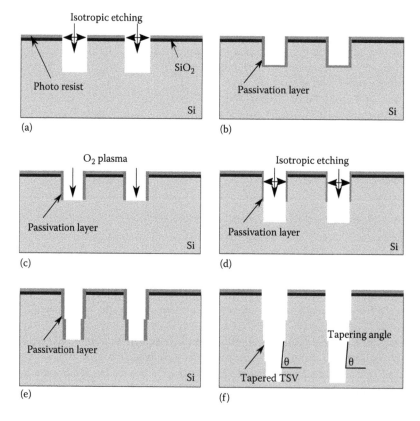

FIGURE 3.1 Illustration of the DRIE process flow, which consists of cycles of isotropic etching (a and d), passivation (b and e), and passivation removal (c). The isotropic etching will inevitably result in a tapered shape (usually around 87°). (a) Thermal oxidation, mask patterning, and the first isotropic etching; (b) deposit passivation layer; (c) passivation removal by oxygen plasma; (d) the second isotropic etching step; (e) deposit passivation layer—new isotropic etching/passivation cycle; and (f) resulting TSV's cross-section with a tapering angle.

Since the TSV's aspect ratio (height:radius) can be as high as 10:1, the variations between two areas of cross sections could lead to different electrical and mechanical behaviors in top and bottom parts.

The dramatic variation in TSV's structure significantly impacts its electrical (i.e., delay) and reliability (i.e., peak current, thermal mechanical stress, and electromigration) properties [9]. For example, current and thermal mechanical stress will not be uniformly distributed across the TSV. This nonuniformity can cause severe reliability problems, for example, (1) nonuniform mechanical distribution and (2) nonuniform electromigration (EM) degradation, since the velocity of the atomic movement strongly depends on the current, thermal mechanical stress, and local temperature.

For the reasons listed earlier, it is necessary to accurately predict the electrical behavior of tapered TSVs as well as to investigate the thermal and reliability properties of the 3D-ICs considering the tapering effect. Specifically in this chapter we make the following contributions:

1. We illustrate that the Elmore delay of tapered TSVs is different when signals travel in different directions, which makes it problematic to use them in bi-directional buses, where designers usually assume bi-directional delays are the same.
2. We set up a detailed simulation framework to quantify the impact of TSV tapering on DC current density distribution within the TSV structure. Our simulation results indicate that tapered TSV, especially at the narrow bottom end, possesses significantly higher peak as well as greater variation in current density.
3. Based on the DC current density simulation and materials' thermal property, we estimate the temperature increase brought by the heat generated by the current. We observed that the two wires (width × height × length = 2 μm × 2 μm × 10 μm) increase the temperature by 25 K, and the temperature is basically uniformly distributed inside the 3D structure.
4. We also set up simulations to investigate the impact of TSV tapering on thermal mechanical stress within and around the TSV structure. The result shows that the tapered TSV, especially at the narrow bottom end, requires much less keep out zone (KOZ) than the cylindrical TSV, which brings opportunities to fit more devices in a 3D structure.
5. Based on the DC current density, thermal, and the thermal mechanical stress simulation, we set up the TSV's EM simulation framework, which accounts for the interdependence between TSV's EM, current density, temperature, thermal mechanical stress, and atomic concentration. The simulation results show that the TSV tapering worsens the TSV's EM trend, mainly because the slim TSV cone at the bottom produces high variations of current density and thermal mechanical stress.

The remainder of this chapter is organized as follows. In Section 3.2, we introduce the mainstream TSV's fabrication method and explain the source of the tapering effect. In Section 3.3, we investigate the impact of TSV tapering on TSV's electrical performance. We use Elmore model to quantify the delay for both cylindrical and tapered TSVs. In Section 3.4, we set up a simulation to investigate the DC current distribution and the thermal distribution within both the cylindrical and the tapered TSV. Then in Section 3.5, TSV's thermal profile is used as an input to obtain the thermal mechanical stress distribution inside the TSV. Section 3.6 presents our simulation framework to combine the DC current, thermal, and stress profile to simulate the TSV's EM, and we show how the tapering effect affects EM trends in TSVs and neighboring wires. Section 3.7 summarizes the implications of TSV tapering on 3D CAD tools. Section 3.8 concludes this chapter.

3.2 TAPERED TSVs: AN INEVITABLE BYPRODUCT OF MANUFACTURING

3.2.1 TSV FABRICATION

TSV's fabrication includes DRIE, via filling by deposition of diffusion barrier and adhesion layers, metallization, wafer thinning and alignment, and bonding [4,12]. A typical via-first (via formation before CMOS process) can be summarized as follows. First of all conventional contact lithography is used to define the etching area. The wafer is subsequently etched (i.e., DRIE), and then thermal oxidation is performed to form the passivation layer. Once the passivation is grown, the seed layer is deposited using physical vapor deposition (PVD) and metal (Copper, Tungsten, etc.) is electroplated to fill the TSV. This is followed by chemical mechanical polishing (CMP), and then Cu is sputtered and patterned on both sides of the wafer as contacts for bonding neighboring layers. Finally bonding is performed under high temperature and pressure. There are other fabrication methods, but the complete literature survey of all different approaches for TSV fabrication has been omitted for brevity.

3.2.2 SOURCE OF TAPERING EFFECT

Much of the conventional researches on modeling and optimization of 3D ICs assume the TSVs to be perfectly cylindrical in shape. However, in reality, the fabricated TSVs end up with a conical tapered structure. This tapered structure comes from DRIE, the most common way to etch vias in silicon. The DRIE alternates repeatedly between etch mode and passivation mode to create vias with high aspect ratios, as shown in Figure 3.1. The DRIE manufacturing process starts with thermal oxidation, photoresist deposition, patterning, and the first isotropic etching (Figure 3.1a). This results in a trench in the silicon substrate where the TSV is to be located. Then a passivation layer is formed on the surface of the TSV trench. This is followed by removing the passivation layer at the bottom to reveal the TSV trench. The bottom of this trench is then etched further toward the silicon substrate. The passivation/etching process is iterated for several hundreds of times till the desired TSV height is reached, as illustrated in Figure 3.1f. During each etching mode, the sidewall is protected by the passivation layer so that the regions, which has already been etched, won't be affected by the following etching process. However, each passivation mode reduces the TSV diameter, therefore, as the TSV trench goes deeper, the effective diameter of the trench becomes less. This is the primary reason that when the aspect ratio (height:radius) is high, a tapered structure is what DRIE process usually ends up with. Another obvious characteristic of the repeated etching–passivation process is that the sidewall is usually rough and scalloped. The roughness of the sidewall will affect the copper filling process, but in this chapter, for simplicity, we assume the copper is perfectly sputtered and there's no void in the TSV cone.

3.2.3 TSV TAPERING IN REAL FABRICATION

Such an iterative manufacturing process inherently results in a tapered TSV structure, which has been reported in many experimental works [2,5,8,14]. The SEM pictures in [2] reveals a tapered TSV with a top diameter of 10.5 μm, a height of 51.4 μm, and a tapering angle of 89°. Reference 5 shows a tapered TSV with top diameter ≈5.8 μm, height ≈22.2 μm, bottom diameter ≈3.2 μm, and tapering angle ≈87°.

Some other works have shown that TSV's degree of taper is more pronounced in thinner TSVs with smaller diameters and higher aspect ratios [10]. The authors of [7,12] indicate that the inherent tapering of TSVs during manufacturing helps improve the manufacturing yields of TSVs since it helps in conformal deposition of isolation dielectric, copper diffusion barrier metallization, and copper seed metallization over the sidewalls of the deep silicon vias. A slight taper also enables a void-free copper via electroplating [7]. According to the simulation results from [13], tapered-shape TSV has the smallest reflection noise and signal loss, compared to some other TSV shapes, such as rectangular, circular, and annular shapes.

In conclusion, the tapering effect causes TSVs to resemble a cone with the bottom region being thinner than the top region, and TSV tapering is a natural, unavoidable consequence of TSV fabrication, and in some circumstances the TSV tapering helps improve the manufacturability and yields.

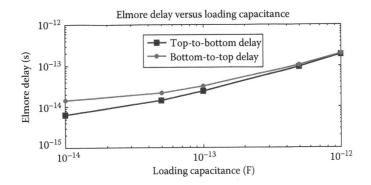

FIGURE 3.2   TSV's Elmore delay in both directions. $\theta = 87°$, $L = 30$ μm, $r_1 = 2$ μm, $r_2 = 0.4$ μm.

3.3   BI-DIRECTIONAL DELAYS OF TAPERED TSV

TSV tapering impacts its delay. Cylindrical TSVs, being symmetric, have the same delay in both directions. This allows us to use them on bi-directional buses with the same clock. However the tapering effect causes delays for both directions to be different. This can be verified using the Elmore delay model as follows:

$$Delay_{a \to b} = \int_0^L \frac{\rho \cdot (c_0(L-x)+C_L)dx}{\pi(r_1 - x\cot(\theta))^2} \tag{3.1}$$

$$Delay_{b \to a} = \int_0^L \frac{\rho \cdot (c_0(L-x)+C_L)dx}{\pi(r_2 + x\cot(\theta))^2} \tag{3.2}$$

Equations 3.1 and 3.2 represent the Elmore delay in the TSV from top (with larger diameter) to narrow bottom and from the narrow bottom to top, respectively. The Elmore delay equals the integral of each infinitesimal TSV segment's resistance times the downstream capacitance. Here a and b are two port of TSV (a is top and b is bottom), ρ is resistivity, L is the length of TSV, r_1 and r_2 are radius of TSV at top and bottom, respectively, ($r_1 > r_2$), c_0 is the unit capacitance per length, and C_L is the loading capacitance.

Figure 3.2 illustrates the delay in different directions. It shows that bottom-to-top delay is consistently greater than top-to-bottom delay. As the TSVs are the crucial signal path for 3D-ICs, the loss of bi-directionality of TSV delay can significantly impact how 3D-ICs are designed and optimized.

3.4   DC CURRENT DENSITY AND THERMAL SIMULATION FOR TAPERED TSV

Excessively high DC current is one of the primary causes for many reliability problems such as overheating and EM. In order to analyze and quantify the DC current density distribution in tapered TSV and cylindrical TSV (as comparison), we use ANSYS WORKBENCH as a simulation tool. The structure of the tapered TSV we used as a test case is shown in Figure 3.3a. This test case consists of three components: (1) A tapered TSV cone (4 μm in top diameter, 0.856 μm in bottom, 87° of tapering angle, and 30 μm in height), (2) two landing pads (length × width × height = 5 μm × 5 μm × 0.2 μm), and (3) two signal wires (length × width × height = 0.2 μm × 0.2 μm × 10 μm). One current source is inserted at the top left corner, feeding 0.4 mA current, and one current sink is inserted at the bottom right corner, absorbing the 0.4 mA current.

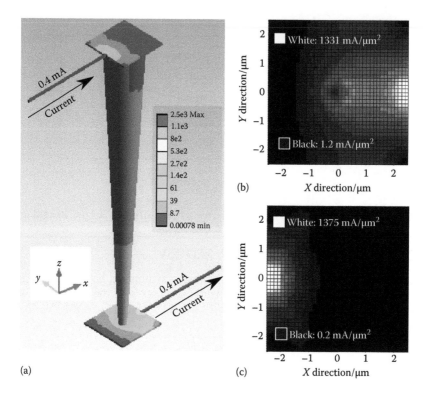

FIGURE 3.3 (a) Current density (mA/μm²) in a tapered TSV and signal wires. (b) The current density distribution at the bottom pad. (c) The current density distribution at the top pad.

Figure 3.3a shows the 3D view of current density distribution in a tapered TSV's structure. Significantly, high current density value is observed approximately 3 μm into the TSV from both the top and bottom interfaces in Z direction. At the bottom pad, as illustrated in Figure 3.3b, the current not only crowds at the transition regions between the pad and the wire, but also at the interface between the TSV and the bottom pad. The current density distribution at the top pad only peaks at the transition region.

A similar cylindrical TSV case is simulated for comparison. As shown in Figure 3.4a, the tapering angle is set to be 90°. The top and bottom diameters are both 4 μm, and height remains to be 30 μm. The size of the landing pads and signal wires is the same as in the tapered case (5 μm × 5 μm × 0.2 μm, and 0.2 μm × 0.2 μm × 10 μm, respectively). Current sources are again inserted at the top left corner and bottom right corner, with 0.4 mA current.

In the cylindrical case, the magnitude of current density is again plotted in 3D view (Figure 3.4a), within the bottom pad (Figure 3.4b) and within the top pad (Figure 3.4c). Both the top pad and the bottom pad have only one current density peak, which is at the transition region between the pad and the wire. We notice that current density distribution at the top pad is very similar to the top pad in a tapered TSV's structure, while the bottom pad isn't. The bottom pad in the tapered case has two current density peaks, leading to higher nonuniformity, which we'll illustrate in later sections, is one of the primary reasons for tapered TSV's having high EM trend.

We also calculate the Joule heat produced by the DC current, and then use that as the heat sources to set up the heat conduction equations, for both tapered and cylindrical TSVs. It turns out so that the TSV tapering has little impact on the temperature increase. The primary Joule heating effect happens in the narrower signal wires. In fact, in both cases, the temperature rises from room temperature (298 K) to 323 K, increased by 25 K. We also notice that the temperature is basically uniformly distributed inside the structure. This is due to the fact that the heat conductivity of the

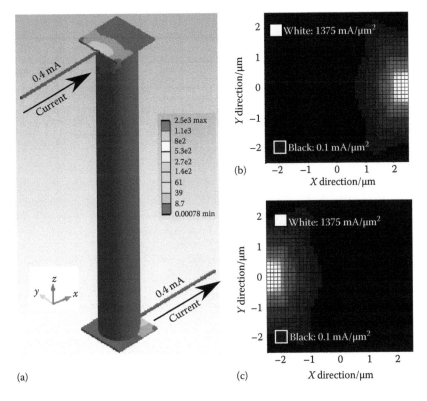

FIGURE 3.4 (a) Current density distribution (mA/μm²) in a cylindrical TSV and signal wires. (b) The current density distribution at the bottom pad. (c) The current density distribution at the top pad.

copper is high enough to spread the Joule heat around. In later sections, we'll use 323 K as the working temperature of TSVs.

3.5 THERMAL MECHANICAL STRESS SIMULATION FOR TAPERED TSV

During the IC fabrication process, TSVs endure many thermal cycles. For example, annealing is one of the common steps to remove the lattice damages caused by ion implantation and activate the implanted ions as well. Another example is the wafer bonding process, where high temperature is needed in order to obtain reliable 3D stack. When the whole chip cools down, because of the significant mismatch between the different materials' coefficients of the thermal expansions (CTEs), significant thermal mechanical stress remains inside the material. These stress accelerate material fatigue, rough the silicon's surface, and shorten the circuit's lifetime. A very simple solution to enhance the device's reliability is to keep the device away from TSV's KOZ. However, large KOZ in the routing area becomes routing obstacles while KOZ in the placement region may potentially enlarge the chip's total area. In this section, we first analyze and quantify the magnitude of the thermal mechanical stress in the tapered TSV, using a finite element analysis tools called COMSOL [1], and then, from the wire's EM point of view, evaluate the impact of TSV's tapering on 3D-IC's reliability properties.

3.5.1 PHYSICAL DIMENSIONS AND SIMULATION SETUP

Figure 3.5a shows the geometry containing the tapered TSV we use for thermal mechanical simulation. It consists of three layers of silicon, a tapered TSV, and bonding materials such as interlayer dielectric (ILD), silicon dioxide, and benzocyclobutene (BCB). Material properties used in stress simulation is summarized in Table 3.1, similar to [11].

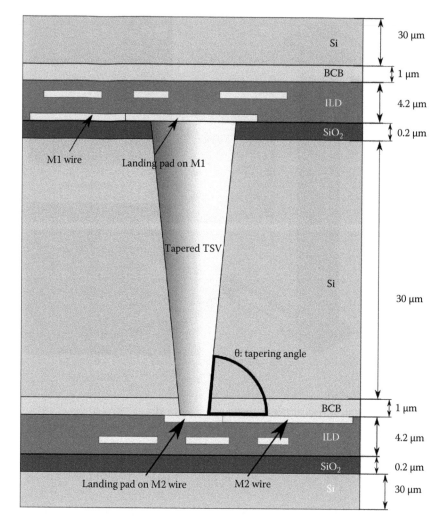

FIGURE 3.5   Tapered TSV's geometry for thermal mechanical stress simulation.

TABLE 3.1

Material Property of TSV Structure

	Elastic Modulus (GPa)	Poisson's Ratio	CTE (K⁻¹)
Silicon	162	0.28	3.05e−6
Copper	111.5	0.343	1.77e−5
Silicon oxide	71.7	0.16	5.1e−7
ILD	9.5	0.3	2.0e−5
BCB	6.1	0.35	3.3e−5

Source: Pak, J.S., Pathak, M., Lim, S.-K., and Pan, D.Z., Modeling of electro-migration in through-silicon-via based 3D IC, *2011 IEEE 61st Electronic Components and Technology Conference (ECTC)*, pp. 1420–1427, Lake Buena Vista, FL, © 2011 IEEE.

3.5.2 Impact of TSV Tapering on KOZ

Mechanical stress can affect devices in many ways. High magnitude of stress may induce cracks or voids in silicon, or change the devices' mobility thus driving current as well. High variations in stress may also lead to significant EM failures in interconnects. The primary purpose of KOZ is to guarantee that the device is free from unpredictable mechanical stress. The size of the KOZ is usually defined by a stress threshold such that stress magnitude outside the KOZ is smaller than that threshold. In this chapter, for simplicity, we use Von Mises stress as a measurement of thermal mechanical stress. We also assume that the KOZ is in the shape of a circle, and we define the threshold to be 200 MPa. Since the early simulation results show that the temperature gradient inside the TSV structure is not remarkable, we assume the temperature distribution is uniform. We set the annealing temperature to be 573 K, and after annealing, the TSV is cooled to a normal working temperature, which is 323 K.

The 3D view of the stress distribution of both tapered and cylindrical TSVs is shown in Figure 3.6. Significant high stress values are observed in the contact regions between the copper cone and the silicon substrate (silicon substrate is not shown for readability). This is due to the fact that the CTE of copper (1.77×10^{-5} K^{-1}) is significantly larger than silicon (3.05×10^{-6} K^{-1}), and when the whole 3D structure is cooled down from the annealing temperature, high compressive stress is formed inside the copper cone.

Now let's quantify the size of the KOZ, using 200 MPa Von Mises Stress as a threshold. In a tapered TSV structure, the radius of KOZs are 3.6 μm at the top pad (Figure 3.7a) and 0.87 μm at the bottom pad (Figure 3.7b). That will require an area of 98.5 and 3.8 μm² at the top and the bottom, respectively.

In a cylindrical TSV structure, however, the radius of KOZs are 3.6 μm at the top pad (Figure 3.8a) and 2.65 μm at the bottom (Figure 3.8b). The area requirement at the top and bottom pad is of 98.5 and 22.1 μm², respectively. From the simulation result we can clearly see that the tapering does not affect the stress distribution at the top pad, but for the bottom pad, it saves 83% of area, which could be used for wire routing and gate placement.

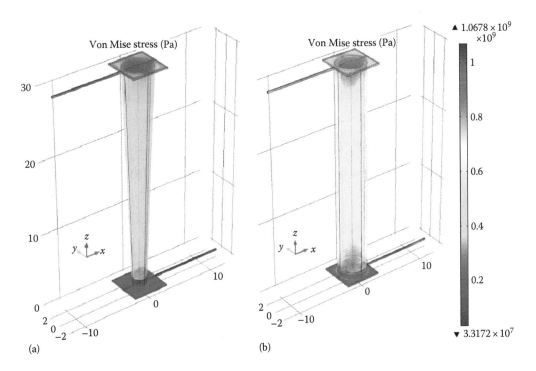

FIGURE 3.6 Thermal mechanical stress distribution in (a) tapered TSV and (b) cylindrical TSV.

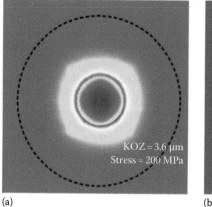
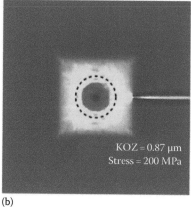

FIGURE 3.7   Thermal mechanical stress distribution in tapered TSV at (a) top pad and (b) bottom pad. KOZ is 3.6 and 0.87 μm, respectively. The area occupied by the TSV is 98.5 μm² at the top and 3.8 μm² at the bottom. The size of both top and bottom pad is 5 μm × 5 μm, respectively. TSV's radius is 2 μm at the top and 0.4 μm at the bottom.

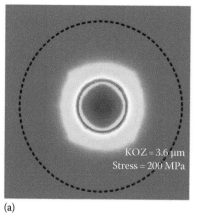
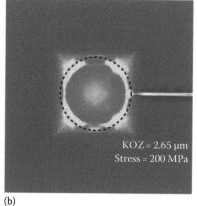

FIGURE 3.8   Thermal mechanical stress distribution in cylindrical TSV at (a) top pad and (b) bottom pad. KOZ is 3.6 and 2.65 μm, respectively. The area occupied by the TSV is 98.5 μm² at the top and 22.1 μm² at the bottom. The size of both top and bottom pad is 5 μm × 5 μm, respectively. TSV's radius is 2 μm at both the top and the bottom.

3.6   ELECTROMIGRATION SIMULATION FOR TAPERED TSV

Electromigration (EM) is the phenomenon whereby atoms migrate overtime, forming hillocks or voids, which eventually results in short circuit or open circuit. The classic Black equation [3] is based on the empirical observation that the interconnect's mean time to failure (MTTF) is related to current density. However, many later works such as [6] have shown other contributing forces to TSV's EM, such as the stress gradient, the thermal gradient, and the atomic concentration gradient. In this section, all these four driving forces are considered. More specifically, we use the following mass transport equation to investigate the EM inside a tapered TSV structure and neighboring wires.

3.6.1 Equations for TSV Electromigration

$$\frac{\partial c}{\partial t} + \nabla \cdot \vec{q} = 0 \tag{3.3}$$

$$\vec{q} = -D\nabla c + \frac{Dc\vec{j}e\rho Z}{kT} + \frac{Dc\Omega}{kT} \cdot (\nabla(\sigma_m)) + \frac{DcQ^\star}{kT} \cdot \frac{\nabla(T)}{T} \tag{3.4}$$

where
  c is atomic concentration
  t is time
  D is diffusivity
  j is current density
  e is electron charge
  Z is effective charge
  Ω is atomic volume
  σ_m is hydrostatic stress
  $Q^\star$ is heat of transport
  ρ is resistivity
  k is Boltzmann's constant
  T is temperature

Equation 3.4 shows all four driving forces of EM: atomic concentration gradient, current density, thermal mechanical stress gradient, and temperature gradient. The unknown variable, c, is very difficult to solve analytically. So we use COMSOL's numerical solver to solve the problem efficiently.

In this chapter, we define the initial atomic concentration to be 1.5×10^{28} cm^{-3}. Again, we assume that the temperature gradient inside the TSV structure is neglectable. We set the environment temperature T to be 323 K. MTTF is defined as the amount of time elapses when the actual atomic concentration has 5% deviation from the initial value.

3.6.2 Electromigration Simulation Results for Tapered TSV

Based on EM equations (Equations 3.3 and 3.4), DC current density simulation results in Section 3.4, and thermal mechanical stress simulation results in Section 3.5.2, we use COMSOL's solver to evaluate MTTF values at each position. Especially, we choose several representative points to illustrate our results, as shown in Figure 3.9. Note that the cylindrical TSV is again shown as a comparison [11].

From Table 3.2 we can see that the TSV tapering has worsened the overall EM trends. Specifically let's first look at the EM trend at the bottom pad. For cylindrical TSV's bottom pad, the position that has the most EM degradation is Position J, which is the center point between the TSV cone and the bottom pad. However, for tapered TSV's bottom pad, the most EM vulnerable position becomes I and K, both of which are located at the transition region between the TSV cone and the bottom pad. This difference is due to the different stress and DC current density distributions at the bottom pad of the TSV structure. At the top pad, Position D, which is the center of the top pad, is always the most unreliable position, in both tapered and cylindrical TSVs. It is worth mentioning that the regular wires, carrying much high current density than the TSV cone, are much more reliable than the TSV, in both tapered and cylindrical cases. The reason is twofold: (1) The speed of atomic migration doesn't solely depend on the magnitude of the DC current density, but also the thermal mechanical stress; (2) The stress inside the TSV, produced by the CTE mismatch between TSV's filling materials and silicon substrate, is much more pronounced than the stress distribution around the wires.

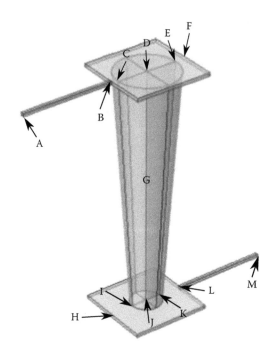

FIGURE 3.9   Points of interest in the EM simulation. Refer to Table 3.2 for MTTF values at points A–H.

TABLE 3.2
Normalized MTTF for Tapered TSV, Comparing with Cylindrical TSV

Position	Normalized Mean Time to Failure						
	A	B	C	D	E	F	G
Tapered TSV	>10	0.45	0.05	0.04	0.08	0.3	0.4
Cylindrical TSV	>100	3.1	3.2	0.9	2.0	1.9	0.6
Position	**H**	**I**	**J**	**K**	**L**	**M**	
Tapered TSV	1.4	0.03	0.05	0.03	2.0	>10	
Cylindrical TSV	2.0	0.7	0.2	17.2	32.5	>100	

Note: Refer to Figure 3.9 for the definition of each position.

3.7   IMPLICATIONS OF TSV TAPERING ON CAD TOOLS

Tapering is an inevitable result of the nonideal fabrication process. It brings lots of challenges and opportunities to the CAD tool designers. We summarize some of the implications as follows:

1. TSV's bi-directional delay is not symmetric. Delay in one direction is significantly larger than the other. When TSV's are used as high-speed bi-directional buses, a more sophisticated model is needed. Even the TSV is used solely as one-way channel, such as a clock tree, the delay of the TSV depends on the bonding strategy, such as Face-to-Face (F2F) and Face-to-Back (F2B) bonding, and the relative position of the clock source and the clock destination.

2. TSV tapering changes the DC current distribution, especially at the bottom pad. Large amount of current is observed when the diameter of the TSV becomes less. Circuit designers may need to come up with methods such as wire sizing to make sure the peak current inside the TSV is under control.

3. TSV tapering also changes the thermal mechanical stress distribution. The amount of impact, in terms of device's performance and reliability, needs to be explored. However, in the sense of the KOZ, our simulation result shows that we don't need as much area when we assume the TSV is perfectly cylindrical. TSV's smaller diameter at the bottom helps shrink the KOZ area, which is beneficial to wire routing and gate placement.

4. Although TSV carries less current density than normal planar wires, high thermal mechanical stress makes them less robust in the sense of EM. The stress distribution of tapered TSV is even less favorable, and our results show that the tapered TSV itself is usually the most vulnerable part in a 3D-IC structure.

3.8 CONCLUSION

In this chapter, we investigated the TSV's tapering effect, which is an inevitable byproduct of the TSV's fabrication process, and explored its significant impact on TSV's electrical, thermal, stress, and reliability properties.

ACKNOWLEDGMENT

The authors acknowledge that this work has been funded by NSF grant 0917057.

REFERENCES

1. Comsol multiphysics. http://www.comsol.com/. Accessed: 2013-08-10.
2. Imec's scientific report 2010, TSV processing. http://www.imec.be/ScientificReport/SR2010/2010/1159081.html. Accessed: 2013-08-10.
3. J.R. Black. Mass transport of aluminum by momentum exchange with conducting electrons. In *Proceedings of Reliability Physics Symposium, 2005*, San Jose, CA, pp. 1–6, 2005.
4. C. Huyghebaert, J. Van Olmen et al. Cu to cu interconnect using 3D-TSV and wafer to wafer thermo-compression bonding. In *2010 International Interconnect Technology Conference (IITC)*, Burlingame, CA, pp. 1–3, 2010.
5. F. Inoue, T. Shimizu, H. Miyake, R. Arima, T. Ito, H. Seki, Y. Shinozaki, T. Yamamoto, and S. Shingubara. Highly adhesive electroless barrier/cu-seed formation for high aspect ratio through-Si vias. *Microelectronic Engineering*, 106:164–167, June 2013.
6. J.P. Jing, L. Liang, and G. Meng. Electromigration simulation for metal lines. *Journal of Electronic Packaging*, 132, 2010.
7. B. Kim, C. Sharbono, T. Ritzdorf, and D. Schmauch. Factors affecting copper filling process within high aspect ratio deep vias for 3D chip stacking. In *Proceedings of 56th Electronic Components and Technology Conference, 2006*, San Diego, CA, 6pp, 2006.
8. A. Klumpp, P. Ramm, and R. Wieland. 3D-integration of silicon devices: A key technology for sophisticated products. In *Design, Automation Test in Europe Conference Exhibition (DATE)*, pp. 1678–1683, Dresden, Germany, 2010.
9. T. Lu and A. Srivastava. Detailed electrical and reliability study of tapered TSVs. In *2013 IEEE International 3D Systems Integration Conference (3DIC)*, San Francisco, CA, pp. 1–7, October 2013.
10. R. Nagarajan, L. Ebin et al. Development of a novel deep silicon tapered via etch process for through-silicon interconnection in 3-D integrated systems. In *Electronic Components and Technology Conference*, San Diego, CA, p. 5, 2006.
11. J.S. Pak, M. Pathak, S.-K. Lim, and D.Z. Pan. Modeling of electro-migration in through-silicon-via based 3D IC. In *2011 IEEE 61st Electronic Components and Technology Conference (ECTC)*, Lake Buena Vista, FL, pp. 1420–1427, 2011.

12. M. Puech, J.M. Thevenoud et al. Fabrication of 3D packaging TSV using DRIE. In *Design, Test, Integration and Packaging of MEMS/MOEMS, 2008*, Nice, France, pp. 109–114, 2008.
13. Z. Xu and J.-Q. Lu. High-speed design and broadband modeling of through-strata-vias (TSVs) in 3D integration. In *IEEE Transactions on Components, Packaging and Manufacturing Technology*, 1:154–162, 2011.
14. X. Zhang, T.C. Chai et al. Development of through silicon via (TSV) interposer technology for large die (21 × 21 mm) fine-pitch cu/low-k fcbga package. In *Electronic Components and Technology Conference*, San Diego, CA, pp. 305–312, 2009.

4 3D Interconnect Extraction

Sung Kyu Lim

CONTENTS

ABSTRACT

Through-silicon-via (TSV) introduces new parasitic components into 3D ICs. They penetrate silicon substrate and are vulnerable to any signal noise through the substrate. They also introduce a large capacitance and affect the timing, power, and noise at the full-chip level. Thus, it is essential to extract TSV-related parasitics in detail to provide accurate signed-off timing and power analysis and verification. Due to the prohibitive time complexity involved in extracting parasitics from thousands of TSVs and millions of nets in a typical design, 3D-IC parasitic extraction has to be highly efficient while maintaining acceptable accuracy.

In this chapter, we model and extract two major types of TSV-related parasitics, namely, TSV-to-TSV and TSV-to-Wire coupling capacitance. TSV-to-TSV coupling, a major coupling element inside silicon substrate, can be modeled and calculated with mathematical equations. We also learn that various silicon-induced phenomena significantly affect the extraction results. TSV-to-Wire coupling, which is strongly depended on geometry structures, cannot be modeled by closed-form formulas easily. Thus, we learn a rule-based extraction, which is extended from the traditional 2D wire extraction. This method solves the computation problem with a good estimation accuracy. We also learn design optimization techniques that alleviate TSV-to-TSV and TSV-to-Wire coupling.

4.1 INTRODUCTION

3D-IC is a promising solution to extend Moore's Law. Through-silicon-via (TSV) is widely used for vertical interconnection in 3D-IC and provides high-density connections between dies. This allows 3D-ICs to obtain ultra-high density and bandwidth at much lower power consumption for data transmission. However, TSVs also introduce new parasitic elements to 3D-ICs. Figure 4.1 shows a 3D-IC structure, where two dies are bonded in face-to-back fashion with via-first technology. Unlike other small metal vias, TSVs are hundred times larger and buried inside the silicon substrate close to transistors and wires. This makes them very sensitive to any noise coupled through the silicon substrate.

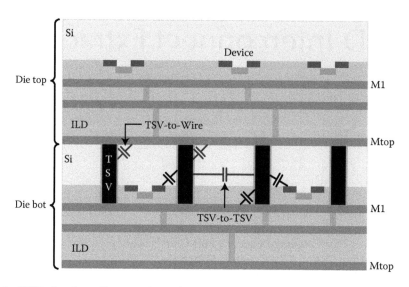

FIGURE 4.1 TSV-related coupling capacitance in a two-die 3D-IC structure.

The TSV coupling not only is a threat to the signal integrity and the logic functionality, but also degrades the delay and power benefits since they introduce extra capacitances and inductances.

TSV parasitics consist of several electrical components: resistance, capacitance, and inductance. Parasitic extraction is essentially an electrical modeling of the electrical and magnetic fields. This requires solving the Maxwell's equations, and it is difficult to find closed-form formulas for all kinds of geometrical structures found in 3D-ICs. However, with a simplified geometry, mathematical solutions are possible. Numerical solutions, which are often used when the geometry is complex, can calculate the resistance quickly. TSV inductance is one critical component when the TSV size is large, and signal switching frequency is high. However, for most digital systems, where the clock frequencies are below 5 GHz, the inductance impact on timing, power, and noise are relatively small compared with capacitive components. Thus, in this chapter, we focus on the extraction of capacitive TSV components.

Two major capacitive coupling types exist that involve TSVs: (1) TSV-to-TSV coupling is originated from the field interactions between two neighbor TSVs. (2) TSV-to-Wire coupling is caused by the field interactions between a TSV and its neighbor wires. Since TSVs are fabricated with a thin liner oxide to isolate from the silicon substrate, this TSV-liner-substrate forms an MOS structure that introduces a depletion region around TSV. It creates a depletion region capacitor between a TSV and substrate on all sides. Also, the doped silicon substrate has a finite resistance and serves as a discharging path when the TSV is switching. These silicon-induced coupling phenomena affect the TSV parasitic significantly and thus need to be considered in detail to obtain an accurate extraction results. On the other hand, electrical field (E-field) interactions also affect the parasitics between TSV and wires. Depending on the technology structure and the layout, TSV-to-Wire coupling capacitance differs significantly. Thus, these E-field-induced effects need to be captured as much as possible to match the extraction results with silicon measurements.

In order to alleviate TSV coupling noise, several noise reduction methods have been proposed. Shielding techniques such as ground TSVs [8,14] and ground plugs [5] are used to protect TSVs from TSV-to-TSV coupling noise. Ground TSV also has a similar impact on TSV-to-Wire coupling due to E-field shielding. Thus, they can be used as a TSV-to-Wire coupling reduction technique as well. Measurement results from [15] show that the H-shaped TSV provides better shielding compared with guard rings. However, these techniques either consume a large silicon area and reduce placement utilization, or require a special fabrication technology and thus increase the chip cost.

In this chapter, we study various modeling and extraction techniques for TSV-to-TSV and TSV-to-Wire extraction. For TSV-to-TSV coupling, we discuss a multi-TSV model [12,14], where the E-field interactions around TSVs are discussed in detail for accurate 3D interconnect extraction. For TSV-to-Wire coupling, we discuss a rule-based pattern-matching algorithm [11] for fast and accurate full-chip extraction. We also study several E-field sharing impacts that affect TSV-to-Wire coupling. Furthermore, we study the TSV parasitics' impact on full-chip timing, power, and noise using GDSII 3D IC layouts. Lastly, we study several design techniques that alleviate TSV-to-TSV and TSV-to-Wire coupling noise.

4.2 TSV-TO-TSV COUPLING EXTRACTION

In this section, we discuss two TSV-to-TSV coupling models and several E-field and silicon impacts on TSV-to-TSV coupling.

4.2.1 Two-TSV vs. Multi-TSV Model

The traditional two-TSV model used in [4,8] is based on a pair of parallel wires shown in Figure 4.2. By modeling TSVs as a cylindrical wire, the two-TSV model consisting a pair of resistor and capacitor, is used to model the E-field coupling between two TSVs. The TSV coupling elements connect to the cylindrical wire through a TSV oxide capacitor. TSV inductance can be ignored for most of the digital applications, but need to be included in the high-frequency operation. In this model, assuming a perfect cylindrical shape of the TSV oxide, the TSV oxide capacitance is calculated by

$$C_{ox} = \frac{\pi \varepsilon_{SiO_2} L_{TSV}}{\ln((R_{TSV} + T_{ox})/R_{TSV})} \tag{4.1}$$

where L_{TSV}, R_{TSV}, and T_{ox} are the TSV height, the TSV radius, and liner thickness, respectively. Since TSVs are buried inside the silicon substrate, the silicon resistance and capacitance have a relationship defined by

$$R_{Si} C_{Si} = \frac{\varepsilon_{Si}}{\sigma_{Si}} \tag{4.2}$$

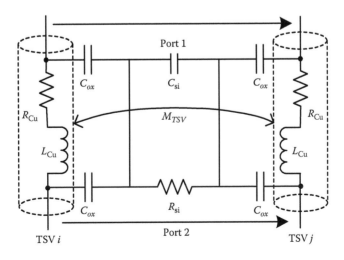

FIGURE 4.2 A traditional circuit model for two-TSV coupling.

assuming a homogenous and a uniformly doped substrate. The capacitance between two parallel cylindrical wires is given by

$$C_{\text{Si},ij} = \frac{\pi \varepsilon_{\text{Si}} L_{TSV}}{\ln\left\{ P_{ij}/(2R_{TSV}+T_{ox}) + \sqrt{(P_{ij}/(2R_{TSV}+T_{ox}))^2 - 1} \right\}} \tag{4.3}$$

where P_{ij} is the distance between TSV i and TSV j. On the full-chip level, by choosing all pairs of TSVs, the TSV-to-TSV coupling is calculated without considering impacts from neighbor TSVs. This model is accurately matching with the measurement results of a TSV-pair and it is a good approximation even if the impact of devices and metal layers are considered since they affect only part of the E-field between TSVs.

The two-TSV model assumes that there are no other objects, which either blocks the coupling path between TSVs or shares the E-field around. This assumption is satisfied only when there are two TSVs buried in the silicon substrate. Even if the impact of devices and metal layers are considered, only part of the E-field between TSVs are affected. Therefore, the two-TSV model is accurate when a small number of TSVs are placed and the TSV pitch is large. However, this assumption is no longer valid if there are other TSVs around, which shares E-fields all the way through silicon substrate. In real 3D designs, there are many TSVs around and all of them affect the E-field distribution within that area. Ignoring the multi-TSV effect, the traditional two-TSV model overestimates the coupling between a TSV pair, and a new model which takes all the neighboring TSVs into account is desired for the full-chip analysis.

Let us look at an example where multiple TSVs are closely placed near each other. From the two-TSV model, the total coupling capacitance on victim TSV increases linearly with number of aggressor TSVs if the coupling path distances are the same for all aggressors. However, from a field-solver simulation results, when there are 4 and 8 aggressor TSVs around the victim, the total coupling capacitance merely increases to 1.94× and 1.99× compared with two-TSV case, respectively. As shown in [14], there is an upper bound of the coupling capacitance calculated by the capacitance of coaxial wire, given by the following formula:

$$C_{\text{max}} = \frac{2\pi \varepsilon_{\text{Si}} L}{\ln(P/r)} \tag{4.4}$$

where P and r are the outer and inner radius of the coaxial wire. According to Equation 4.4, even when there are many aggressor TSVs, the total coupling capacitance cannot be more than 2.26× of the coupling between a TSV pair whose distance is P. But if two-TSV model is used, an 8-TSV-aggressor structure results in 8× coupling capacitance to the victim in total. Thus, if a two-TSV model is used for multiple TSVs, the coupling capacitance is overestimated significantly.

To consider the E-field sharing from multiple TSVs, a multi-TSV model is presented in [14]. By assuming the aggressor TSV array as the multi-conductor transmission line within silicon substrate and the victim TSV as the ground signal [3], the TSV-array inductance matrix $[L_{\text{Si}}]$ is computed by applying the following formula:

$$L_{\text{Si},ij} = \begin{cases} \dfrac{\mu_{\text{Si}} L_{TSV}}{\pi} \ln\left[\dfrac{P_{i0}}{R_{TSV}+T_{ox}} \right] & \text{when } i = j \\[3mm] \dfrac{\mu_{\text{Si}} L_{TSV}}{2\pi} \ln\left[\dfrac{P_{i0} P_{j0}}{P_{ij}(R_{TSV}+T_{ox})} \right] & \text{when } i \neq j \end{cases} \tag{4.5}$$

where P_{ij} is the distance between aggressor TSV i and j and the victim is labeled 0. Note in this formula, unlike the two-TSV model, not only the distances between aggressor and the victim are considered, but also the distances between aggressors are considered. This makes it useful for any TSV placement style even when TSVs are not placed on a regular grid. By using the relation of homogeneous material between the capacitance matrix and the inductance matrix [10], the capacitance matrix for TSV array is calculated by

$$[C_{Si}] = \mu_0 \varepsilon_{Si} L_{TSV}^2 [L_{Si}]^{-1} \tag{4.6}$$

Since we focus on the coupling on victim TSV, only the coupling components between aggressor TSV i and the victim is used, which is given by

$$C_{Si,i0} = \sum_{k=1}^{N} C_{Si,ik} \tag{4.7}$$

And the silicon coupling resistance can be obtained from Equation 4.2. The original model from [3] used all of the coupling components between TSVs, which is not a feasible solution in full-chip level. For example, an array of 100 TSVs leads to more than 20,000 RC components in the model. Therefore, to reduce the number of electrical components, the coupling paths between aggressor TSVs, which is given by $C_{Si,ij}$ ($i \neq j$), is ignored for the full-chip timing and noise calculation. This is because the model is applied in the full-chip level by choosing each TSV as a victim. The coupling paths between aggressor TSVs have a smaller impact since they do not directly affect the voltage waveform on the victim TSV. To verify our model, many test cases are generated containing up to 8 TSVs and transient SPICE simulations are performed. Ten layouts are generated for each sample cases. Model calculated using our equations is compared with extraction results from field solver in frequency domain, and the maximum error on coupling S-parameter is reported in Table 4.1. The results show that for all tested layouts, the coupling parameter error of our extracted model is less than 0.02 dB, and we conclude that our multi-TSV model accurately handles multi-TSV effects and is scalable with different TSV dimensions.

TABLE 4.1
Coupling S-Parameter Comparison between Our Model and 3D Solver

TSV Radius	TSV Height	TSV Liner Width	Max Error
2	30	0.2	0.016
		0.5	0.011
2	60	0.2	0.017
		0.5	0.012
4	30	0.2	0.015
		0.5	0.014
4	60	0.2	0.018
		0.5	0.013

Note: TSV dimensions are in μm and error in dB.

4.2.2 Silicon and E-Field Distribution Impacts

In this section, the effect of silicon depletion region on TSV coupling is discussed. TSV, usually made of copper or tungsten, is insulated from the silicon substrate with an oxide liner, which together form an MOS structure. Due to the nonlinearity of the MOS capacitance, many previous works [8,14] ignore the depletion region around the TSV and assume that the oxide capacitance is the only part that contributes to the TSV capacitance. If the depletion region is considered, the oxide thickness T_{ox} in Equation 4.5 should be replaced by $(T_{ox} + W_{dep})$. The following equation is used to calculate TSV MOS capacitance, which is the serious combination of oxide capacitance (C_{ox}) and depletion capacitance (C_{dep}):

$$C_{MOS} = \frac{C_{ox}C_{dep}}{C_{ox} + C_{dep}} = \frac{\pi\varepsilon_{ox}L_{TSV}}{\ln((R_{TSV} + T_{ox} + W_{dep})/R_{TSV})} \tag{4.8}$$

where W_{dep} needs to be calculated based on the MOS structure, substrate doping, and voltage. From Figure 4.3, the MOS capacitance can be as low as 36% of the oxide capacitance, thus it is

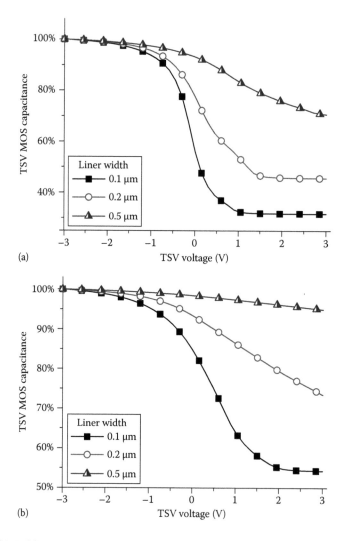

FIGURE 4.3 TSV MOS capacitance with substrate doping of (a) 1015/cm³ and (b) 1016/cm³.

overestimated if the depletion region is ignored. This results in a pessimistic estimation on TSV-induced delay and noise. Another observation is that the MOS capacitance becomes smallest when TSV voltage is tied at VDD while it reaches maximum value when TSV is grounded.

Another impact comes from the conductive silicon substrate. Since TSV is buried in doped silicon substrate, the substrate impact needs to be considered. Previous model assumed the silicon substrate is a floating net. This assumption is not appropriate since most designs ground the substrate using substrate contacts. Even though each TSV has a KOZ, there is a finite impedance from substrate around the TSV to the ground node. Whenever a victim TSV is affected by the aggressor, charges will accumulate at silicon–oxide interface. With a finite silicon impedance, the MOS capacitance can be discharged through the discharging path of the substrate. Therefore, the coupling noise on the victim TSV is reduced. Especially when the RC time constant of the discharging path is small and aggressor switching frequency is low, the accumulated charges can be quickly discharged even when the aggressor signal is still switching. Therefore, the peak noise voltage on the victim reduces due to fewer charges. The traditional model assumes a floating net at silicon substrate and therefore, overestimates the coupling noise on victim TSV since there is no discharging path. But it underestimates TSV-induced delay and power as the capacitance of the discharging path is also ignored. Therefore, the discharging path needs to be modeled using substrate resistors and capacitors. Figure 4.4 illustrates our proposed multi-TSV model with components to model silicon and E-field effects, where C_{Sig} and R_{Sig} represent the silicon substrate capacitance and resistance, respectively, between TSV and the substrate contact to model the charging path.

In previous works, all of the coupling components connecting other TSVs share a single node around victim TSV, which is connected to the TSV net by the MOS capacitor. This model assumes that all sides of the TSV have electrically the same node. However, in real case, since the silicon

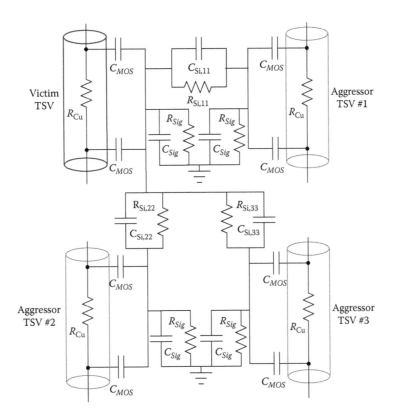

FIGURE 4.4 A multi-TSV coupling model with the depletion capacitance and body resistance modeled.

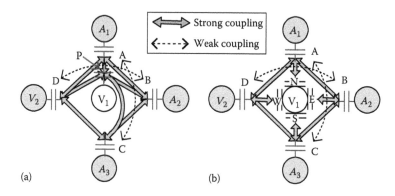

FIGURE 4.5 A circuit model for 5-TSV case. (a) Original and (b) E-field distribution-aware model.

substrate is not a perfect conductor, the electrical properties on different sides of the victim TSV are not the same. Moreover, the coupling between TSVs is mostly between two sides, which is directly facing each other and there is a few coupling on other sides. Especially in multi-TSV case, where each victim TSV is facing many aggressor TSVs in multiple directions, the E-field will be shared heavily.

Consider a 5-TSV case, which is shown in Figure 4.5, where there are four TSVs placed on each side of TSV V_1 and the E-field around the victim TSV is distributed among each aggressor. In this case, only neighbor TSVs are strongly coupled and there is only a weak coupling between TSV A_2 and TSV V_2 due to the E-field blocking effect of TSV V_1. Shown in Figure 4.5, the traditional model uses a common node P to connect all the coupling path from other aggressors. This creates a direct coupling path between TSVs that are weakly coupled. With the common node P, aggressor A_2 is directly coupling with victim V_2 through path $B–P–D$, which results pessimistically in TSV-coupling estimation. Figure 4.6a illustrates the HFSS simulation on E-field distribution of this structure. It is clearly seen from the plot that the coupling from each aggressor is mainly through one of the four sides of the TSV V_1, and there is fewer coupling between the other sides of the victim and the aggressor because of the distributed E-field. Therefore, the traditional model overestimates the coupling noise on the victim TSV.

To model the impact of the E-field distribution, four nodes around victim TSV are used to connect the coupling parameters to aggressors, shown in Figure 4.5b. Depending on the relative location of aggressor TSVs, the coupling path will be connected using the facing node of the victim TSV. Therefore, the direct coupling path between weakly coupled TSV is eliminated in the new model. Figure 4.6b shows the coupling parameters of the circuit model compared with the results extracted using HFSS field solver. The result indicates overall both model match well with the field solver results on the coupling noise. But there is a 1.1 dB over-estimation in coupling noise due to the direct path between TSVs in the original model. Our model shows smaller errors up to 15 GHz not only in noise magnitude but also in noise phase compared with the original one. Therefore, we conclude that our model is more accurate to reflect the E-field distribution impact in TSV-to-TSV coupling.

4.2.3 TSV-TO-TSV COUPLING IMPACT ON FULL-CHIP NOISE

For full-chip analysis, we first extract TSV locations and 2D parasitics for each die separately from Cadence Encounter. Then an RC parasitic network is generated for all the TSVs using our multi-TSV model. After the TSV coupling model is calculated for each TSV, SPICE netlists as well as a top-level SPEF file are generated containing TSV parasitic information. For full-chip noise analysis, HSPICE simulation is performed and the coupling noises on victim nets are extracted. Only the maximum noise that appears on a single net is measured so that the noise value is not counted many times. This procedure is performed on every TSV net in the design and the sum of maximum noise on all TSVs

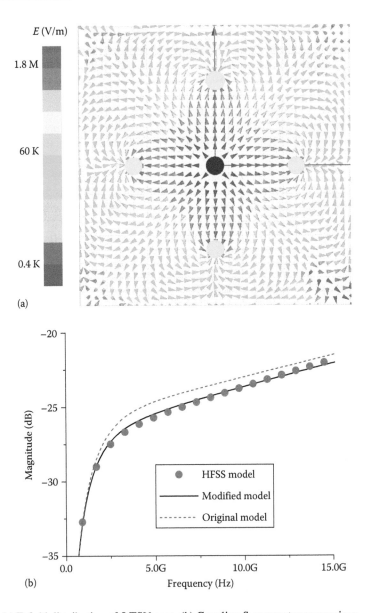

FIGURE 4.6 (a) E-field distribution of 5-TSV case. (b) Coupling S-parameter comparison.

is used as the total noise. Primetime is used to read the parasitic information for each die as well as TSV-coupling information altogether and then perform full-chip static timing and power analysis.

Since TSV parasitics depend on TSV voltage, and it is difficult to estimate the signal arriving time for all possible cases, different strategies are used for worst-case and average-case analysis. For worst case analysis, it is assumed that all the aggressive signals are arrived at the same time and they all have the same switching waveform from 0 V to VDD. In this case, charges due to TSV coupling accumulate around the victim TSV and introduce a large voltage spike at the victim node. Then, the maximum voltage on the victim net is measured. Note that it is only theoretically possible that all aggressors have the same waveform, and the victim would see such a large noise, however, it is a good indicator of how severe is the coupling in the full-chip level and the result is only related to the design itself. We use TSV MOS capacitance measured when the TSV voltage is 0 V since the depletion region width is minimum and TSVs are strongly coupled through the substrate.

For average case study, a time window is chosen which is no larger than the target clock period. We use the TSV MOS capacitance values measured at half of the VDD. Moreover, some aggressors may not even switch during the same clock cycle. Since not all aggressor nets are switching at the same time, the arrival times of the aggressor signals are randomly located within the time window. A switching activity factor, which is less than 1, is used to determine the possibility of signal switching. Note in worst case analysis, since all of the aggressors are switching, the switching factor is 1. Also, different from worst case analysis where all the signals are switching in the same direction, the peak-to-peak voltage difference on the victim TSV net is measured as the noise value.

A 64-point FFT design is used to demonstrate the full-chip impact of TSV-to-TSV coupling. It has 47 K gates and 330 TSVs. The target clock frequency is 200 MHz. We implement this design on a two-die 3D IC using 45 nm technology. The TSV landing pad size is 5 µM and the TSV liner thickness is 0.5 µM. Each TSV has a small KOZ to ensure all the logic cells are outside of the TSV depletion region so that their threshold voltage and performance will not be affected by the depleted substrate. The total footprint area of the design is 380 µM × 380 µM, and the total TSV area is 16,170 µM², which is 11.2% of the total area. We apply different TSV placement strategies and obtain two kinds of designs. During regular placement, TSVs are placed on regular grid with a pitch of 18 µM. TSVs are distributed all over the design space and TSV placement density is about the same everywhere. For irregular placement, TSVs are treated the same way as other logic cells and try to minimize the total wirelength. The minimum TSV-to-TSV pitch in irregular placement is 11 µM so that it can be manufactured. Figure 4.7 shows the die shots with TSV landing pads highlighted.

From Figure 4.8, the largest coupling noise is measured on the TSV net rather than 2D nets and the average noise value on TSVs is much larger than that of 2D nets. Also, compared with 2D nets, 3D TSV nets heavily suffer from coupling noise and delay. This is because of the following reasons: (1) It is difficult for current technology to fabricate TSVs with very small dimensions and large aspect ratio. Therefore, TSV has large MOS capacitance due to long shape and large radius; (2) In future technology nodes, more TSVs and higher TSV placement density are allowed to increase die-to-die bandwidth, therefore there will be larger coupling between TSVs; (3) The permittivity of the inter-layer dielectric (ILD) between 2D interconnections is very low if low-K material is used (2–$3\varepsilon_0$). However, the silicon substrate that is covering around the TSV has a very high permittivity ($11.9\varepsilon_0$), which results in large TSV coupling capacitance.

Table 4.2 compares the two analyses in various metrics. In all cases, the average case shows much smaller total TSV coupling noise than the worst case. In average case, the victim TSV has

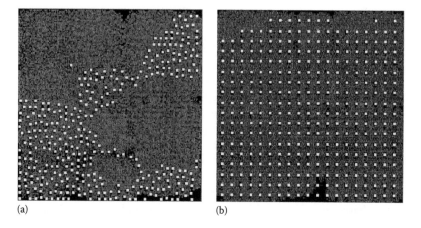

(a)   (b)

FIGURE 4.7   3D-IC layouts. (a) Irregular TSV placement and (b) regular TSV placement. The squares denote TSVs.

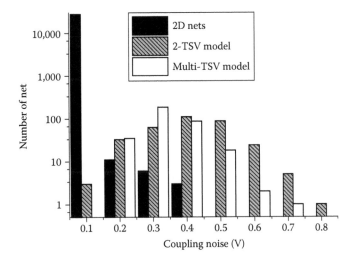

FIGURE 4.8 Full-chip noise comparison. Multi-TSV model is fixing the overestimation issue of the pairwise two-TSV model.

TABLE 4.2
Average and Worst-Case Comparison of the Total TSV Net Noise (V)

	Activity	Slew (ns)	Irregular	Regular
Average-case	0.2	0.1	26.51	24.65
	0.5	0.1	39.61	35.37
	0.2	0.5	14.04	14.62
Worst-case	1.0	0.1	139.01	132.44

much smaller peak-to-peak noise due to the following reasons: (1) Not all the aggressors switch in one clock period, and those switching aggressors do not start voltage transition at the same time. Smaller aggressor signal activity results in smaller coupling noise on victim TSV. (2) Due to the load capacitance, many aggressor nets have longer transition time, especially for nets with weak driver. Slower transition time on aggressor introduces fewer charges through the coupling path thus it reduces the coupling noise on victim TSV. The average-case analysis provides an estimation on average noise level on TSV nets when multiple aggressors with different voltage waveforms are considered. The results show that both the switching activity and the signal slew have a large impact on the noise results on the TSV nets. Larger switching activity and smaller signal slew increase the TSV coupling noise significantly and they should be considered in noise analysis.

Moreover, compared with regular placement design, irregular placement design is showing 5% larger coupling noise. This is because, in irregular placement design, minimum distance between TSVs is smaller, and TSVs are placed with higher density. Therefore, irregular placement suffers more TSV coupling that results in a larger timing degradation. However, since the regular placement is a special case of irregular placement, it is possible to find a better irregular TSV placement which has smaller noise coupling.

To study the impact of field and substrate effects, we disable each field and silicon effect one by one while keeping other effects the same and perform noise analysis on the full-chip level. The worst case analysis flow is used because its result only depends on the circuit itself which makes it a fair comparison. Table 4.3 details chip-level E-field and silicon effects comparison. Without considering the depletion region, TSV MOS capacitance is overestimated, especially when TSV liner

TABLE 4.3

Silicon and E-Field Impact on the Total TSV Net Noise (V), TSV-Induced Delay (ns), and Power Consumption (µW) Increase

Irregular Design	Total TSV Noise	TSV-Induced Delay	TSV Net Power
No depletion region	153.7 (+10.5%)	0.85 (+7.6%)	13.53 (+6.7%)
No body resistance	144.9 (+4.2%)	0.78 (−1.2%)	12.54 (−1.1%)
No E-field distribution	146.3 (+5.2%)	0.79 (0%)	12.68 (0%)
All-effects-included	139.0	0.79	12.68
Regular Design	**Total TSV Noise**	**TSV-Induced Delay**	**TSV Net Power**
No depletion region	145.9 (+10.2%)	0.98 (+7.7%)	13.66 (+7.0%)
No body resistance	138.9 (+4.9%)	0.90 (−1.1%)	12.63 (−1.1%)
No E-field distribution	138.9 (+4.9%)	0.91 (0%)	12.77 (0%)
All-effects-included	132.4	0.91	12.77

thickness is thin and the substrate doping concentration is low. Since our design runs at 200 MHz, a complete depletion around TSV is assumed. If depletion region is ignored, the result show a 10.5% and 10.2% increase in total TSV net noise for irregular and regular designs, respectively. This is because the MOS capacitance is overestimated by 17%.

Moreover, ignoring substrate resistors and capacitors is also a pessimistic estimation on coupling noise. The discharging path through a substrate is critical to limit the peak noise on the victim and it also affects delay and power consumption. Also, without considering the electrical field distribution, the noise is over-estimated because every aggressor sees the whole TSV MOS capacitance around victim TSV, even though it only faces to one side of the victim TSV. Since the electrical field distribution effects do not change any capacitance value, the calculated delay and power is the same using Primetime. Overall, the depletion region impact has the largest impact on full-chip metrics as the MOS capacitance is the dominating component in TSV-to-TSV coupling.

4.2.4 FULL-CHIP TSV-TO-TSV COUPLING MITIGATION

Since the silicon substrate provides a discharging path to the ground, it can be used to reduce the coupling noise on TSVs by making the discharging easier and reducing substrate-to-ground resistors (R_{Sig}). We use a grounded guard ring proposed in [4] in the active layer with $P+$ doping to build a short discharging path for the victim TSV. The ring is connected with grounded rings in Metal 1. Therefore, the TSV is protected by ground ring in active layer and landing pad is protected by ring on Metal 1. To reduce the model complexity, we propose a new guard ring model with a few added components to multi-TSV model. The proposed guard ring structure is shown in Figure 4.9a. The discharging path through the grounded ring contains two components C_{Sig} and R_{Sig}, and we use Synopsys Raphael to extract the substrate capacitance to the ground. Detailed extraction results are listed in Figure 4.9b, with various edge-to-edge distance and guard ring width. Small ground resistance leads to a strong connection between the substrate and the ground net, thus it can help shielding coupling noise introduced by TSV-to-TSV coupling. The ring width shows a large impact on the ground resistance. Thus, the coupling noise reduces further if the width of the guard ring is increased. However, the distance between TSVs and the guard ring does not affect much on the ground resistance. Longer edge-to-edge distance between TSV and guard ring results in a larger guard ring but the coupling E-field strength is reduced. The drawbacks of this method include a slight timing degradation on TSV nets due to the increased ground capacitance and a small area overhead. Wider guard ring shows larger noise reduction but they introduce longer delay. On other hand, while the silicon around TSV is depleted and cannot be used for devices, the guard ring in

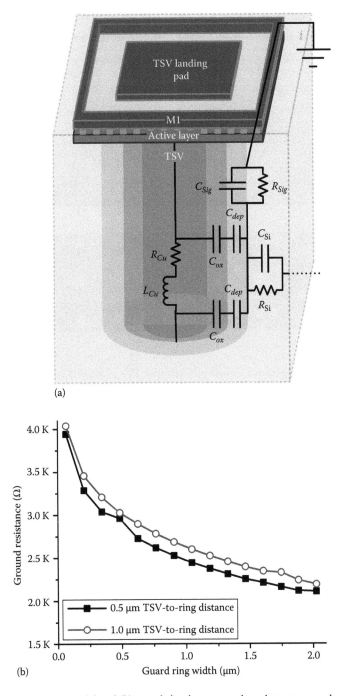

(a)

(b)

FIGURE 4.9 (a) A guard ring model and (b) guard ring impact on the substrate ground resistance.

the active area can make use of this area and help reducing noise. This makes the guard ring more appropriate for designs with large KOZ and increase the silicon utilization. Transient analysis is performed on the 3-TSV test structure with our multi-TSV model and the guard ring shows 47.5% noise reduction on victim TSV net.

To efficiently find TSVs which need noise protection, the following strategy is utilized to perform the noise optimization. First, a worst case noise analysis is performed on the full-chip

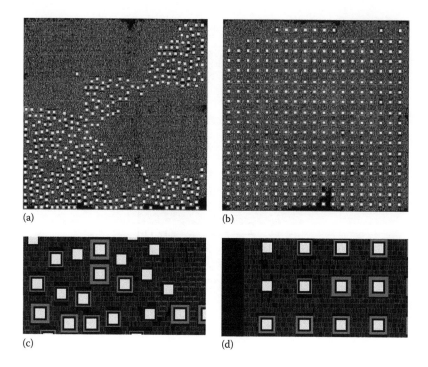

FIGURE 4.10 Guard ring optimized layouts. (a) Irregular TSV placement, (b) regular TSV placement, (c) zoom-in of (a), and (d) zoom-in of (b).

design and obtained the noise levels on each TSV. Then, TSVs are sorted according to the noise levels and guard rings with different widths are added around TSVs. To minimize the area overhead, a noise threshold is used below which no guard ring will be added. Above the threshold, TSVs that suffer larger coupling noise is protected with a wider guard ring and vice versa. Worst case analysis is used here as it is not random seed dependent. Figure 4.10 shows the layout with TSV and guard ring highlighted after the optimization is performed on our regular placement and irregular placement designs. Shown in the layout, TSVs with large coupling noise are mostly located in the center of the die where TSVs are surrounded by more aggressor TSVs as well as standard cells.

After the guard rings are added to the design, the overlapping in the layout is fixed using incremental placement and routing and then perform worst case analysis on the new layout. Table 4.4

TABLE 4.4

Full-Chip Coupling Optimization with Guard Rings for Irregular and Regular TSV Placement Designs

	Irregular	Regular
Total TSV noise without guard ring (V)	139.0	132.4
Total TSV noise with guard ring (V)	101.1	96.5
Noise reduction	27.3%	27.1%
TSV-induced delay (ns)	0.81	0.93
TSV-induced power (μW)	12.75	12.86

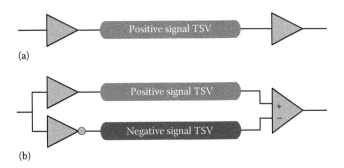

FIGURE 4.11   Signal transmission using TSV. (a) Single-ended and (b) differential pair.

shows the noise optimization results. There is a 27.3% reduction in total TSV net noise with only 7.65% area overhead from guard rings. The delay of the design also increases a little due to the increased substrate ground capacitance. Our results show that guard ring protection is very effective in TSV noise reduction with minimum area overhead.

Another method to enhance the signal transmission reliability is using differential TSV pairs [7,9]. Figure 4.11 compares the single-end TSV and the differential TSV transmission. In differential TSV transmission case, voltages on a pair of TSVs are compared and the difference is used to determine the output level. To analyze the full-chip impact of differential TSV, our multi-TSV model is used in the following discussion.

For a full-chip implementation, a simple digital comparator proposed in [1] is used. For regular design, the worst case noise analysis resulting from original design is used to set a noise threshold. TSVs with noise above the threshold will be replaced by differential TSV pairs. However, for irregular design, since TSVs are placed closer, it is possible that when a single TSV is replaced by a differential pair, the inserted TSV overlaps with existing TSVs. Therefore, for irregular design, starting from the TSV with largest noise, we try to replace TSVs with differential pair, unless the new inserted TSV will cause overlapping in the layout. Compared with regular design, a slightly lower noise threshold is used if same number of TSVs is protected. After the differential TSV insertion, a refine placement is performed to fix TSV overlapping with standard cells and to insert new cells such as comparators. Then incremental routing is performed so that no major redesign is needed when applying the differential TSV insertion.

The differential TSV pair impact comes from two aspects. Once a victim TSV is replaced by a differential pair, it has better noise immunity. On the other hand, when other TSV is considered as the victim, the coupling noise from each member of the differential pair cancels each other and it results in a smaller noise on the victim TSV. To consider both effects, full-chip analysis flow needs to be modified for differential-TSV-awareness. First, differential pair is divided into positive TSV and negative TSV where positive TSV has the same voltage switching direction as other aggressors and the negative TSV has the opposite switching direction. Moreover, for differential TSV pair, noises are compared at both the TSVs, and the absolute value of voltage subtraction is taken as the noise of a differential pair of TSVs rather than the peak-to-peak voltage in single-end TSV case.

Layouts of designs with differential TSV pairs are shown in Figure 4.12 and full-chip analysis results are shown in Table 4.5. For the regular design, the TSV on the critical path is replaced by a differential pair so there is a small increase in the longest path delay due to slower comparators and longer signal transition time. However, for the irregular design, such TSV is not protected. Therefore, only minor change exists on the longest path delay. From the results, we conclude that differential TSV transmission is very efficient in TSV coupling noise reduction with a small overhead in timing and area.

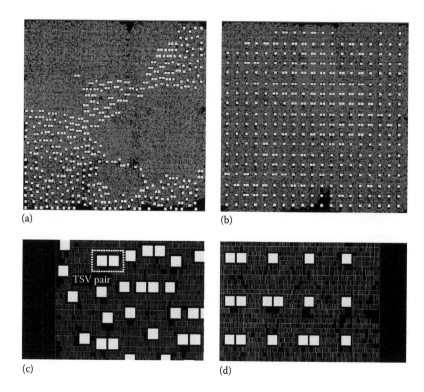

FIGURE 4.12 Coupling mitigation with differential TSV pairs. (a) Irregular TSV placement, (b) regular TSV placement, (c) zoom-in of (a), and (d) zoom-in of (b).

TABLE 4.5
Full-Chip Coupling Optimization with Differential TSV Pairs for Irregular and Regular TSV Placement Designs

Design Style	Irregular		Regular	
With differential TSV?	No	Yes	No	Yes
Protected TSV#	0	100	0	100
Area increase	—	3.4%	—	3.4%
LPD (ns)	4.62	4.64	4.36	5.02
Total TSV noise (V)	139.0	76.5	132.4	83.1
TSV noise reduction	—	44.9%	—	37.2%
Total TSV coupling cap (pF)	4.32	8.10	3.27	6.33
Total TSV MOS cap (pF)	21.9	28.5	21.9	28.5

4.3 TSV-TO-WIRE COUPLING EXTRACTION

In this section, we discuss several extraction techniques and E-field sharing impacts on TSV-to-Wire coupling.

4.3.1 E-FIELD SHARING IMPACT

Kim and Yetal propose an analytical solution in [6] and extend the empirical wire coupling model to handle TSV structures assuming square-shaped TSVs. This math model consists of many empirical formulas to calculate both overlap, corner and fringe capacitance between a pair of TSV and wire considering various wire locations. Since this model is based on closed-form formulas, calculation takes only a negligible runtime. However, the model is not scalable because empirical equations handle a fixed fabrication technology, and a curve fitting needs to be applied for various interconnect dimensions. Moreover, this traditional empirical TSV-to-Wire model considers a pair of TSV and wire at one time [6] and ignores the E-field sharing from other interconnect components. Though a careful curve fitting can accurately model simple strictures, the extraction error on a complicated structure can be large. This is due to E-field sharing among multiple wires located around TSV.

To study the E-field sharing among wires, we build a structure with four metal rings in HFSS [2] shown in Figure 4.13a and extract their E-field interactions with the TSV. Figure 4.13b shows the cross-section E-field distribution map simulated with the coupling capacitance extracted. As results shown, the strongest coupling E-field is formed between the TSV and the nearest wire, and the coupling capacitance is the largest. The outermost wire also shows large capacitance because there is no neighbor conductor outside to share the coupling E-field. However, for middle rings that have neighbor conductors on both inner and outer sides, only small coupling capacitance is formed, which is because of strong E-field sharing. As results shown, without considering the E-field sharing, using a TSV-wire pair-based formula to extract all wire capacitance results in a large overestimation. And it is difficult to come up with a compact model considering various complicated geometries.

Moreover, the E-field sharing effect is also observed even for a single wire. If a wire is divided into several segments, coupling E-fields are not the same for every segments. A regular segment has neighbors on both sides while a corner segment has only one neighbor. We build a single wire structure, which is divided into 0.5 µM segments, and extract the capacitance of each segment using Raphael. Figure 4.14 shows an unbalanced coupling distribution between regular segments and corner segments. All regular segments have similar coupling capacitances to the TSV but corner segments show 80% larger capacitances even though they are located further from TSV. This is because sidewalls of corner segments also contribute to the fringe capacitance and there is no outside neighbor, which shares the coupling E-field. Therefore, for accurate extraction, corner segments need to be considered separately. Treating corner segments as regular segments underestimates the TSV-to-Wire capacitance especially for short wires. For designs with mostly long wires, the corner segment impact is small.

If there are multiple wires surrounding a TSV, another E-field sharing impact also affects TSV-to-Wire extraction results. As shown in Figure 4.15a, if the TSV is only facing to wires on one side, a few E-field sharing exists around TSV and the total coupling for a single wire is 2.15 fF. However, if the TSV is facing to wires in more directions as shown in Figure 4.15b and c, the single wire capacitance reduces to 1.31 fF. This is because TSV-to-Wire coupling is evenly distributed to

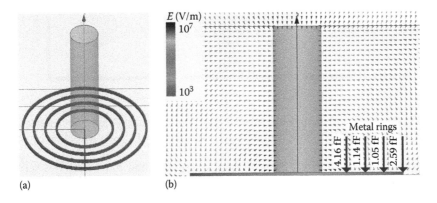

(a) (b)

FIGURE 4.13 TSV-to-Wire coupling model. (a) HFSS structure with a TSV and four rings, (b) cross-section E-field distribution around the TSV in (a).

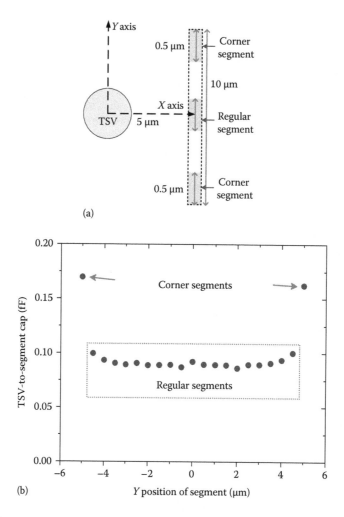

FIGURE 4.14 Corner segment impact. (a) Simulation structure with 0.5 µM wire segments and (b) extraction results for each segment.

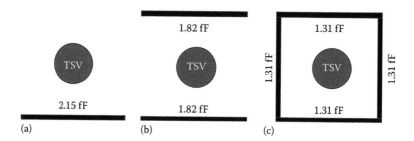

FIGURE 4.15 Impact of wire coverage around a TSV on TSV-to-Wire coupling capacitance. (a) A TSV with one neighboring wire, (b) a TSV with two neighboring wires, and (c) a TSV surrounded by 4 wires.

all four neighbors. We use a wire coverage factor to represent how much a TSV is surrounded by wires. A wire coverage calculation example is shown in Figure 4.16a and wire coverage factors for structures in Figure 4.15a, b, and c are 25%, 50%, and 100%, respectively. Larger wire coverage results in stronger E-field sharing and smaller capacitance per unit length. However, the total TSV-to-Wire capacitance increases since more conductors are around the TSV. Therefore, for accurate

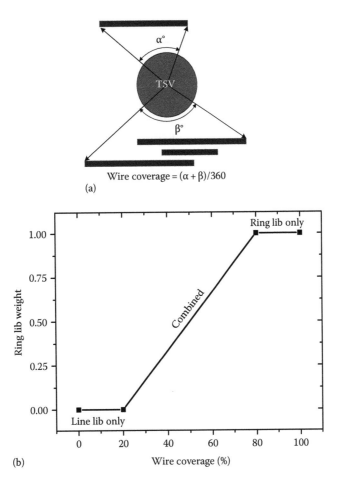

Wire coverage = $(\alpha + \beta)/360$

(a)

(b)

FIGURE 4.16 Our combined method. (a) Calculation of wire coverage and (b) calculation of weighted average based on the line and ring libraries.

TSV-to-Wire capacitance extraction, the wire coverage effect needs to be considered carefully especially when there are many wires in the full-chip level.

4.3.2 PATTERN-MATCHING TECHNIQUE

To handle 3D full-chip TSV-to-Wire extraction, we propose a pattern-matching technique. This technique is similar to traditional 2D full-chip extraction tools which correlate closely with silicon measurements. But our technique considers all special TSV-related impacts which traditional tools cannot handle. The first stage of a pattern-matching extraction is using a general extraction engine such as a field solver, to perform extraction on various pre-defined structures. Results are saved into a database called library. Then, during extraction stage, full-chip interconnects are compared to the library and extraction results of pre-calculated structures closest to the layout are used for capacitance calculation. Though generating the library and extraction rules takes a long time as thousands of layout geometries need to be simulated, a common library or a set of rules can be used for a certain technology on various designs. Therefore, they are provided in the Process Design Kit (PDK) by the foundry. Since only library look-up and math calculation are performed during extraction, the pattern-matching extraction can extract parasitics of a large-scale circuit within minutes and they are suitable for extraction of next generation 3D-ICs with billions of transistors and thousands of TSVs.

However, traditional pattern-matching engines can only handle 2D designs, where interconnect extraction is limited to several neighbor wires. Unlike metal vias, TSVs are hundreds of times larger and they interact with many surrounding neighbors due to their large influence regions. E-field sharing from multiple conductors also introduces new challenges which must be considered during extraction. To consider all aforementioned effects, we build three special libraries for TSV-to-Wire coupling extraction. Two libraries (Yie, a line library, and a ring library) are used for regular segments while a third corner library is used specifically for corner segments. These libraries enable detailed considerations of E-field sharing among wires. During extraction, wires are divided into segments and the capacitance is calculated for each segment. The geometry information of the segment and its context is used to match patterns in the library, and a linear interpolation of closest structures is used when there is no exact matching structure.

We build the line library for TSVs with a low-wire coverage. As shown in Figure 4.17a, the line library is built by placing straight wires on only one side of the TSV. All wires are segmented and a single structure is able to produce results for many segments with various locations. This increases the extraction parallelism and reduces the generation time. The length of each wire segment depends on its relative location to the TSV. Each segment always has a facing angle of 5° to the TSV and wire segments far from TSV are longer. This is because that wires far from the TSV have weaker coupling and smaller capacitance per unit length. Our segmenting method takes advantage of the cylindrical shape of TSV so that the capacitance of each wire segment is in a similar order to prevent small error accumulations. Similar to Finite Element Analysis (FEA), finer segmenting is used on areas where E-field is strongest and rapidly changing while coarser segmenting is used on less critical areas. This enables a best tradeoff between simulation time and accuracy.

The line library assumes that only one side of a TSV is surrounded by metal wires thus only weak E-field sharing exists around the TSV and the unit length capacitance of wires are high. Therefore, this library is suitable for layouts where TSVs are covered by a few wires around. For a general case where the TSV is surrounded by wires on multiple sides, the line library gives overestimated capacitances since the line library always assumes a weak E-field sharing. In terms of library generation time, since the line library consists of less complicated geometries structures such as straight wires, it is faster to generate.

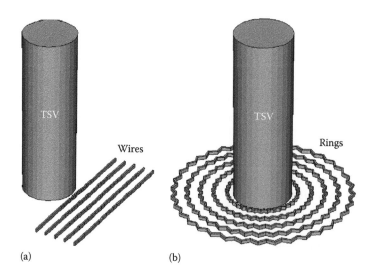

(a)   (b)

FIGURE 4.17   Test structures for library generation. (a) Line library and (b) ring library.

To consider layouts where TSVs are surrounded by many wires on all sides, another ring library is built. As shown in Figure 4.17b, we duplicate wire segments with various locations to form a ring around TSV. In this structure, as E-field of the TSV are evenly distributed to all directions, we extract the total capacitance of the ring and divide it by the total number of ring segments. Different number of rings is added around the TSV to simulate multiple wires. As the wire coverage for a ring structure is 100%, E-field sharing around TSV is high and the unit length capacitance for ring library is small. Unlike the line library, the ring library always assumes strong E-field sharing around TSV, thus they are suitable for designs with congested routing around TSVs. For a general case where the TSV is surrounded by few wires and the wire coverage is low, the ring library underestimates the TSV-to-Wire capacitance. Thus, the ring library is complementary to the line library to provide accurate extraction for weak E-field sharing cases. However, as the ring structures are built with many segments, the complicated geometry needs a longer extraction time for field solving.

As in previous discussions, corner segments with a single neighbor have larger coupling capacitances due to sidewall capacitances and less E-field sharing from neighbors. Therefore, based on the line library, a special corner library is built to extract capacitances of corner segments at various locations. The corner library structure is similar to the one for the line library. However, only the capacitance of the corner segment is extracted and saved. Compared to line and ring libraries, the unit length capacitance of surrounding wires is the highest and geometry complexity for corner library is low. However, since there are not many corner segments in the full-chip level especially for top metal layers, its impact on system performance and noise metrics is small. However, for short wires, the extraction error is reduced significantly considering corner segment effects.

We develop an algorithm for pattern-matching-based TSV-to-Wire coupling capacitance extraction. Extraction is performed on each TSV. Areas around the TSV are divided into 72 circular sectors, each with 5° in central angle and the same radius as the TSV influence region. These sectors are numbered clockwise. Therefore, wires closer to the victim TSV have finer segments and only segments within the TSV influence region are considered. Wires are segmented similarly as in the line library at the sector boundary and all wire segments in the same sector are gathered into a list. Wire dimension and location, number of wires, and average pitch of wires are used as indexes to search through the library. The look-up procedure takes place on each list and compares the layout structure to the pre-generated libraries. Linear interpolation is used when the library index does not exactly match.

For corner segments, results from the corner library are used. For regular segments, we combine both line library and ring library based on the wire coverage around the TSV. As shown in Figure 4.16b, if the wire coverage is above 80%, we only use the ring library as the coupling capacitance per unit length is small. On the other hand, if the coverage is below 20%, we only use the line library assuming a weak E-field sharing. Otherwise, results from both libraries are combined and a weighted average is calculated depending on the wire coverage. This enables wire coverage consideration during full-chip extraction. After all lists are parsed, TSV-to-Wire parasitics are exported into a Standard Parasitic Exchange Format (SPEF) which can be integrated into standard full-chip CAD flow for further timing and noise analysis.

4.3.3 TSV-to-Wire Coupling Impact on Full-Chip Noise

As a case study of the impact of TSV-to-Wire coupling in the full-chip level, we use a 64-point FFT design based on a 45-nm technology similar to the design in Section 4.2.3. However, since our libraries are built assuming a regular TSV pitch, we apply our pattern-matching method to FFT designs for TSV-to-Wire extraction on a regular placement design. For TSV-to-TSV coupling extraction, the multi-TSV coupling model is used and the same full-chip analysis flow from [12] is used where full-chip static timing and power analysis is performed with Primetime and worst-case noise analysis is performed with Hspice to measure TSV noises with an accurate multi-TSV model.

To validate our extraction method in the full-chip level, we apply our pattern-matching algorithm and compare it with a field solver (Raphael). For each TSV, we build a Raphael structure exactly as the layout around the TSV. We set the TSV influence region as 10 μm to save Raphael simulation time but most coupling around TSV is still captured. As discussed before, using a single ring library gives a 8.3% underestimated total capacitance, while using a single line library gives a 5.3% overestimated total capacitance. However, both of them are accurate enough: an average error of 0.171 fF for the ring library only case and 0.163 fF for the line library only case. If we consider the wire coverage effect and combine both libraries in calculation, the total capacitance error is only 1.9% and the average error drops to only 0.112 fF. The correlation coefficients are 0.971, 0.966, and 0.981 for ring library, line library, and combined method, respectively. Full-chip extraction result is shown in Figure 4.18 and the library comparison is listed in Table 4.6. Although our method needs longer time to build the libraries, once all libraries are ready, it can be applied to any designs. Compared with field solver, which takes 7.5 h and 615 MB memory space to compute all TSVs, our pattern-matching algorithm takes only 2.4 s and 21 MB memory space. Also, field solver needs much longer time and memory on larger design with more TSVs. The runtime and memory usage are negligible

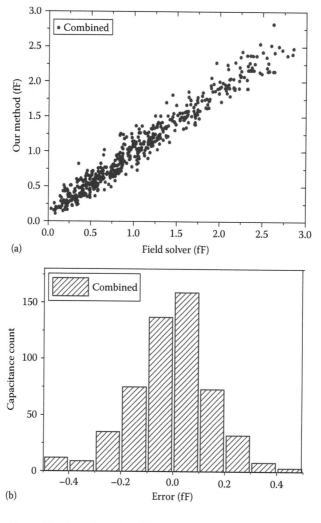

FIGURE 4.18 Full-chip verification of TSV-to-Wire extraction. (a) Comparison with a field solver and (b) error distribution.

TABLE 4.6
Extraction with Various Pattern Libraries

	Ring Lib	Line Lib	Combined
Total Cap (fF)	538	618	579
Total Cap error	−8.3%	+5.3%	−1.9%
Correlation coefficient	0.971	0.966	0.981
Average error (fF)	0.171	0.163	0.112

Note: The total capacitance from Raphael simulation is 568 fF.

compared with the field solver case. Therefore, we conclude that both ring and line libraries are accurate for multiple wires, and our combined method considering wire coverage effect is very fast and highly accurate.

Since the partition of this design is to minimize the TSV count and the system performance is not taken into consideration, a 3D path has the longest delay. Critical paths of both designs are the same. It starts from a register in the top die, goes through the bottom die, and ends on another register of the top die. As a result, both TSV-related parasitics and top metal routing parasitics affect the full-chip timing results. Table 4.7 summarizes full-chip TSV-to-Wire impacts on timing, power, and noise. Note that the longest path delay (LPD) change only comes from TSV nets as we assume clock network is ideal. If a real clock tree network is included, then the LPD is further affected since the clock signal needs to be delivered to the top die using TSVs as well, and the clock skew is also affected by TSV-to-Wire parasitics. Therefore, it also calls for fast and accurate TSV-to-Wire coupling extraction during clock tree synthesis.

For the FFT64 design, if the TSV-to-Wire coupling capacitance is ignored, the timing and noise analysis is inaccurate. From the result, without TSV-to-Wire coupling, the LDP is only 4.48 ns. This is underestimated since the load capacitance are not fully captured. After TSV-to-Wire capacitances are annotated from SPEF, the LPD increases to 5.08 ns due to increased capacitances on TSVs. Many other 3D nets are affected by TSV-to-Wire coupling as well. Even if the critical path of the original design is not a 3D path, with TSV-to-Wire coupling considered, the critical path may change and timing impacts are noticeable. Note that since 2D routers and timing optimization engines are not aware of the TSV-to-Wire capacitance, not enough buffers are inserted to TSV and wires on top metal layer. This results in a large delay increase after TSV-to-Wire extraction is applied.

Also, ignoring the TSV-to-Wire coupling results in a significant under-estimation in TSV net noise. This is because traditional 2D extraction only extract parasitics from TSV landing pads, thus the coupling capacitance is much smaller. Moreover, the influence region of a TSV landing pad is significantly smaller than that of a TSV. As a result, many aggressors are ignored with 2D extraction simply because wires are too far away. Even though TSV-to-TSV coupling contributes to TSV

TABLE 4.7
Full-Chip Impact of TSV-to-Wire Coupling

Is TSV-to-Wire included?	No	Yes
Longest path delay (ns)	4.48	5.08 (+13.4%)
Total power on TSV net (mW)	0.303	0.356 (+17.6%)
Total net switching power (mW)	2.42	2.50 (+3.3%)
Total noise on TSV net (V)	32.5	78.2 (+104%)

noise significantly, its impact is large if the TSV is tall and TSVs are placed closely. Without considering TSV-to-Wire coupling, total TSV net noise is underestimated. In terms of power, though a large increase in TSV net power is observed, there is only a negligible impact on total power from TSV-to-Wire coupling. This is because the major portion of total power is consumed by transistors, and 3D nets are only a small portion of all nets. Note that the FFT64 is a small circuit, if a design has larger footprint with more TSVs and longer wirelength, the TSV-to-Wire impact on the power will increase.

4.3.4 FULL-CHIP TSV-TO-WIRE COUPLING MITIGATION

Since TSV-to-Wire coupling is majorly between TSVs and the nearest metal layer. Therefore, to reduce TSV-to-Wire coupling, a simple technique is increasing the minimum distance between TSV and the nearest wire routing. To implement this method, a routing blockage on top metal layer is placed around each TSV. This blockage region is the routing keep-out-zone (KOZ). To study the impacts of various KOZ sizes, we implement two designs with KOZ sizes of 2.5 and 5 μm, respectively. Figure 4.19 shows these design layouts. The original design has KOZ of 0.5 μm. With a larger KOZ, the capacitance between a TSV and its nearest wire further reduces due to a weaker coupling E-field. In addition, a larger KOZ reduces routing resources available on top metal layer. This results in a reduction in top metal layer wire length and total aggressors around TSV. Note that increasing routing KOZ does not have any silicon area overhead, therefore, placement results are kept the same as the original design, and only incremental routing is performed to correct any routing violations. Thus, layouts have minimum changes and it is a fair comparison for all designs.

However, one drawback for increasing KOZ is that heavier routing congestion on other layer is observed due to reduced routing tracks on the top metal layer. Potentially, this may result in negative impacts on design quality and coupling increase between 2D wires. For the FFT64 design up to M4, since top metal wires are reduced by 43.7% with 5 μm KOZ, wire congestion on other metal layers

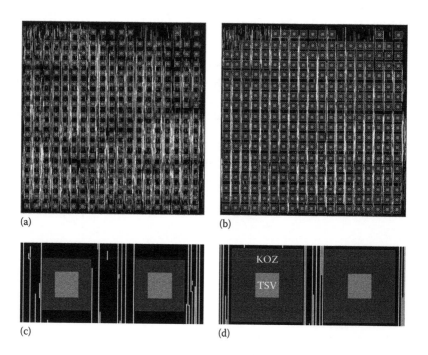

FIGURE 4.19 TSV-to-Wire coupling mitigation with keep-out-zone (KOZ) resizing. (a) 2.5-μm KOZ, (b) 5-μm KOZ, (c) zoom-in of (a), and (d) zoom-in of (b).

TABLE 4.8
Keep-Out-Zone Impact on Full-Chip Design Metrics

KOZ size (µm)	0.5	2.5	5
Longest path delay (ns)	5.08	4.95 (−2.6%)	4.77 (−6.1%)
Total power on TSV net	0.356	0.342 (−3.9%)	0.327 (−8.1%)
Total net switching power (mW)	2.50	2.47 (−1.2%)	2.45 (−2.0%)
Total noise on TSV net (V)	78.2	67.0 (−14.3%)	42.9 (−45.1%)

Note: The percentage values are with respect to 0.5 µm KOZ.

is more severe. Wirelength on M2 increases by 29.9% and the total wirelength increases from 373.9 to 376.3 mm. Therefore, for some designs which have limited routing resources, increasing KOZ may not be beneficial with increased routing congestion.

We perform full-chip analysis with updated extraction results. With the same TSV location, the TSV-to-TSV coupling capacitance remains the same. Full-chip analysis result is shown in Table 4.8. With larger KOZ, total wire length on top metal layer is reduced. Therefore, the TSV-to-Wire coupling also shows smaller impact on the full-chip timing and noise. Compared with the original design, the LPD decreases by 2.6% and 6.1% for 2.5 and 5 µm KOZ, respectively. For some designs, it may introduce timing degradation if there's not enough routing resources. The KOZ impact on full-chip power result is much smaller since TSV-to-Wire capacitance decreases but the wire-to-wire capacitance increases on other layer. Larger KOZ also helps reducing total noise on TSV net. The total worst-case noise on TSV net can be reduced by 14.3% and 45.1% for 2.5 and 5 µm KOZ, respectively. From the result, we conclude that increased KOZ is effective in reducing TSV-to-Wire coupling.

Another way to protect the victim TSV is to provide E-field shielding using grounded conductors around TSVs. Similar technologies are widely used to increase SNR in communication area. We use a physical design optimization technique specifically designed to reduce TSV-to-Wire coupling. Fabricated using the standard technology, grounded wire guard rings on top metal layer are inserted around TSV and they do not have any overhead on silicon areas. However, the guard ring consumes some routing tracks on the top metal layer and needs additional routing to connect the ring to the ground. The grounded guard ring shields some E-field around TSV, and introduces a ground capacitor on the TSV. As a result, there is a small delay and power overhead on TSV nets, but it reduces coupling noise on the TSV. Moreover, the guard ring now becomes the nearest wire around TSV which makes all other wires have neighbors on both sides. Therefore, the coupling capacitance between the victim TSV and signal wires is reduced and the TSV is better shielded with coupling noise.

To study the full-chip impact of wire guard rings, we build three multi-shielded-TSV libraries, Yie, line library, ring library and corner library, where TSVs are surrounded by grounded guard rings. To study the guard ring width impact, libraries are built with both 0.5 and 1.5 µm guard rings. In addition, we implement FFT64 designs with 0.5 and 1.5 µm guard rings on the top metal layer. Figure 4.20 shows top die layouts. These two designs are based on our previous FFT64 design with 2.5 µm KOZ. We insert the guard ring into the KOZ so that the placement and signal routing are kept the same. Similarly to KOZ insertion, guard rings also consume routing resources on the top metal layer. However, they provide better protections to TSVs.

After re-extraction on TSV-to-Wire and wire-to-wire coupling capacitance, we perform full-chip analysis and the results are shown in Table 4.9. The longest path will increase by 0.6% and 1.2% for 0.5 and 1.5 µm guard rings, respectively. This impact comes from two aspects: The ground capacitance on TSV increases but TSV-to-Wire capacitance decreases. The total capacitance on TSV will always increase with wider guard ring since the metal coverage increased.

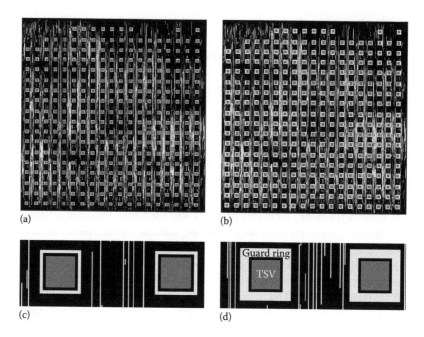

FIGURE 4.20 TSV-to-Wire coupling mitigation with guard ring. (a) 0.5-μm guard ring, (b) 1.5-μm guard ring, (c) zoom-in of (a), and (d) zoom-in of (b).

TABLE 4.9
Guard Ring Impact on Full-Chip Design Metrics

Guard ring width (μm)	0	0.5	1.5
Longest path delay (ns)	4.95	4.98 (+0.6%)	5.01 (−1.2%)
Total power on TSV net (mW)	0.342	0.351 (+2.6%)	0.358 (+4.7%)
Total net switching power (mW)	2.47	2.475 (+0.2%)	2.479 (+0.4%)
Total noise on TSV net (V)	67.0	58.0 (−13.4%)	53.6 (−20.0%)

Note: The percentage values are with respect to no guard ring case. KOZ is set to 2.5 μm.

Although there is a small increase on TSV switching power, guard ring impact on total power is negligible. Our results show that guard ring is very effective in TSV net noise reduction and the total TSV net noise reduces by 13.4% and 20.0% for 0.5 and 1.5 μm guard rings, respectively. Since the KOZ of this design is 2.5 μm, compared with the original design with 0.5 μm KOZ, a combination of increasing KOZ and guard ring protection can reduce up to 31.4% total TSV noise. Since the TSV-to-TSV coupling elements remain the same and it also contributes to the TSV net noise, increasing KOZ and adding guard ring cannot reduce TSV net noise too much. From the results, we conclude that the guard ring is effective in TSV net noise reduction with small delay and power overhead.

4.4 CONCLUSIONS

In this chapter, we studied several fast and accurate full-chip TSV parasitic extraction methods and learned full-chip TSV parasitics' impacts on timing, power, and noise. For TSV-to-TSV coupling, a compact multi-TSV model that considers E-field and substrate effects provided better accuracy compared with traditional two-TSV model. Depletion region, substrate impedance, and E-field distribution effects were critical in TSV modeling. Both worst-case and average-case full-chip analysis showed that TSV-to-TSV coupling has a large impact on full-chip timing and noise. To alleviate the TSV-to-TSV coupling noise, we studied grounded guard rings and differential TSVs for signal protection. These optimization methods were very effective in noise reduction, easy to implement, and area efficient.

For TSV-to-Wire coupling, we studied various factors affecting TSV influence region and TSV-to-Wire capacitance. We also studied a fast and accurate pattern matching algorithm, which considered multiple wire effects, corner segment effects, and wire coverage effects. We verified the method against a field-solver and studied TSV-to-Wire impacts on full-chip level. Full-chip analysis results showed that TSV-to-Wire coupling had a large impact on full-chip delay and TSV net noise. We also increased the keep-out-zone around TSV in top routing layer and used ground guard rings for TSV-to-Wire coupling reduction. Both methods were effective in TSV net noise reduction with a small overhead on routing congestion and full-chip timing and power.

REFERENCES

1. R. Anglada and A. Rubio. A digital differential-line receiver for CMOS VLSI currents. *IEEE Transactions on Circuits and Systems*, 38(6):673–675, June 1991.
2. ANSYS. High Frequency Structural Simulator. www.ansys.com.
3. Y.-J. Chang et al. Novel crosstalk modeling for multiple through-silicon-vias (TSV) on 3-D IC: Experimental validation and application to Faraday cage design. In *Proceedings of IEEE Electrical Performance of Electronic Packaging*, Tempe, AZ, pp. 232–235, 2012.
4. J. Cho et al. Modeling and analysis of through-silicon via (TSV) noise coupling and suppression using a guard ring. *IEEE Transactions on Components, Packaging, and Manufacturing Technology*, 1(2): 220–233, 2011.
5. N.H. Khan, S.M. Alam, and S. Hassoun. GND plugs: A superior technology to mitigate TSV-induced substrate noise. *IEEE Transactions on Components, Packaging, and Manufacturing Technology*, 3(5):849–857, May 2013.
6. D.H. Kim, S. Mukhopadhyay, and S.K. Lim. Fast and accurate analytical modeling of through-silicon-via capacitive coupling. *IEEE Transactions on Components, Packaging, and Manufacturing Technology*, 1(2):168–180, February 2011.
7. J. Kim et al. Modeling and analysis of differential signal Through Silicon Via (TSV) in 3D IC. *IEEE CPMT Symposium*, Tokyo, Japan, pp. 1–4, August 2010.
8. C. Liu, T. Song, J. Cho, J. Kim, J. Kim, and S.K. Lim. Full-chip TSV-to-TSV coupling analysis and optimization in 3D IC. In *Proceedings of ACM Design Automation Conference*, San Diego, CA, pp. 783–788, June 2011.
9. K.-C. Lu et al. Wideband and scalable equivalent-circuit model for differential through silicon vias with measurement verification. In *IEEE Electronic Components and Technology Conference*, Las Vegas, NV, pp. 1186–1189, May 2013.
10. C.R. Paul. *Analysis of Multiconductor Transmission Lines*. John Wiley & Sons, Lexington, KY, 1994.
11. Y. Peng, D. Petranovic, and S.K. Lim. Fast and accurate full-chip extraction and optimization of TSV-to-wire coupling. In *Proceedings of ACM Design Automation Conference*, San Francisco, CA, pp. 1–6, June 2014.
12. Y. Peng, T. Song, D. Petranovic, and S.K. Lim. On accurate full-chip extraction and optimization of TSV-to-TSV coupling elements in 3D ICs. In *Proceedings of IEEE International Conference on Computer-Aided Design*, San Jose, CA, pp. 281–288, November 2013.
13. C. Serafy and A. Srivastava. TSV replacement and shield insertion for TSV–TSV coupling reduction in 3D global placement. *IEEE Transactions on Computer-Aided Design of Integrated Circuits and Systems*, PP(99):1–1, 2014.
14. T. Song, C. Liu, Y. Peng, and S.K. Lim. Full-chip multiple TSV-to-TSV coupling extraction and optimization in 3D ICs. In *Proceedings of ACM Design Automation Conference*, Austin, TX, pp. 1–7, May 2013.
15. S. Uemura et al. Isolation techniques against substrate noise coupling utilizing through silicon via (TSV) process for RF/mixed-signal SoCs. *IEEE Journal of Solid-State Circuits*, 47(4):810–816, April 2012.

Section II

Physical Design Methods
for 3D Integration

5 3D Placement and Routing

Pingqiang Zhou and Sachin S. Sapatnekar

CONTENTS

ABSTRACT

A critical part of building a successful 3D system lies in the ability to physically arrange the circuitry in such a way that manages myriad complexities related to thermal management, wire length optimization, and communication overheads. This chapter first presents an overview of methods for 3D placement, both under TSV-based and monolithic integration models. Next, it discusses algorithms for 3D routing for gate-level design after placement. Finally, methods for optimized chip-level communication using 3D NoCs are discussed.

5.1 INTRODUCTION

The physical arrangement of logic blocks plays a crucial role in exploiting the advantages of short interconnects in 3D technologies and in averting the disadvantages of thermal hot spots. This chapter presents an overview of techniques used for physical planning, placement, and routing of 3D-integrated circuits (3D-ICs).

The benefits that can be leveraged from physical design depend on the specific flavor of 3D-IC technology that is employed. Although 3D-ICs potentially offer increased system integration over 2D planar ICs, their integration densities vary for different technologies [8,16]. At the coarsest level, wire bonding may be used to connect multiple separately fabricated layers through connections at the periphery; for such technologies, the vertical interconnect pitch is of the order of 35 µm. A larger number of inter-layer connections are made possible by microbump technology, which uses bumps on die surface for vertical interconnect and can achieve smaller pitches of the order of 10 µm. The through-silicon-via (TSV) approach (Figure 5.1a) can reduce the vertical pitch down to 1–5 µm by etching a hole right through the wafer to build a via that connects the layers. At the finest level, monolithic 3D integration (Figure 5.1b) can make 100 nm pitch vertical interconnections possible and enables the 3D stacking of transistors on the same wafer. It is generally accepted that true 3D integration is based on the use of TSVs or monolithic integration, and other technologies are sometimes lumped into the category of 2.5D integration, that is, somewhere in between 2D and 3D

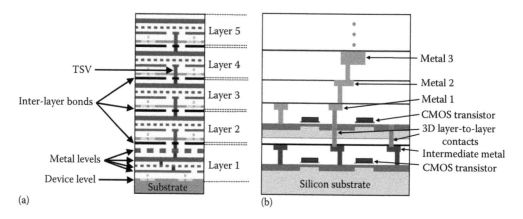

FIGURE 5.1 Two 3D schemes showing (a) TSV-based and (b) monolithic integration (Adapted from Bobba, S. et al., CELONCEL: Effective design technique for 3-D monolithic integration targeting high performance integrated Circuits, In *Proceedings of the Asia-South Pacific Design Automation Conference*, Yokohama, Japan, pp. 336–343, 2011.).

technologies. Of the true 3D technologies, TSV-based integration is the closest to reality because it provides the right balance between its potential to achieve high packing density and its ability to reuse traditional 2D-IC fabrication processes by bonding prefabricated dies.

Circuit design using 3D integration techniques can theoretically stack circuit components at various levels of granularity [9]. Individual units in each 3D stack may correspond to cores and/or memory, or functional blocks, or gates, or even transistors. The first two of these map on to TSV-based integration, where the footprint of a TSV is relatively large, while the latter two can only be achieved through monolithic integration. Aside from being farther from economic feasibility, monolithic integration is also more disruptive as it requires changes to the entire traditional design flow: from the standard cell library design to technology mapping, physical design, and up to GDSII file generation.

The move from 2D to 3D implies larger challenges to circuit design in many aspects. First, the larger design space in 3D results in increased design complexity. For example, Li et al. [31] showed that, given a floorplanning problem with n blocks, the solution space of 3D floorplanning with L layers increases by $n^{L-1}/(L-1)!$ times compared to the 2D case. Second, in 3D circuits, a large number of active devices are packed into a much smaller area, so that the power density is much higher than in a corresponding 2D circuit. As a result, the thermal issue is a key factor in 3D design [18]. Third, due to the significant mismatch of the coefficients of thermal expansion (CTEs) between TSV fill material (e.g., the CTE of copper is 17.7 ppm/°C) and silicon substrate (with a CTE of 3.05 ppm/°C), thermal stress builds up in the region surrounding TSVs during 3D-IC fabrication process [25,35]. This thermally induced mechanical stress can affect both chip performance and reliability. Therefore, it is necessary to develop analysis tool for thermal stress and develop design optimization techniques to alleviate its impact on 3D-ICs.

Fundamentally, the problem of 3D design is related to topological arrangements of blocks, and therefore, physical design plays a natural role in determining the success of 3D design strategies. Compared to its 2D counterpart, new challenges arise in 3D physical design:

1. The vertical space in the 3D design is not continuous; in other words, in the legalized layout, the circuit components (blocks or cells) are only allowed to be placed at a given set of discrete layers. Therefore, although circuit components can be freely moved about in a continuous 3D space at intermediate steps of floorplanning [51] or placement [18] algorithms, a layer-assignment step is required to assign a circuit component to a certain layer finally. The transition from a continuous solution to a discrete one incurs a mismatch in the solution quality, and a smooth transition is necessary for quality control.

2. Inter-layer connections are unique in 3D-IC design. Inter-layer vias can be explicitly inserted into the circuit before global placement [26], or after global placement but before the legalization step [23]. The number and location of inter-layer connections have a great impact on the performance of a 3D circuit. Considering that the inter-layer connections typically occupy significant silicon area, it is necessary to consider the planning of inter-layer connections at the early stage of 3D physical design, including reserving necessary white space for inter-layer connections in the floorplanning or placement stage.

3. Thermal considerations are particularly important in 3D chips. Thermal hot spots can be overcome through a number of strategies, such as reducing the amount of power dissipated on chip by lower power design techniques, controlling the thermal profile across the chip by altering the distribution of the heat sources, and improving the heat sinking pathways on the chip by the insertion of thermal vias [49]. Thermal vias are inter-layer metal connections that have no electrical function, but instead, constitute a passive cooling technology that draws heat from the problem areas to the heat sink. The thermal conductivity for epoxy is $k_{epoxy} = 0.05$ W/mK, while the corresponding values for silicon and copper are $k_{silicon} = 150$ W/mK and $k_{copper} = 401$ W/mK. Thermal via insertion can be built into floorplanning, placement and routing, or performed as an independent post-processing step, depending on the design methodology.

Therefore, physical design in the 3D realm requires a fresh approach. A new design flow must be devised to handle 3D-specific problems such as layer assignment and vertical interconnect (e.g., TSV) assignment, new cost functions become important such as thermal and stress, and ordinary extensions of 2D approaches are unequal to the task of solving these problems.

5.2 3D PLACEMENT

The 3D and 2D placement problems have many common features: they both arrange standard cells into rows within the layout, under objective functions related to parameters such as wirelength, congestion, and chip area; they both must consider the special case of the mixed-size placement problem that has large functional blocks (e.g., embedded memories, IP blocks, etc.) in addition to standard cells [10]; they can both leverage a common set of approaches that were developed for 2D placement problem, such as force-directed placement, quadratic placement, and partitioning-based placement [18,20].

However, 3D placement has its specific concerns. First, inter-layer vias occupy significant amount of layout area, and therefore, during the placement step, it is necessary to reserve white space for their insertion. Second, the number and location of inter-layer vias must be carefully planned, so as to achieve a good balance between performance and chip area. Third, inter-layer vias can serve as good heat conduction paths because of their high thermal conductivity, and judiciously placing the signal and thermal inter-layer vias helps to achieve a better temperature distribution in the 3D stack. Fourth, 3D placement can also optimize the TSV distribution to relieve the impact of TSV-induced thermal stress on circuit performance. We will discuss 3D-specific placement issues in detail in Sections 5.2.1 and 5.2.2.

5.2.1 TSV-BASED INTEGRATION

In this section, we discuss the 3D-specific placement problems in TSV-based 3D circuits.

5.2.1.1 Thermally Driven 3D Placement

The result of a typical 3D thermally aware placement [18] is displayed in Figure 5.2, which also shows a layout for the benchmark circuit, ibm01, in a four-layer 3D process. The cells are positioned in ordered rows on each layer, and the layout in each individual layer looks similar to a 2D standard cell layout. The heat sink is placed at the bottom of the 3D chip, and the lighter shaded regions are cooler than the darker shaded regions. The coolest cells are those in the bottom layer, next to the heat sink, and the temperature increases as we move to higher layers.

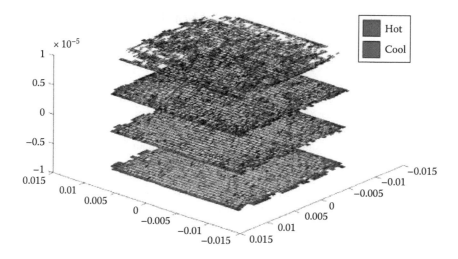

FIGURE 5.2 A placement for the benchmark ibm01 in a four-layer 3D technology.

Figure 5.2 shows that the on-chip thermal effects can be alleviated in 3D placement stage by

1. *Altering the distribution of the heat sources*: This can be achieved by making the upper layers sparser than the lower layers, in terms of the percentage of area populated by the cells [18], or by moving hot blocks/cells closer to the heat sink or apart away from each other in each layer [13,51].
2. *Improving the heat sinking pathways on the chip*: Considering that the inter-layer metal interconnects are the main channels for heat flow across the layers, we can either
   a. *Optimize the distribution of signal TSVs*: The work in [11] shows that if we assume the dielectric layer is an ideal heat insulator, then the peak temperature of the 3D chip is minimized when the number of TSVs in each bin is proportional to the lumped total power consumption in that bin, together with the bins in all the layers directly above it, or
   b. *Deliberately insert thermal vias into 3D circuits*: A key observation in work [19] is that the insertion of thermal vias is most useful in areas with a high thermal gradient, rather than areas with a high temperature. Effectively, the thermal via acts as a pipe that allows the heat to be conducted from the higher temperature region to the lower temperature region; this, in turn, leads to temperature reductions in areas of high temperature. This is illustrated in Figure 5.3, which shows the 3D layout of the benchmark *struct*, before and after the addition of thermal vias, respectively. The hottest region is the center of the uppermost layer, and a major reason for its elevated temperature is because the layer below it is hot. Adding thermal vias to remove heat from the second layer, therefore, also significantly reduces the temperature of the top layer. For this reason, the regions where the insertion of thermal vias is most effective are those that have high thermal gradients.

5.2.1.2 Stress-Driven 3D Placement

TSVs may be made of copper, tungsten, or polysilicon; of these, copper is the primary choice because of its low resistivity. During the manufacturing process for 3D-ICs, the TSV and silicon wafer undergo several thermal cycles before a final annealing step that embeds the TSV in the wafer. During annealing and the subsequent cooling, the structure is subject to a thermal ramp from about 250°C down to room temperature. Because of the mismatch in the CTE between copper TSV (17.7 ppm/°C) and the silicon (3.05 ppm/°C), copper contracts faster than silicon. This creates

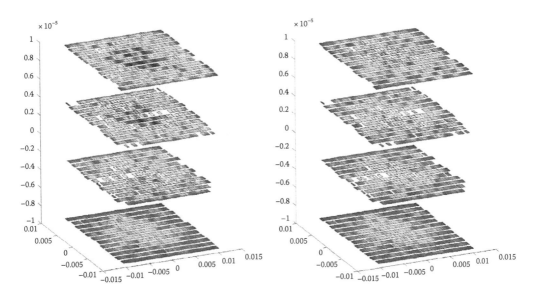

FIGURE 5.3 Thermal profile of struct before (a) and after (b) thermal via insertion. The four layers of the right figure correspond to the four layers in the left figure.

a tensile stress in the region surrounding the TSV as the surrounding silicon is pulled in due to the CTE differential [32]. According to the piezoresistive theory of materials, this stress alters the electrical conductivity of devices and hence the mobility of the transistors [35].

Figure 5.4 shows the color maps of the delay changes in PMOS and NMOS transistors in the gates for the benchmark circuit *spi*. The square white portions correspond to the TSV locations and represent the keep-out zones (KOZs) around each TSV, where the stress is too high to allow circuitry to be placed therein. From the figure we can see that the amount of path delay variation depends on the relative location of the gates with the TSVs, and the maximum delay changes are observed in the regions horizontal and vertical to the TSVs where the stresses are concentrated. Moreover, by comparing the scales, we can conclude that the change in the PMOS delays are dominant in the circuit compared to the NMOS delays.

Given the fact that the TSV stress may degrade the performance of a PMOS/NMOS, the work in [7] proposes a stress-driven timing optimization flow by exploiting the TSV stress-dependent mobility variation. The approach first uses min-cut partitioning to assign cells to layers, and then adds TSVs to the netlist and performs placement. The stress maps associated with the TSVs are

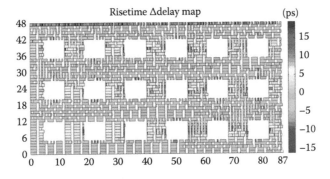

FIGURE 5.4 Rise delay changes in the gates for the *spi* circuit.

then computed on each layer, and these stress values are used to perform static timing analysis and determine the critical paths. The cells on the critical paths are moved using stress-aware placement, which is an extension of the force-directed 3D placement [18] that captures stress effects on timing. Stress maps are updated between placement iterations and special measures are taken to ensure the stability of the placer between iterations. Results on IWLS 2005 benchmarks [2] and industrial circuits show that stress-driven placement can achieve reduction of 21.6% in worst negative slack and 28% in total negative slack compared to wirelength-driven placement.

5.2.1.3 TSV Planning

Vertical interconnects in TSV-based 3D-ICs can be enabled by TSVs (Figure 5.1a). According to [26], the area of a minimum TSV cell at 45 nm is 6.1 μm², while the area of a minimum two-input NAND gate of the NCSU 45-nm library is 1.88 μm² [1], and the area of a minimum D flip-flop gate in the Nangate 45-nm library [3] is 4.52 μm². Therefore, TSVs occupy significant area and their layout impact should be considered in the placement stage. If we insert TSVs during the routing stage, then the distribution of the white space available for these TSVs greatly affects the quality of the routing solution. This is because the available locations for TSV insertion tend to be located along the chip periphery, if the white space is not reserved for TSVs during placement. As a result, the placement solution would incur longer wirelength since the available TSVs may be far from the cells that they are connected to [23].

Figure 5.5 shows how white space can be reserved during global placement. Given a net and its connected pins, we can first find its bounding box, that is, the range spanned by this net. Assuming that the communication between neighboring layers of a net requires one TSV, the net in this example spans three layers, so it requires two TSVs. Then the required space for these two TSVs is distributed into density cubic cubes overlapping with the net-box evenly, where cubes are used in the density model of the global placement stage to evenly distribute blocks and TSVs over a 3D chip.

Considering that the TSVs in 3D circuits are competing for limited white space resources with other circuit blocks such as CMOS decaps [52], it is necessary to optimize the TSV count during placement. At the macro block level, researchers have addressed this during the floorplanning step [51,53]. At the placement stage, the work in [20] seeks to minimize the wirelength and inter-layer via (or TSV) counts in all the three steps of placement stage, namely global placement, coarse legalization, and detailed legalization, by applying the following objective function in all steps:

$$\sum_{\text{each net } i} [WL_i + \alpha_{ILV} \cdot ILV_i] \tag{5.1}$$

where
  WL_i is the bounding box wirelength
  ILV_i is the number of inter-layer vias for net i

They further explore the tradeoff between inter-layer via count and wirelength as shown in Figure 5.6. It is clearly seen that as the inter-layer via coefficient, α_{ILV}, increases, inter-layer via counts decrease and wirelength increases. Wirelength reductions within 2% of the maximum can be achieved using 46% fewer inter-layer vias.

In addition to controlling the total number of TSVs during the placement of a 3D chip, it is also beneficial to control the TSV density in a local region during placement. Lim et al. [27] point out that the topography variation in metal layers is becoming more serious as technology advances below 45 nm, and semiconductor manufacturers require strict and tight metal density rules, including the range of metal density, because topography is determined mainly by underlying feature density, and in TSV-based 3D circuits, the landing pads are large enough to cause significant metal density mismatches. Therefore, it is necessary to control the density of TSVs across the chip area

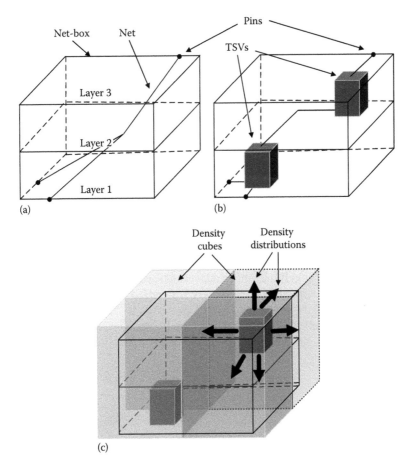

FIGURE 5.5 White space reservation during 3D global placement: (a) a net and its corresponding bounding box, (b) the two TSVs required for this net, and (c) the required spaces for TSVs are distributed into cubic cubes overlapping with the bounding box. (Adapted from Hsu, M.-K., Balabanov, V., and Chang, Y.-W., TSV-aware analytical placement for 3-D IC designs based on a novel weighted-average wirelength model, *IEEE Transactions on Computer-Aided Design of Integrated Circuits and Systems*, 32(4), 497–509, © 2013 IEEE.)

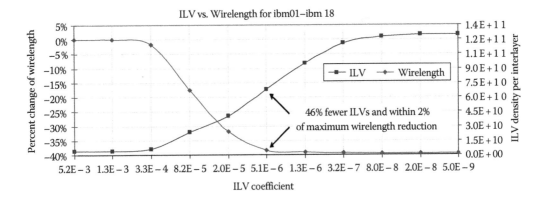

FIGURE 5.6 Average wirelength vs. ILV tradeoff for 18 IBM benchmarks. (From Goplen, B. and Sapatnekar, S.S., Placement of 3D ICs with thermal and interlayer via considerations, In *Proceedings of the ACM/IEEE Design Automation Conference*, San Diego, CA, pp. 626–631.)

and Lim et al. [27] propose to optimize the distribution of TSVs in global placement, by adding TSV density forces in force-directed 3D placement. Their results show that TSV-density-driven placement can achieve more uniform TSV distribution and therefore smaller metal density variation.

5.2.2 3D Monolithic Integration

Early work [18] has explored the problem of moving standard cells in a 3D continuous space during the placement stage. This idea is easily extended to monolithic integration as shown in Figure 5.1b where the active layer of one standard cell can be stacked above that of another cell. 3D monolithic integration starts with the design of 3D monolithic library. Researchers have proposed two possible solutions recently [8,9]. The first solution is to fold the layout of one *single* cell in multiple device layers as shown in Figure 5.7 [8]. In this example, the p-type and n-type devices are realized on two active layers, with p-type device sitting on top of n-type device. In this kind of transformation, the area of a 3D cell is determined by the p-type device because it is typically larger than the n-type device. In addition, there could be a slight increase in the space due to the I/O pins in the 3D layout, so as to obey the design rules if we take the close proximity of wider power rails into account. One of the main advantages of intra-cell stacking is that it can be easily integrated with the conventional design flow: only the design effort of the 3D library is required, and no changes are needed to be made to the existing physical design tools.

The other solution is to perform cell-on-cell stacking [9], by stacking cells on top of each other, as shown in Figure 5.8. The pin access of the bottom cell is maintained by allocating some extra

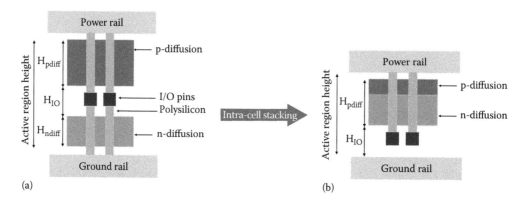

FIGURE 5.7 (a) The layout of a typical cell (inverter). (b) Intra-cell stacking by realizing p-type and n-type devices, respectively, on two adjacent active layers. (Adapted from Bobba, S. et al., CELONCEL: Effective design technique for 3-D monolithic integration targeting high-performance integrated circuits, In *Proceedings of the Asia-South Pacific Design Automation Conference*, Yokohama, Japan, pp. 336–343, 2011.)

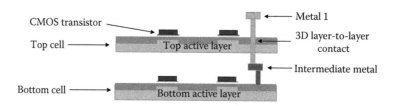

FIGURE 5.8 Cell-on-cell stacking, with two cells sitting at adjacent layers. (Adapted from Bobba, S. et al., CELONCEL: Effective design technique for 3-D monolithic integration targeting high-performance integrated circuits, In *Proceedings of the Asia-South Pacific Design Automation Conference*, Yokohama, Japan, pp. 336–343, 2011.)

space at the top active layer. An intermediate metal layer between the two active layers is used for the intra-cell routing of the bottom cell. Compared to existing 2D standard cell libraries, intra-cell stacking is shown to achieve 29% reduction on average in the cell height, while cell-on-cell stacking leads to 25% increase in the cell height.

The work [9] also proposes a physical design tool, named CELONCEL, for the placement of 3D circuit using cell-on-cell stacking technology. The design flow consists of four main steps:

1. *Deflation*: The standard cell library is transformed, so that the width of each cell is shrunk by half, together with the *x*-coordinates of the pin geometry defined for each cell. Next, any commercial placement tool can be used to generate a virtual seed placement without any overlap among the transformed cells.
2. *Inflation*: The width of each cell is transformed back to its original size, with the center of the cell remaining fixed in the seed placement. As a result of this operation, it is likely that overlaps will be generated between cells.
3. *Active layer assignment*: The active layer of each cell is assigned, with the objective of minimizing the overlap with the neighboring cells. Note that at this step the cells are fixed in the horizontal active area plane. A 0–1 linear program is formulated and solved for each cell row for this step.
4. *Legalization*: After the layer assignment step, there may be still some overlap that remains. Therefore, a legalization step is performed to remove such overlap, with the objective function of minimizing the total displacement of all the cells in their own active layer from the optimal location determined by the placement tool. This can be achieved by solving a linear program problem for each cell row.

5.3 3D ROUTING

3D routing involves the routing for power, signal, and clock nets [29]. In this section, we mainly focus on signal routing, and the reader may refer to Chapters 6, 8, and 9 for issues related to power and clock networks in 3D-ICs.

3D signal routing solution can be classified into two scales: at the gate/cell level and at the core level. At the gate level, routing involves finding the optimal path to connect all the pins of each net. At the chip level, the 3D routing problem is related to the generation of Network-on-Chip (NoC) architecture, which involves (1) the synthesis of NoC topology, that is, to determine the number of used routers and their connections to the cores/other routers, and (2) the routing of traffic flows in the synthesized NoC topology.

5.3.1 GATE-LEVEL ROUTING

In 2D routing, it is important to optimize the wire length, the delay, and the congestion. As in 3D routing, several 3D-specific issues come into play. First, the delay of a wire increases with its temperature, so that more critical wires should avoid the hottest regions, as far as possible. Second, inter-layer vias are a valuable resource that must be optimally allocated among the nets. Third, congestion management and blockage avoidance is more complex with the addition of a third dimension. For instance, a signal via or thermal via that spans two or more layers constitutes a blockage that wires must navigate around.

Consider the problem of routing in a three-layer technology. The layout is gridded into rectangular tiles, each with a horizontal and vertical capacity that determines the number of wires that can traverse the tile, and an inter-layer via capacity that determines the number of free vias available in that tile. These capacities account for the resources allocated for nonsignal wires (e.g., power and clock wires) as well as the resources used by thermal vias. For a single net, the degrees of freedom that are available are in choosing the locations of the inter-layer vias, and selecting the precise routes within each layer. The locations of inter-layer vias will depend on the resource contention for vias within each grid. Moreover, critical wires should avoid the high-temperature tiles, as far as possible.

The work in [12] presents a thermally conscious router, using a multilevel routing paradigm, with integrated interlayer via planning and incorporating thermal considerations. An initial routing solution is constructed by building a 3D minimum spanning tree (MST) for each multipin net, and using maze routing to avoid obstacles. At each level of the multilevel scheme, the inter-layer via planning problem assigns vias in a given region at level $k - 1$ of the multilevel hierarchy to tiles at level k. The problem is formulated as a min-cost max-flow problem, which has the form of a transportation problem. The flow graph is constructed as follows:

1. The source node of the flow graph is connected through directed edges to a set of nodes v_i, representing candidate thermal vias; the edges have capacity 1 and cost 0.
2. Directed edges connect a second set of nodes, T_j, from each tile to the sink node, with capacity equaling the number of vias that the tile can contain, and cost zero. The capacity is computed using a heuristic approach that takes into account the temperature difference between the tile and the one directly in the layer below it (under the assumption that heat flows downwards toward the sink); the thermal analysis is based on a commercial FEA solver.
3. The source and sink both have cost m, which equals the number of inter-layer vias in the entire region.
4. Finally, a node v_i is connected to a tile T_j through an arc with infinite capacity and cost equaling the estimated wirelength of assigning an inter-layer via v_i to tile T_j.

The 3D routing approach presented in [50] combines the problem of 3D routing with heat removal by inserting thermal vias in the z direction, and introduces the concept of thermal wires. Like a thermal via, a thermal wire is a dummy object; it has no electrical function, but is used to spread heat in the lateral direction. Each layer is tiled into a set of regions.

The global routing scheme goes through two phases. In Phase I, an initial routing solution is constructed. A 3D MST is built for each multipin net, and based on the corresponding two-pin decomposition, the routing congestion is statistically estimated over each lateral routing edge using the method in [46]. This congestion model is extended to 3D by assuming that a two-pin net with pins on different layers has an equal probability of utilizing any inter-layer via position within the bounding box defined by the pins.

A recursive bipartitioning scheme is then used to assign inter-layer vias. This is also formulated as a transportation problem, but the formulation is different from the multilevel method described earlier. Signal inter-layer via assignment is then performed across the cut in each recursive bipartition. Figure 5.9a shows an example of signal inter-layer via assignment for a decomposed two-pin signal net in a four-layer circuit with two levels of hierarchy. The signal inter-layer via assignment is first performed at the boundary between regions of group 0 and group 1 at topmost level, and then it is processed for layer boundary within each group. At each level of the hierarchy, the problem of signal inter-layer via assignment is formulated as a min-cost network flow.

Figure 5.9b shows the network flow graph for assigning signal inter-layer vias of five inter-layer nets to four possible inter-layer via positions. The idea is to assign each net that crosses the cut to an inter-layer via. Each inter-layer net is represented by a node N_i in the network flow graph; each possible inter-layer via position is indicated by a node C_j. If C_j is within the bounding box of the two-pin inter-layer net N_i, we build a directed edge from N_i to C_j, and set the capacity to be 1, the cost of the edge to be cost (N_i, C_j). The cost (N_i, C_j) is evaluated as the shortest path cost for assigning inter-layer via position C_j to net N_i when both pins of N_i are on the two neighboring layers; otherwise it is evaluated as the average shortest path cost over all possible unassigned signal inter-layer via positions in lower levels of the hierarchy. The shortest path cost is obtained with Dijkstras algorithm in the 2D

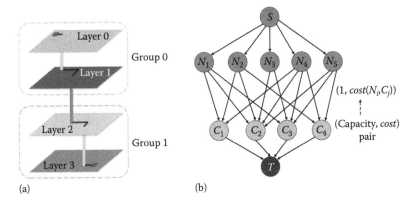

FIGURE 5.9 (a) Example of hierarchical signal via assignment for a four-layer circuit. (b) Example of min-cost network flow heuristics to solve signal via assignment problem at each level of hierarchy. (Adapted from Zhang, T. et al., Temperature-aware routing in 3D ICs, In *Proceedings of the Asia-South Pacific Design Automation Conference*, Yokohama, Japan, pp. 309–314, 2006.)

congestion map generated from the previous estimation step, and the cost function for crossing a lateral routing edge is a combination of edge length and an overflow cost function similar to that in [21]. The supply at the source, equaling the demand at the sinks, is N, the number of nets.

Finally, once the inter-layer vias are fixed, the problem reduces to a 2D routing problem in each layer, and maze routing is used to route the design.

Next, in Phase II, a linear programming approach is used to assign thermal vias and thermal wires. A thermal analysis is performed, and fast sensitivity analysis is carried out using the adjoint network method, which has the cost of a single thermal simulation. The benefit of adding thermal vias, for relatively small perturbations in the via density, is given by a product of the sensitivity and the via density, a linear function. The objective function is a sum of via densities and is also linear. Additional constraints are added in the formulation to permit overflows, and a sequence of linear programs is solved to arrive at the solution.

5.3.2 CHIP-LEVEL ROUTING

At the chip level, the communication among the circuit components, also called processing elements (PEs), for example, general or special purpose processors, embedded memories, application-specific components, mixed-signal I/O cores, etc., are enabled by NoCs [14]. Figure 5.10 shows one such example. The traffic through the NoC depends on how the applications are partitioned and mapped onto the PEs [33].

The 3D routing problem at the chip level is equivalent to the generation of NoC architecture, which involves

1. The synthesis of NoC topology, that is, to determine the number of used routers and their connections to the cores/other routers. The optimal network topology depends on the actual applications. For instance, a 2D mesh network (Figure 5.11a) is widely used in regular tile-based chip multiprocessors (CMPs), which are designed for general purpose of applications with various kinds of traffic patterns. Because the diameter of a mesh grows linearly with the dimension, it results in long communication distance, hence, poor performance for large networks. On the other hand, fully customized networks, such as in Figure 5.11b, are preferred in application-specific system-on-chip (SoC) applications because such applications have predictable traffic patterns; a customized topology can

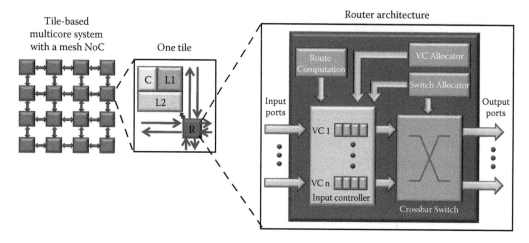

FIGURE 5.10 A general NoC architecture in a tile-based multicore system. Each tile consists of a processing core and caches, and is connected to other tiles by a router. The NoC in a tile-based system usually applies a mesh structure.

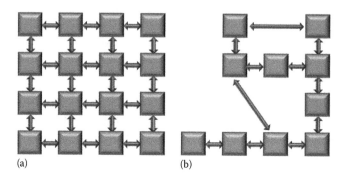

FIGURE 5.11 Regular and fully customized communication architectures. Regular structures (a) provide greater scalability but often worse performance, while fully customized ones (b) require more design effort but yield better performance results. (a) Regular topology and (b) customized topology.

provide significant improvements in terms of performance, power consumption, and area compared to the regular topologies such as mesh.

2. The routing of traffic flows in the synthesized NoC topology, that is, to find the paths in the NoC topology to forward the packets in the traffic flows between a source node and a destination node. Figure 5.12 shows an example of traffic routing in a 3D chip with two layers. A core graph shown in Figure 5.12a specifies the communication demands between the PEs in the chip, while a routing graph as shown in Figure 5.12b represents the topology of the NoC. In this example, PE v_1 in layer 1 needs to communicate with PE v_2 in layer 2. The corresponding routing graph consists of five routers, three of them are in layer 1 and the other two are in layer 2. When we route a traffic flow in the given core graph ($v_1 \rightarrow v_2$ in this example), we must find at least one routing path from the source node v_1 to the destination node v_2 in the given routing graph. In this example, the path $v_1 \rightarrow v_{1,1}^r \rightarrow v_{1,2}^r \rightarrow v_2$ is one solution to the problem.

In the area of designing NoC architectures for 3D-ICs, most of the literature has focused on regular 3D NoC topologies such as 3D meshes [5,17,28,36,42], which are appropriate for regular 3D designs

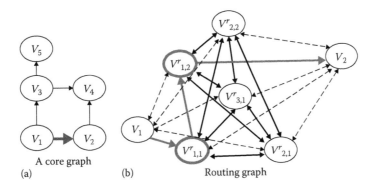

(a) A core graph

(b) Routing graph

FIGURE 5.12 Example of traffic routing in a 3D chip with 2 layers.

[30,37], such as the CMPs. The main task of the 3D routing problem in a regular 3D NoC is to route the traffic flows in the given NoC topology, by applying a proper routing algorithm at the routers in the network. The routing algorithm consists of two parts: one that determines which output ports a packet can be sent to so that it may reach its destination, and another that selects one of these ports. In deterministic routing, only one output port is considered in the first step and further selection is not needed. In oblivious routing, multiple choices are available, and one of these is selected based solely on information in the header flit, while adaptive routing uses information on network status, such as buffer and link utilizations, in selecting the output port. Routing algorithms not only affect the network transmission latency but also can impact the power/energy consumption and throughput of the communication network. For details about the routing algorithms, the reader is referred to [15,41].

Some work [47] also considers partially regular scheme in regular 3D design, that is, the topology in the horizontal direction (x and y) is regular, but the interconnects in the vertical (z) direction can be customized. This is because inter-layer vias in 3D chips are scarce resources, and both their numbers and locations on chip should be optimized so as to find a balance between performance and inter-layer via cost.

As discussed before, most modern SoC architectures consist of heterogeneous cores such as CPU or DSP modules, video processors, and embedded memory blocks, and the traffic requirements among the cores can vary widely. Therefore, regular topologies such as meshes may have significant area and power overhead [39,48], and tuning the topology for application-specific solutions can provide immense benefits.

The synthesis of an application-specific NoC architecture includes finding the optimal number and size of switches, establishing the connectivity between the switches and with the cores, and finding deadlock-free [15] routing paths for all the traffic flows. For 2D systems, the problem of designing application-specific NoC topologies has been explored by several researchers [6,22,38,45]. Srinivasan et al. [45] present a three-phase NoC synthesis technique consisting of sequential steps that floorplan the cores, next perform core-to-router mapping, and then generate the network topology. In [38], Murali et al. present an NoC synthesis method that incorporates the floorplanning process to estimate link power consumption and detect timing violations. Several topologies are explored, starting from one where all the cores are connected to one switch, to one where each core is connected to a separate switch. The traffic flows are ordered so that larger flows are routed first.

In the 3D domain, Yan and Lin [48] present an application-specific 3D NoC synthesis algorithm that is based on a rip-up-and-reroute procedure for routing flows, where the traffic flows are ordered in the order of increasing rate requirements so that smaller flows are routed first, followed by a router merging procedure. Murali et al. [39] propose a 3D NoC topology synthesis algorithm, which is an extension to their previous 2D work, described earlier. The 3D NoC synthesis problem has been shown to be NP-hard in [44].

However, there are several issues with the aforementioned work:

- First, the final results of application-specific NoC topology synthesis depend on the order in which the traffic flows are routed. In some cases, routing larger flows first provides better results [22,38], while in others, routing the smaller flows first may yield better results [48]. A strategy is required to reduce the dependency of the results on flow ordering.
- Second, in all of the works mentioned previously, the average hop count is used to approximate the average packet latency in NoCs. This ignores the queuing delays in switch ports and the contention among different packets for network resources such as switch ports and physical links, and cannot reflect the impact of physical core-to-switch or switch-to-switch distances on network latency. More accurate delay models that include the effects of queuing delay and network contention, and better delay metrics, should be applied for NoC performance analysis.
- Last, the delays and power dissipation for physical links in NoCs are closely linked to the physical floorplan and topology of cores and switches. However, none of them consider the optimization of floorplan of the cores and switches in NoC topology synthesis.

Zhou et al. address these important problems in their application-specific NoC topology synthesis work [53]. Their design flow is shown in Figure 5.13. The input to their algorithm is a directed graph, called the core graph, $G(V, E, \lambda)$. Each node $v_i \in V$ represents a core (either a processing element or a memory unit) and each directed edge $e_{v_i,v_j} \in E$ denotes a traffic flow from source v_i to destination v_j. The bandwidth of traffic flow from core v_i to v_j is given by $\lambda(e_{v_i,v_j})$ in *MB/s*. In addition, NoC architectural parameters such as the NoC operating frequency, f, and the data link width, W, are also assumed to be provided as inputs. The operating frequency is usually specified by the design, and data link width is dictated by the IP interface standards. The 3D NoC synthesis framework in [53] permits a variety of objectives and constraints, including considerations that are particularly important in 3D, such as power dissipation, temperature, and the number of TSVs, and NoC-specific

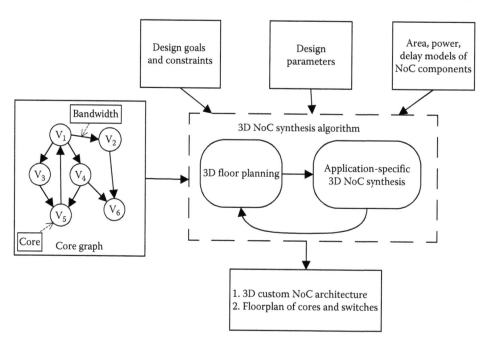

FIGURE 5.13 Application-specific 3D NoC synthesis flow. (From Zhou, P. et al., Optimized 3D network-on-chip design using simulated allocation. *ACM Transactions on Design Automation of Electronic Systems*, 17(2):1–19, April 2012.)

issues such as minimizing the average/maximum network latency, limitations on the maximum bandwidth, as well as general factors such as the design area.

Given a core graph, an initial floorplan of the cores is first obtained using a thermally aware floorplanner. This precedes the 3D NoC synthesis step, and is important because the core locations significantly influence the NoC architecture.

Then the 3D NoC synthesis algorithm is performed on a directed routing graph $G'(V', E')$: V' is the vertex set, which is the union of core set V in the input core graph $G(V, E, \lambda)$ and the set of added switches, V_s. It is assumed that the maximum number of switches that can be used in each 3D tier l equals to the number of cores in that tier, although it is easy to relax this restriction. The edge set E' is constructed as follows: the cores in a tier l are connected only to the switches in the same tier l, and adjacent tiers $l - 1, l + 1$, and the switches from all the 3D tiers form a complete graph. A custom NoC topology is a subgraph of the routing graph, G'.

The solution proposed in [53] to overcoming the ordering problem is based on the use of a multi-commodity flow (MCF) network formulation for the NoC synthesis problem; the advantage of such an approach is that it takes a global view of the problem and eliminates the aforementioned problem of finding the best order in which to route the traffic flows. For a core graph $G(V, E, \lambda)$ and a cor-responding routing graph $G'(V', E')$ (corresponding to a flow network), let $c(u, v)$ be the capacity of edge $(u, v) \in E'$. The capacity $c(u, v)$ equals to the product of the operating frequency f and data link width W. Each commodity $K_i = (s_i, t_i, d_i)$, $i = 1,...,k$ corresponds to the weight (traffic flow) along edge e_{s_i,t_i} in the core graph from source s_i to destination t_i, and $d_i = \lambda(e_{s_i,t_i})$ is the demand for com-modity i. Therefore, there are $k = |E|$ commodities in the core graph. Let the flow of commodity i along edge (u, v) be $f_i(u, v)$. Then the MCF problem is to find the optimal assignment of flow which satisfies the constraints:

$$\text{Capacity constraints:} \quad \sum_{i=1}^{k} f_i(u,v) \le c(u,v)$$

$$\text{Flow conservation:} \quad \sum_{\omega \in V', u \ne s_i, t_i} f_i(u, \omega) = 0$$

$$\text{where } \forall v, u \; f_i(u, v) = -f_i(v, u)$$

$$\text{Demand satisfaction:} \quad \sum_{\omega \in V'} f_i(s_i, \omega) = \sum_{\omega \in V'} f_i(\omega, t_i) = d_i$$

Superficially, this idea seems similar to [24], where an MCF formulation is proposed. However, that work is directed to 2D NoC synthesis with a single objective of minimizing NoC power, modeled as a linear function of the flow variables $f_i(u, v)$. The corresponding linear programming problem is solved using an approximation algorithm. The more general formulation in [53] integrates more objectives and more accurate modeling for NoC components. In fact, most components of the objective func-tion in [53] are nonlinear or, as in case of network latency, unavailable in closed form, rendering an LP-based approach impossible. Therefore, a stochastic simulated allocation (SAL) approach [43], that is particularly suitable for solving the MCF problems where the objective function is in such a form, is adapted to efficiently solve the MCF problem. The work in [53] also uses an accurate delay model for routers in NoCs [40], which considers the queuing delay and network contention. The SAL procedure yields the NoC topology and the paths for all the traffic flows in the core graph.

After the 3D NoC synthesis step, the actual switches and links in the synthesized 3D NoC archi-tecture are fed back to the floorplanner to update the floorplan of the cores and used switches, and the refined floorplan information is used to obtain more accurate power and delay estimates. The process continues iteratively: with the refined floorplan, a new SAL-based 3D NoC synthesis proce-dure is invoked to find a better synthesis solution, and so on.

Experimental results on a set of benchmarks show that the work in [53] can produce greatly improved solutions for various design objectives such as NoC power (average savings of 34%), network latency (average reduction of 35%), and chip temperature (average reduction of 20%), compared to the baseline algorithm with fixed-order flow routing, simple delay model and without feedback from floorplanning step.

5.4 SUMMARY AND CONCLUSIONS

In this chapter, we discussed different 3D integration technologies and their application in integrating different granularity of circuit components. We then discuss the 3D-specific design issues such as inter-layer vias, thermal profile, and thermal-induced stress, and then present solutions in placement and routing stages to overcome these issues. We also discuss the problems of physical design for 3D monolithic integration and 3D on-chip communication network, together with their CAD solutions.

REFERENCES

1. NCSU EDA Group, FreePDK45. Available online at: http://www.eda.ncsu.edu/wiki/FreePDK. Accessed August 31, 2015.
2. C. Albrecht, IWLS 2005 Benchmarks, 2005. Available online at http://www.iwls.org/iwls2005/benchmarks.html. Accessed August 31, 2015.
3. Nangate 45 nm Open Cell Library. Available online at http://www.nangate.com. Accessed August 31, 2015.
4. C. Ababei, Y. Feng, B. Goplen, H. Mogal, T. Zhang, K. Bazargan, and S. S. Sapatnekar. Placement and routing in 3D integrated circuits. *IEEE Design and Test*, 22(6):520–531, November 2005.
5. C. Addo-Quaye. Thermal-aware mapping and placement for 3-D NoC designs. In *Proceedings of the IEEE International SoC Conference*, Herndon, VA, pp. 25–28, 2005.
6. T. Ahonen, H. Bin, and J. Nurmi. Topology optimization for application-specific networks-on-chip. In *System Level Interconnect Prediction Workshop*, Paris, France, pp. 53–60, 2004.
7. K. Athikulwongse, A. Chakraborty, J.-S. Yang, D. Z. Pan, and S. K. Lim. Stress-driven 3D-IC placement with TSV keep-out zone and regularity study. In *Proceedings of the IEEE/ACM International Conference on Computer-Aided Design*, San Jose, CA, pp. 669–674, 2010.
8. P. Batude, M. Vinet, A. Pouydebasque, C. Le Royer, B. Previtali, C. Tabone, L. Clavelier et al. GeOI and SOI 3D monolithic cell integrations for high density applications. In *Proceedings of the IEEE Symposium on VLSI Technology*, Kyoto, Japan, pp. 166–167, 2009.
9. S. Bobba, A. Chakraborty, O. Thomas, P. Batude, T. Ernst, O. Faynot, D. Z. Pan, and G. De Micheli. CELONCEL: Effective design technique for 3-D monolithic integration targeting high performance integrated circuits. In *Proceedings of the Asia-South Pacific Design Automation Conference*, Yokohama, Japan, pp. 336–343, 2011.
10. J. Cong and G. Luo. An analytical placer for mixed-size 3D placement. In *Proceedings of the International Symposium on Physical Design*, San Francisco, CA, pp. 61–66, 2010.
11. J. Cong, G. Luo, and Y. Shi. Thermal-aware cell and through-silicon-via co-placement for 3D ICs. In *Proceedings of the ACM/EDAC/IEEE Design Automation Conference*, San Diego, CA, pp. 670–675, 2011.
12. J. Cong and Y. Zhang. Thermal-driven multilevel routing for 3D ICs. In *Proceedings of the Asia-South Pacific Design Automation Conference*, Shanghai, China, pp. 121–126, 2005.
13. J. Cong and Y. Zhang. Thermal-aware physical design flow for 3-D ICs. In *Proceedings of the International VLSI Multilevel Interconnection Conference*, pp. 73–80, 2006.
14. W. J. Dally and B. Towles. Route packets, not wires: On-chip interconnection networks. In *Proceedings of the ACM/IEEE Design Automation Conference*, Las Vegas, NV, pp. 684–689, 2001.
15. W. J. Dally and B. Towles. *Principles and Practices of Interconnection Networks*. Morgan Kaufmann, 2003.
16. W. R. Davis, J. Wilson, S. Mick, J. Xu, H. Hua, C. Mineo, A. M. Sule, M. Steer, and P. D. Franzon. Demystifying 3D ICs: The pros and cons of going vertical. *IEEE Design and Test*, 22(6):498–510, November 2005.

17. B. Feero and P. P. Pande. Performance evaluation for three-dimensional networks-on-chip. In *Proceedings of IEEE Computer Society Annual Symposium on VLSI*, Porto Alegre, Brazil, pp. 305–310, 2007.

18. B. Goplen and S. S. Sapatnekar. Efficient thermal placement of standard cells in 3D ICs using a force directed approach. In *Proceedings of the IEEE/ACM International Conference on Computer-Aided Design*, San Jose, CA, pp. 86–89, 2003.

19. B. Goplen and S. S. Sapatnekar. Placement of thermal vias in 3-D ICs using various thermal objectives. *IEEE Transactions on Computer-Aided Design of Integrated Circuits and Systems*, 25(4):692–709, April 2006.

20. B. Goplen and S. S. Sapatnekar. Placement of 3D ICs with thermal and interlayer via considerations. In *Proceedings of the ACM/IEEE Design Automation Conference*, San Diego, CA, pp. 626–631, 2007.

21. R. T. Hadsell and P. H. Madden. Improved global routing through congestion estimation. In *Proceedings of the ACM/IEEE Design Automation Conference*, Anaheim, CA, pp. 28–31, 2003.

22. A. Hansson, K. Goossens, and A. Rădulescu. A unified approach to constrained mapping and routing on network-on-chip architectures. In *Proceedings of the International Conference on Hardware-Software Codesign and System Synthesis*, Jersey city, NJ, pp. 75–80, 2005.

23. M.-K. Hsu, V. Balabanov, and Y.-W. Chang. TSV-aware analytical placement for 3-D IC designs based on a novel weighted-average wirelength model. *IEEE Transactions on Computer-Aided Design of Integrated Circuits and Systems*, 32(4):497–509, April 2013.

24. Y. Hu, Y. Zhu, H. Chen, R. Graham, and C.-K. Cheng. Communication latency aware low power NoC synthesis. In *Proceedings of the ACM/IEEE Design Automation Conference*, San Francisco, CA, pp. 574–579, 2006.

25. M. Jung, J. Mitra, D. Z. Pan, and S. K. Lim. TSV stress-aware full-chip mechanical reliability analysis and optimization for 3D IC. *Communications of the ACM*, 57(1):107–115, January 2014.

26. D. H. Kim, K. Athikulwongse, and S. K. Lim. Study of through-silicon-via impact on the 3-D stacked IC layout. *IEEE Transactions on VLSI Systems*, 21(5):862–874, May 2013.

27. D. H. Kim, R. O. Topaloglu, and S. K. Lim. TSV density-driven global placement for 3D stacked ICs. In *Proceedings of the International SoC Design Conference*, Taipei, Taiwan, pp. 135–138, 2011.

28. J. Kim, C. Nicopoulos, D. Park, R. Das, Y. Xie, V. Narayanan, M. S. Yousif, and C. R. Das. A novel dimensionally-decomposed router for on-chip communication in 3D architectures. In *Proceedings of the ACM/IEEE International Symposium on Computer Architecture*, San Diego, CA, pp. 138–149, 2007.

29. Y.-J. Lee and S. K. Lim. Co-optimization and analysis of signal, power, and thermal interconnects in 3-D ICs. *IEEE Transactions on Computer-Aided Design of Integrated Circuits and Systems*, 30(11):1635–1648, November 2011.

30. F. Li, C. Nicopoulos, T. Richardson, Y. Xie, V. Narayanan, and M. Kandemir. Design and management of 3D chip multiprocessors using network-in-memory. In *Proceedings of the ACM/IEEE International Symposium on Computer Architecture*, Boston, MA, pp. 130–141, 2006.

31. Z. Li, X. Hong, Q. Zhou, Y. Cai, J. Bian, H. H. Yang, V. Pitchumani, and C.-K. Cheng. Hierarchical 3-D floorplanning algorithm for wirelength optimization. *IEEE Transactions on Circuits and Systems I*, 53(12):2637–2646, December 2006.

32. K. H. Lu, X. Zhang, S.-K. Ryu, J. Im, R. Huang, and P. S. Ho. Thermomechanical reliability of 3-D ICs containing through silicon vias. In *Proceedings of the IEEE Electronic Components and Technology Conference*, San Diego, CA, pp. 630–634, 2009.

33. R. Marculescu, U. Y. Ogras, L.-S. Peh, N. E. Jerger, and Y. Hoskote. Outstanding research problems in NoC design: System, microarchitecture, and circuit perspectives. *IEEE Transactions on Computer-Aided Design of Integrated Circuits and Systems*, 28(1):3–21, January 2009.

34. R. Marculescu, U. Y. Ogras, and N. H. Zamora. Computation and communication refinement for multiprocessor SoC design: A system-level perspective. *ACM Transactions on Design Automation of Electronic Systems*, 11(3):564–592, June 2006.

35. S. K. Marella, S. V. Kumar, and S. S. Sapatnekar. A holistic analysis of circuit timing variations in 3D-ICs with thermal and TSV-induced stress considerations. In *Proceedings of the IEEE/ACM International Conference on Computer-Aided Design*, San Jose, CA, pp. 317–324, 2012.

36. H. Matsutani, M. Koibuchi, and H. Amano. Tightly-coupled multi-layer topologies for 3-D NoCs. In *Proceedings of the IEEE International Conference on Parallel Processing*, Xi'an, China, pp. 75–75, 2007.

37. P. Morrow, B. Black, M. J. Kobrinsky, S. Muthukumar, D. Nelson, C.-M. Park, and C. Webb. Design and fabrication of 3D microprocessors. In *Proceedings of the Materials Research Society Symposium*, Boston, MA, 2007.

38. S. Murali, P. Meloni, F. Angiolini, D. Atienza, S. Carta, L. Benini, G. De Micheli, and L. Raffo. Designing application-specific networks on chips with floorplan information. In *Proceedings of the IEEE/ACM International Conference on Computer-Aided Design*, San Jose, CA, pp. 355–362, 2006.

39. S. Murali, C. Seiculescu, L. Benini, and G. De Micheli. Synthesis of networks on chips for 3D systems on chips. In *Proceedings of the Asia-South Pacific Design Automation Conference*, Yokohama, Japan, pp. 242–247, 2009.

40. U. Y. Ogras and R. Marculescu. Analytical router modeling for networks-on-chip performance analysis. In *Proceedings of the Design, Automation and Test in Europe*, Nice Acropolis, France, pp. 1096–1101, 2007.

41. M. Palesi and M. Daneshtalab. *Routing Algorithms in Networks-on-Chip*. Springer, Boston, MA, 2013.

42. V. F. Pavlidis and E. G. Friedman. 3-D topologies for networks-on-chip. *IEEE Transactions on Very Large Scale Integration Systems*, 15(10):1081–1090, 2007.

43. M. Pióro and D. Medhi. *Routing, Flow, and Capacity Design in Communication and Computer Networks*. Morgan Kaufmann Publishers, San Francisco, CA, 2004.

44. C. Seiculescu, S. Murali, L. Benini, and G. De Micheli. SunFloor 3D: A tool for networks on chip topology synthesis for 3D systems on chip. In *Proceedings of the Design, Automation and Test in Europe*, Nice, France, pp. 9–14, 2009.

45. K. Srinivasan, K. S. Chatha, and G. Konjevod. An automated technique for topology and route generation of application specific on-chip interconnection networks. In *Proceedings of the IEEE/ACM International Conference on Computer-Aided Design*, San Jose, CA, pp. 231–237, 2005.

46. J. Westra, C. Bartels, and P. Groeneveld. Probabilistic congestion prediction. In *Proceedings of the International Symposium on Physical Design*, Phoenix, AZ, pp. 204–209, 2004.

47. T. C. Xu, G. Schley, P. Liljeberg, M. Radetzki, J. Plosila, and H. Tenhunen. Optimal placement of vertical connections in 3D network-on-chip. *Journal of Systems Architecture*, 5(7):441–454, August 2013.

48. S. Yan and B. Lin. Design of application-specific 3D networks-on-chip architectures. In *Proceedings of the IEEE International Conference on Computer Design*, San Jose, CA, pp. 142–149, 2008.

49. Y. Zhan, S. V. Kumar, and S. S. Sapatnekar. Thermally aware design. *Foundations and Trends in Electronic Design Automation*, 2(3):255–370, March 2008.

50. T. Zhang, Y. Zhan, and S. S. Sapatnekar. Temperature-aware routing in 3D ICs. In *Proceedings of the Asia-South Pacific Design Automation Conference*, Yokohama, Japan, pp. 309–314, 2006.

51. P. Zhou, Y. Ma, Z. Li, R. P. Dick, L. Shang, H. Zhou, X. Hong, and Q. Zhou. 3D-STAF: Scalable temperature and leakage aware floorplanning for three-dimensional integrated circuits. In *Proceedings of the IEEE/ACM International Conference on Computer-Aided Design*, San Jose, CA, pp. 590–597, 2007.

52. P. Zhou, K. Sridharan, and S. S. Sapatnekar. Congestion-aware power grid optimization for 3D circuits using MIM and CMOS decoupling capacitors. In *Proceedings of the Asia-South Pacific Design Automation Conference*, Yokohama, Japan, pp. 179–184, 2009.

53. P. Zhou, P.-H. Yuh, and S. S. Sapatnekar. Optimized 3D network-on-chip design using simulated allocation. *ACM Transactions on Design Automation of Electronic Systems*, 17(2):1–19, April 2012.

6 Power and Signal Integrity Challenges in 3D Systems-on-Chip

Emre Salman

CONTENTS

ABSTRACT

A significant physical design challenge in both high-performance 3D integrated circuits and low-power 3D systems-on-chip is to guarantee system-wide power and signal integrity. This chapter provides an overview of these challenges with emphasis on through silicon via (TSV)-based 3D ICs. Different TSV types and their implications to power/signal integrity are first discussed. In the next section, power distribution methodology for a nine-plane processor-memory stack is described. Different decoupling capacitor topologies are also investigated for power-gated 3D ICs. The following section focuses on TSV-to-transistor noise coupling as an important signal integrity issue. A compact model is also proposed to achieve efficient noise coupling analysis in 3D ICs.

Finally, a case study is described to analyze signal integrity in a 3D integrated SoC with application to bioelectronics.

6.1 INTRODUCTION

Three-dimensional (3D) integration is a promising technology that maintains the benefits of miniaturization by utilizing the vertical dimension rather than decreasing the size of the devices in two dimensions [1–7]. Utilizing the vertical dimension not only increases the integration density (thereby reducing cost), but also reduces the length and number of global interconnects. This reduction enhances system performance (due to reduced interconnect delay) while lowering the power consumption (due to reduced switching capacitance). These advantages have attracted significant attention in the past decade to develop high-performance computing systems such as many-core processors with embedded memory, as illustrated in Figure 6.1a [8–11]. 3D technologies alleviate the existing gap between logic blocks and memory units in high-performance microprocessors by utilizing vertically embedded dynamic random access memory (DRAM), thereby significantly increasing the memory bandwidth and reducing memory access time.

In addition to facilitating high-performance computing systems, 3D technologies provide a unique opportunity for heterogeneous integration of diverse functions onto a monolithic chip, as depicted in Figure 6.1b. Heterogeneous integration expands the application domain of 3D-integrated circuits from high-performance computing to relatively lower power systems-on-chips (SoCs) consisting of, for example, sensors, analog interface circuit, memory, digital processing blocks, and RF wireless transmission circuitry. The ability to merge disparate circuits and technologies is essential for multifunctional systems since each plane can be optimized according to the requirements of that particular function. This advantage also supports the concept of *More-than-Moore* that targets the scaling of a system by incorporating nondigital portions into the same monolithic IC [12,13]. 3D integration of diverse functions is expected to enable exciting opportunities in a variety of fields including energy-efficient mobile computing, health care, and environmental control. For example, The Semiconductors Electronics Division of the National Institute of Standards and Technology

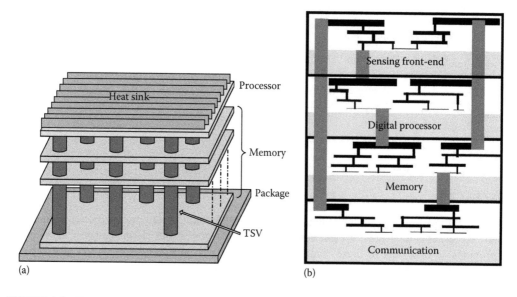

(a)   (b)

FIGURE 6.1 Two primary applications of TSV-based 3D integration technology: (a) many core processors with embedded memory for high-performance computing and (b) heterogeneous 3D SoC with application to low-power mobile computing, health care, and environmental control.

(NIST) identifies 3D technology as one of the fundamental enabling technologies for "extremely scaled intelligent bioelectronic microsystems for *in vivo* operations" [14].

A significant circuit- and physical-level challenge in both high-performance 3D computing systems and relatively low-power 3D SoCs is to ensure *system-wide power and signal integrity*, as explored in this chapter. Power integrity refers to the quality of the power supply voltage delivered to each transistor within a 3D stack, whereas signal integrity refers to the quality of both analog and high-speed digital data signals that are subject to different types of noise. These issues are investigated in the following sections with particular emphasis on wafer-level 3D integration technologies where multiple wafers are thinned, aligned, and vertically bonded. Communication among the dies is achieved by high-density TSVs. Global interconnect length is therefore reduced, lowering the overall power dissipation and latency. TSVs achieve significantly higher interconnection density among the planes of a 3D stack as compared to other vertical interconnects such as through hole vias in a system-in-package (SiP) [3].

The rest of the chapter is organized as follows. The primary characteristics of the three TSV fabrication technologies are summarized in Section 6.2. The issues of power integrity and signal integrity in TSV-based 3D systems are discussed in Sections 6.3 and 6.4, respectively. The chapter is concluded in Section 6.5.

6.2 TSV TECHNOLOGIES AND IMPLICATIONS TO POWER/SIGNAL INTEGRITY

As illustrated in Figure 6.2, existing efforts in 3D-ICs can be broadly classified under three primary categories: (1) transistor-level 3D integration, (2) system-in-package (SiP) and system-on-package (SoP), and (3) wafer-level TSV-based 3D integration. In transistor-level 3D integration [15], active devices within a logic gate are fabricated on different layers, but only a single substrate exists. The upper layer devices are formed on a recrystallized silicon obtained by laser heating or rapid thermal annealing [16]. Polysilicon and single crystal silicon films are also used to host upper layer transistors. Despite the highest density of vertical interconnects, transistor-level 3D integration suffers significantly from thermal challenges and high-defect density [17]. The upper layer silicon should be formed with sufficiently low temperatures to not damage the metal interconnects and devices at the bottom layers. Due to this limitation, the existing research in transistor-level 3D integration primarily focuses on material development and novel fabrication methods.

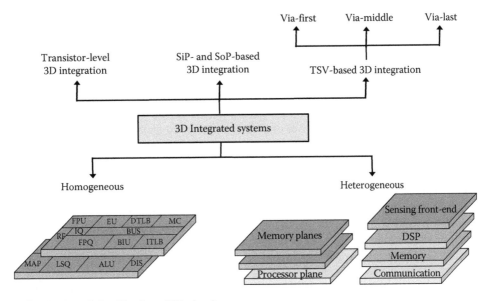

FIGURE 6.2 Broad classification of 3D circuits.

SiP- and SoP-based 3D integration [18–20] are relatively more mature, and commonly encountered in commercial electronic devices where multiple dies are packaged by utilizing (1) wire bonding, (2) low-density vertical interconnects, and (3) area–array-based C4 (controlled collapse chip connection) bumps. Comprising a primary practical concern, the length of these interconnects is typically in the range of several millimeters, significantly degrading the reliability and packaging efficiency in addition to preventing dense integration and causing high power consumption.

In wafer-level 3D integration [21–23], as depicted in Figure 6.1, entire wafers are bonded to produce a 3D stack where TSVs are utilized for interplane communication. This method not only achieves higher interconnect density, but also lower cost per connection due to greater alignment accuracy and enhanced surface planarization [24]. Furthermore, greater flexibility is provided to implement *in-stack* passive devices that are highly important in analog and RF functions. Thus, wafer-level 3D integration is considered to be one of the most promising approaches to realize dense and heterogeneous ICs in the nanoscale era [25–27].

Despite the ongoing research to further optimize 3D fabrication steps, no fundamental limitation exists. For example, alignment accuracy of one micrometer has already been demonstrated [26]. Multiple bonding techniques have also been developed such as adhesive [28,29], oxide [30], and metal bonding [31]. The wafer thinning capability varies from hundreds of nanometers to several hundred micrometers, depending upon whether bulk silicon or silicon-on-insulator is utilized [32,33]. Finally, several distinct TSV fabrication methods exist depending upon when the TSVs are formed [34]. These methods are (1) via-first [35], (2) via-middle [36,37], and (3) via-last TSVs [38]. Note that each TSV type exhibits significantly different electrical characteristics, and therefore require distinct design guidelines, as described in the following sections.

6.2.1 Via-First TSV Technology

In a via-first method, TSVs are fabricated before the transistors are patterned in silicon, that is, prior to front-end-of-line (FEOL) [35,39–41]. Thus, TSVs fabricated with the via-first technique do not pass through the metallization layers, as depicted in Figure 6.3a. The TSV of a plane is connected between the first metal layer of the current plane and the top most metal layer of the previous plane. Polysilicon is typically used as the filling material due to its ability to withstand high temperatures [35,39–41]. Via-first TSVs are less sensitive to contamination since both the filling and substrate materials are the same [42]. The physical dimensions of via-first TSVs are smaller than via-last TSVs [43]. Via-first TSVs, however, are highly resistive and have a lower filling throughput due to the use of polysilicon as the filling material [42,44].

6.2.2 Via-Middle TSV Technology

In a via-middle process, TSVs are fabricated after FEOL, but before the metallization layers are patterned, that is, prior to back-end-of-line (BEOL) [36–38,42,45]. Similar to via-first TSVs, via-middle TSVs connect the first metal layer of a plane with the last metal layer of the previous plane, as illustrated in Figure 6.3b. Since the high-temperature FEOL process precedes TSV fabrication steps, via-middle process permits the use of tungsten (or possibly copper) as the filling material, which is significantly less resistive as compared to doped polysilicon. Tungsten can be used as the filling material due to low thermal expansion coefficient (4.6 ppm/K) as compared to copper (17 ppm/K) [37]. A material with low sensitivity to temperature is required since the TSV fabrication step is followed by a moderately high temperature BEOL process. Note however that copper-filled via-middle TSVs have also been demonstrated [46].

Via-middle TSVs require a relatively large (12:1 or higher) aspect ratio [36–38,45]. Thus, the width of via-middle TSVs is comparable to the width of via-first TSVs, whereas the height is comparable to via-last TSVs. This characteristic produces a relatively high inductive behavior. Also note that the fabrication process of via-middle TSVs is relatively more challenging. For example, a conformal barrier

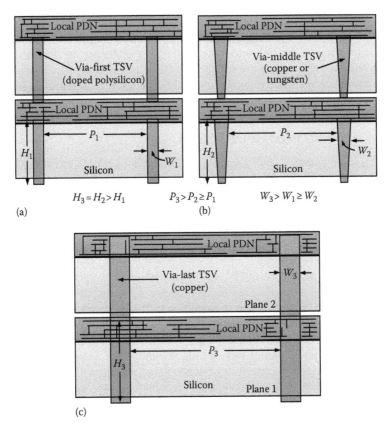

FIGURE 6.3 Illustration of the three primary TSV technologies: (a) via-first, (b) via-middle, and (c) via-last.

process is required for the tungsten to adhere to the dielectric in the cavity. A 20-nm titanium nitride (TiN) layer is typically deposited using metal organic chemical vapor deposition (MOCVD) [36]. TiN is a hard, dense, refractory material with sufficiently low electrical resistivity (22 $\mu\Omega$ cm) [47].

Another challenge is the high level of stress during the deposition of the oxide layer, which is exacerbated when a conformal layer is required. This challenge is partially negated with the use of a tapered TSV structure that progressively shrinks in size [36,37]. This structure is illustrated in Figure 6.3b. Other challenges include sensitivity to contamination and the requirement to maintain the temperature within 500°C [42]. Novel deposition techniques such as atomic layer deposition (ALD) and time-modulated deposition alleviate some of these issues [36,48].

From the design perspective, via-middle technology is an interesting compromise between a highly resistive via-first and a low resistive, but highly inductive via-last TSVs. Furthermore, similar to via-first TSVs, via-middle TSVs do not cause metal routing blockages. The Semiconductor Manufacturing Technology (SEMATECH) consortium has chosen via-middle TSVs as a primary focus area [49].

6.2.3 Via-Last TSV Technology

In the via-last approach, TSV formation occurs after the metallization layers are fabricated, that is, after BEOL [39,42,43]. Thus, as opposed to via-first and via-middle TSVs, via-last TSVs pass through the metal layers, causing metal routing blockages, as depicted in Figure 6.3c [43,50]. A lower resistivity filling material such as copper is used since high temperature FEOL and BEOL processes are performed before the via formation [39,42]. The use of copper as a filling material makes the process sensitive to both temperature (should be maintained less than 230°C) and contamination [42]. Despite exhibiting relatively low resistance, the inductive characteristics of via-last TSVs are relatively more significant than via-first TSVs due to greater dimensions [43]. The physical connection between the TSV and metal layers is typically achieved at the top most metal layer. Note that process technologies to manufacture TSVs are currently under investigation based on recent research results on wafer thinning, alignment accuracy, mechanical stress, and bonding methods. Thus, TSV geometries for a certain TSV type may vary depending upon the foundry.

6.3 SYSTEM-WIDE 3D POWER INTEGRITY

A significant circuit- and physical-level challenge in a TSV-based 3D-IC is to design a robust power distribution network that achieves reliable power delivery to each die. Maintaining the power network impedance smaller than a target impedance (i.e., satisfying the power supply noise) is a difficult task due to reduced operating voltages, increased current magnitudes, and the existence of multiple dies and TSVs. A conservatively large number of power/ground TSVs can significantly increase the area overhead in addition to producing highly inductive characteristics due to a smaller damping factor.

A case study is described in Section 6.3.1 where the power integrity characteristics of a nine-plane processor-DRAM system are investigated. A design space is developed to satisfy power supply noise while minimizing the physical area consumed by the TSVs and decoupling capacitors [51].

3D-ICs are expected to be heavily power-gated due to the importance of subthreshold leakage current in nanoscale bulk CMOS technologies. Furthermore, with heterogeneous integration, the switching activity factor of different blocks/planes is expected to significantly vary, emphasizing the need for power gating. Power gating, however, significantly affects the system-wide power integrity of a 3D-IC, particularly in the presence of decoupling capacitors [52,53]. Various decoupling capacitor topologies are discussed in Section 6.3.2 to enhance power integrity in the power-gated 3D-ICs. Advantages of a reconfigurable decoupling capacitor topology are emphasized [54].

6.3.1 CASE STUDY: POWER DISTRIBUTION IN A NINE-PLANE 3D PROCESSOR-DRAM STACK

6.3.1.1 Architecture

A 3D system in a 32-nm CMOS technology consisting of eight memory planes and one plane for the processor is considered. Each plane contains nine metal layers where the metal thickness and aspect ratio are determined according to 32 nm technology parameters [55]. The power supply voltage is equal to 1 V. Each plane occupies an area of 120 mm², excluding the TSVs and the intentional decoupling capacitance. The system has 1 GB of DRAM spread uniformly across eight memory planes. Each memory plane has 1 gigabit (GB) DRAM divided into 32 modules of equal size, where each module has 32 MB memory. Similarly, the processor plane is also divided into 32 modules. Each of these modules consume an area of 1500 × 2500 μm². This topology is depicted in Figure 6.4. For via-first and via-middle technologies, the TSVs are placed beneath the active circuit, as illustrated in Figure 6.4a, whereas in via-last technology, the power and ground TSVs are distributed on both sides of each module, as depicted in Figure 6.4b.

6.3.1.2 Power Distribution Network Model

An equivalent electrical model corresponding to the power distribution network of one of the 32 modules is illustrated in Figure 6.5. This model consists of the TSVs (R_{TSV}, C_{TSV}, and L_{TSV}), substrate (R_{Si} and C_{Si}), power distribution network within a plane ($R_{Network}$, $R_{Vertical}$, and R_{M1}), switching load circuit, and decoupling capacitance C_{decap}. Note that in addition to these on-chip impedances, the parasitic package resistance and inductance are, respectively, 3 mΩ and 100 pH at both the power and ground supplies, assuming an organic flip-chip package [56]. The primary characteristics of the three TSV technologies are listed in Table 6.1.

The DRAM consumes 3 W, uniformly distributed across the eight stacks [57]. Alternatively, the processor consumes 90 W. Thirty percent of the overall power is due to static power dissipation,

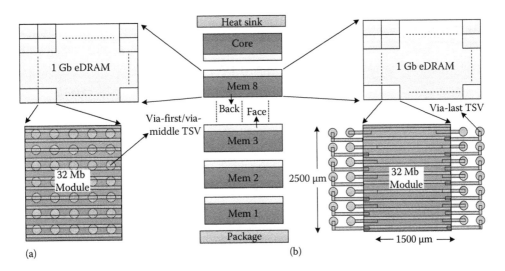

FIGURE 6.4 3D processor-memory stack: (a) via-first and via-middle TSVs and (b) via-last TSVs.

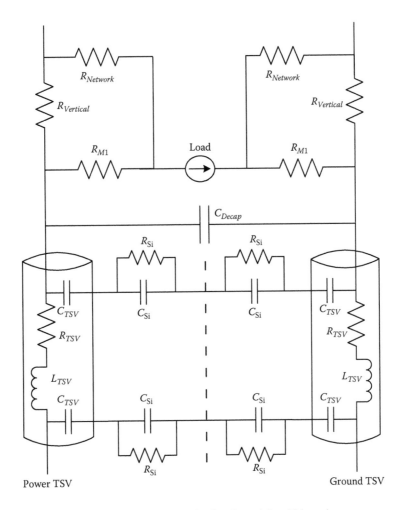

FIGURE 6.5 Equivalent power distribution network of each module within a plane.

TABLE 6.1

Primary Characteristics of Via-First (Filled with Polysilicon), Via-Middle (Filled with Tungsten), and Via-Last (Filled with Copper) TSVs

Parameter	Via-First	Via-Middle	Via-Last
Diameter, W	4 µm	4–2.66 µm	10 µm
Height, H	10 µm	60 µm	60 µm
Pitch, P	8 µm	8–9.34 µm	20 µm
Oxide thickness, t_{ox}	0.2 µm	0.2 µm	0.2 µm
TSV resistance, R_{TSV}	5.7 Ω	858.36 mΩ	20 mΩ
TSV inductance, L_{TSV}	4.18 pH	49.76 pH	34.94 pH
TSV capacitance, C_{TSV}	23 fF	117.81 fF	283 fF

Sources: Agarwal, A., Murthy, R.B., Lee, V., and Viswanadam, G., Polysilicon interconnections (FEOL): Fabrication and characterization, In *Proceedings of the IEEE Electronics Packaging Technology Conference*, pp. 317–320, © 2009 IEEE; Katti, G., Stucchi, M., De Meyer, K., and Dehaene, W., Electrical modeling and characterization of through silicon via for three-dimensional ICs. *IEEE Transactions on Electron Devices*, 57(1), 256–262, © 2010 IEEE; Pares, G., Bresson, N., Minoret, S., Lapras, V., Brianceau, P., Lugand, J.F., Anciant, R., Sillon, N., Through silicon via technology using tungsten metallization, *Proceedings of the IEEE International Conference on IC Design and Technology*, pp. 1–4, © 2011 IEEE.

whereas the remaining portion is due to dynamic power consumption [57]. A triangular current waveform is assumed with 400 ps period, 100 ps rise time, and 150 ps fall time. The DC current and peak current are determined based on, respectively, the static and dynamic power consumption.

6.3.1.3 Results

In 3D power distribution networks, it is important to determine the appropriate number of TSVs and decoupling capacitance that satisfy the constraint on power supply noise. From this design space, a valid pair that minimizes the physical area overhead is chosen. Note that this specific design point is dependent upon the implementation of the decoupling capacitance. Three methods are considered: (1) metal-oxide-semiconductor (MOS) capacitance in a 32 nm technology node with a capacitance density of 39.35 fF/µm² as determined from the equivalent oxide thickness (EOT) of the technology [55], (2) deep trench capacitance with a density of 140 fF/µm² [58], and metal–insulator–metal (MIM) capacitance with a density of 8 fF/µm² [59,60].

As an illustrative example, results for via-middle TSVs (filled with tungsten) are shown. The peak noise surface as a function of decoupling capacitance and number of TSVs is plotted in Figure 6.6a. As illustrated in this figure, the power supply noise monotonically decreases as the number of TSVs and decoupling capacitance increase. The noise contour at 100 mV (power supply noise constraint) is illustrated in Figure 6.6b. Any point on (and above) this curve satisfies the noise constraint. The physical area overhead for each of these points is depicted in Figure 6.7 for three different decoupling capacitance densities, as mentioned earlier. A specific design point exists where the physical area overhead is minimized. For example, if MOS capacitance is used, this design point is at 620 TSVs and 0.6 nF of decoupling capacitance.

A similar procedure is followed for via-first and via-last TSVs. The design points that minimize the physical area overhead while satisfying the power supply noise constraint are summarized in Table 6.2. The following important conclusions can be made based on these results:

• Highly resistive via-first TSVs (filled with polysilicon) can be used to deliver power at the expense of approximately 9% area overhead as compared to less than 2% area overhead in via-middle (filled with tungsten) and via-last (filled with copper) technologies.

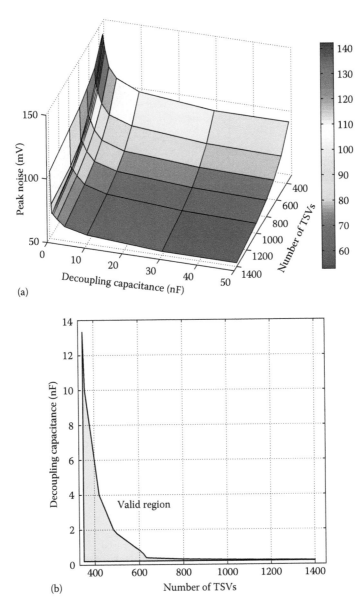

FIGURE 6.6 Peak noise characteristics as a function of number of TSVs and decoupling capacitance in via-middle TSV technology: (a) surface plot and (b) contour plot at 100 mV.

- Despite the higher area requirement, a power distribution network with via-first TSVs is typically overdamped and the issue of resonance is alleviated.
- For via-middle and via-last TSV technologies, the impedance at the resonant frequency should be sufficiently small. This issue is exacerbated for a via-last based power distribution network since the TSV resistance is significantly lower, and therefore the network is typically underdamped.

6.3.2 DECOUPLING CAPACITORS FOR POWER-GATED 3D-ICS

In traditional topologies, the decoupling capacitance of a power-gated block (or plane) cannot provide charge to neighboring planes since these capacitors are typically connected to a virtual power

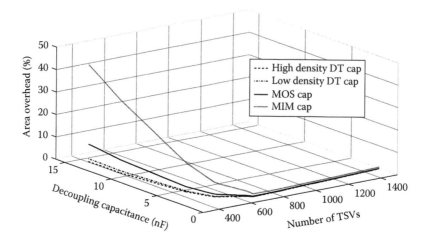

FIGURE 6.7   Area overhead in via-middle technology. Note that each point on the curve satisfies the target power supply noise.

TABLE 6.2

Valid Design Points That Satisfy the Peak Power Supply Noise While Minimizing the Area Overhead for Each TSV Technology

Cap Type	Cap Density (fF/μm²)	Area Overhead (%)	Number of TSVs	Decoupling Capacitance (nF)	Power Loss (%)	Peak Noise (mV)
Via-First						
MOS-C	39.35	9.06	2750	2.7	8.3	33.42
MIM	8	11.83	3437	0.8	7.3	23.5
Via-Middle						
MOS-C	39.35	1.96	620	0.60	6.7	63.56
DT	140	1.68	484	2	7.6	45.13
MIM	8	3.01	631	0.4	6.6	67.5
Via-Last						
MOS-C	39.35	1.54	76	0.385	4	134.72
DT	140	1.21	30	3.74	4.9	134.67
MIM	8	2.56	76	0.385	4	134.72

Note: DT refers to deep trench.

network that is closer to the switching circuit [53]. Placing the decoupling capacitors sufficiently close to the switching load is helpful in reducing the power supply noise [61,62]. However, if the block or plane is power-gated, those capacitors that provide charge to the block (or plane) are disconnected from the global power network, making these capacitors ineffective for the neighboring planes. Since system-wide power integrity is a critical challenge in 3D-ICs, effective use of intentional decoupling capacitance is crucial, even when power gating is adopted.

It has been observed that the effective range of a decoupling capacitor exceeds single plane in 3D-ICs with low-resistance TSVs [63]. This phenomenon is important particularly when block- or plane-level power gating is performed due to significant decoupling capacitance that cannot provide charge to the remaining, active planes. Thus, when one or more number of planes is power-gated, power supply noise in one of the remaining active planes may increase and violate the constraint despite a reduction in the overall switching current drawn from the power supply. It is therefore highly critical to ensure that the decoupling capacitors within the power-gated planes are not ineffective for the remaining planes. A reconfigurable decoupling capacitor topology is described in the following section to alleviate this issue [53].

6.3.2.1 Reconfigurable Topology

In the reconfigurable topology, two switches are used to form a configurable decoupling capacitor, as conceptually illustrated in Figure 6.8 [64]. If a certain plane is active, the decoupling capacitors on that plane are connected to the virtual V_{DD} grid through switch 2, thereby reducing the power supply noise on that plane. Alternatively, if the plane is power-gated (sleep transistors are turned off), the decoupling capacitors are connected to the global V_{DD} grid, bypassing the sleep transistors. Thus, even if the plane is power-gated, the decoupling capacitors are effective for the remaining planes. The overhead of this topology includes the reconfigurable switches, metal resources required to route the related control signals, and a possible increase in overall power consumption depending upon how the capacitors are implemented, as discussed in the following section.

6.3.2.2 Evaluation Setup

A power distribution network for a three-plane 3D-IC with via-last TSVs is developed, as conceptually illustrated in Figure 6.9 for a single plane [53]. A 45 nm CMOS technology with 10 available metal layers in each plane is adopted [65]. A portion of the power network with an area of 1 mm × 1 mm is analyzed. Each plane consists of a global power network, virtual power network, distributed PMOS sleep transistors, distributed decoupling capacitance (implemented as MOS capacitors), and distributed switching load circuit consisting of inverter gates, as depicted in Figure 6.9. Top two metal layers (9 and 10) on each plane are dedicated to global power distribution network with an interdigitated grid of 11 × 11. Metal layers 8 and 7 are used as the virtual power network represented by an interdigitated grid of 21 × 21 [66]. Power gating is achieved using a distributed method where the sleep transistors that control a plane are placed within that plane [67]. Note that the top plane also consists of C4 bumps to connect the on-chip grid with the flip-chip substrate.

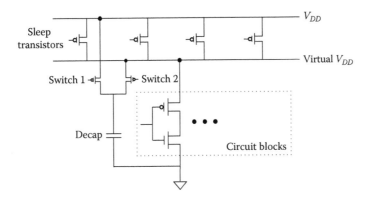

FIGURE 6.8 Conceptual representation of the reconfigurable decoupling capacitor topology with power gating.

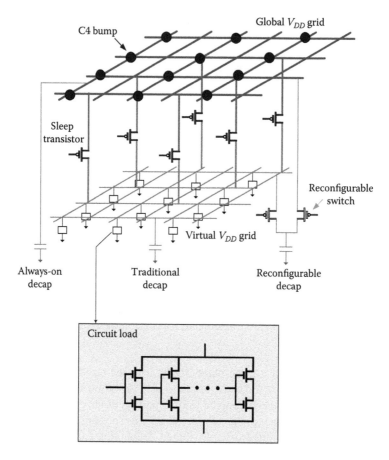

FIGURE 6.9 Plane-level power network illustrating distributed sleep transistors, decoupling capacitors (traditional and proposed topologies), switching load circuits (gates with active devices), and the C4 bumps (for the top plane only).

Similar to [68], inverter pairs with varying number and size are used to model the switching load circuit. The overall area is divided into 30 sub-areas and a switching circuit is connected to each sub-area to consider the spatial heterogeneity of the current loads. The spatial load current distribution and power densities are based on [69] where the peak power density reaches 40W/cm^2, comparable to the power density in modern processors [70]. All of the decoupling capacitors are implemented as MOS capacitors in the 45 nm technology with an oxide thickness of 1 nm [65].

6.3.2.3 Results

The efficacy of the reconfigurable decoupling capacitor placement topology is evaluated by comparing the methodology with the traditional topology where the capacitors are connected to a virtual power grid. Both power supply noise and power gating noise are analyzed. Three power gating scenarios are considered:

- *Scenario 1*: All of the three planes are active, representing the greatest workload.
- *Scenario 2*: The top and bottom planes are active, while the middle plane is power-gated.
- *Scenario 3*: Only the bottom plane is active, while the middle and top planes are power-gated.

Power supply noise results are listed in Table 6.3 for each scenario. As listed in this table, when some of the planes are power-gated (scenarios 2 and 3), the reconfigurable topology achieves less power

TABLE 6.3

Power Supply Noise Obtained from Each Scenario and Noise Reduction Achieved by Reconfigurable Topology

| | Power Status | | | Power Supply Noise (mV) | | | | | |
| | | | | Traditional | | Reconfigurable | | | |
	Top	Mid	Btm	Peak	RMS	Peak	Redtn. (%)	RMS	Redtn. (%)
Scenario 1	On	On	On	50	30.27	50	N/A	28.33	6.4
Scenario 2	On	Off	On	48.94	23.41	42.90	12.3	16.79	28.3
Scenario 3	Off	Off	On	52.86	17.53	38.34	27.4	9.39	46.4

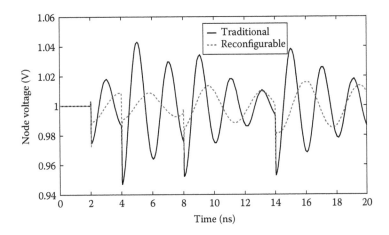

FIGURE 6.10 Transient behavior of voltage noise at a specific node within the bottom plane for each topology for scenario 3.

supply noise by exploiting the capacitors of the power-gated plane(s). For example, in scenario 3 where two planes are power-gated, the reduction in peak noise is approximately 27%, whereas the reduction in RMS noise is approximately 46%. Note that in the traditional topology, the peak noise in scenario 3 exceeds 50 mV despite the reduction in overall switching current due to power gating. This characteristic is due to less decoupling in the power network since the decoupling capacitors in the power-gated planes cannot behave as charge reservoirs for the remaining, active planes. This observation justifies the need for reconfigurable capacitors. Transient behavior of voltage noise at a specific node within the bottom plane is also depicted in Figure 6.10 for scenario 3, demonstrating the reduction in peak and RMS noise.

   To investigate power gating noise, the power-gated middle plane transitions from inactive to active state in scenarios 2 and 3, and the voltage fluctuation due to in-rush current during the wake-up process is analyzed. Peak power gating noise is observed at the bottom plane. Results are listed in Table 6.4. The reconfigurable topology achieves more than 80% reduction in peak and RMS power gating noise. In traditional topology, a significant amount of in-rush current flows not only for the activated circuit, but also to charge the associated decoupling capacitors. Alternatively, in the reconfigurable topology, the decoupling capacitors are connected to the global V_{DD} grid when the plane is power-gated. Thus, even if the plane is power-gated, these capacitors remain charged (significantly reducing in-rush current) and can behave as charge reservoir once the plane transitions to active state. The transient behavior of the power gating noise is illustrated in Figure 6.11 for scenario 3 where the transition happens at 1 ns.

TABLE 6.4

Power Gating Noise Obtained from Each Scenario and Noise Reduction Achieved by Reconfigurable Topology

| | Power Status | | | Power Gating Noise (mV) | | | | | |
| | | | | Traditional | | Reconfigurable | | | |
	Top	Mid	Btm	Peak	RMS	Peak	Redtn. (%)	RMS	Redtn. (%)
Scenario 1	On	On	On	N/A			N/A		
Scenario 2	On	Off	On	90.73	72.83	15.55	82.8	11.46	84.2
Scenario 3	Off	Off	On	103.2	78.76	15.66	84.8	11.57	85.3

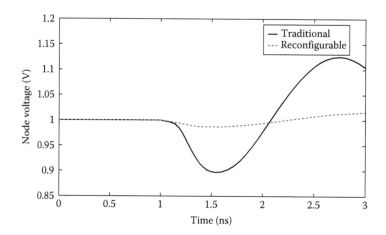

FIGURE 6.11 Transient behavior of power gating noise at a specific node within the bottom plane for each topology for scenario 3.

In the reconfigurable topology, the physical area overhead increases by only 1.55% due to reconfigurable switches and higher decoupling capacitance required to compensate for the shielding effect of the reconfigurable switches. This slight increase in the physical area significantly enhances power integrity when one or more planes are power-gated, as described earlier.

To quantify the power overhead of the reconfigurable topology, both topologies are simulated for each scenario and the overall average power consumption is determined. All of the decoupling capacitors are implemented as MOS capacitors to consider MOS-C leakage. It is observed that the maximum power (drawn in scenario 1) increases by only 1.36% due to increased capacitance and switches. This power overhead can be further reduced if low leakage decoupling capacitor implementations are adopted, such as MIM capacitors.

6.4 SYSTEM-WIDE 3D SIGNAL INTEGRITY

In addition to traditional noise coupling and propagation mechanisms such as crosstalk, power supply noise, and substrate coupling, 3D-ICs suffer from TSV-induced noise coupling [71–73]. Specifically, during a signal transition within a TSV, noise couples from TSV into the substrate due to both dielectric and depletion capacitances. The coupling noise propagates throughout the substrate and affects the reliability of nearby transistors [74]. This issue is exacerbated for TSVs that carry signals with high switching activity factors and fast transitions such as clock signals.

Analog/RF blocks and memory cells are among the most sensitive circuits to substrate noise coupling [75–78]. For example, in [79], experimental data demonstrate that the signal-to-noise-plus-distortion ratio (SNDR) of a delta-sigma modulator is reduced by more than 20 dB due to substrate noise. Note that in heterogeneous 3D systems, the front-end circuitry consisting of analog/RF blocks is typically located at the top plane (closer to the I/O pads) to reduce the overall impedance between the pads and analog inputs. In this floorplan, TSVs are required to transmit the digital signals (including the clock signal) to the data processing plane. Thus, TSV-induced noise becomes an important issue for the reliability of the analog/RF blocks. Digital transistors are also affected by TSV-induced noise if the physical distance between the TSV and device is sufficiently short [80]. TSV-induced noise changes the drain current characteristics of both an on and off transistor, as observed in [81].

A methodology is described in Section 6.4.1 to efficiently analyze TSV-to-transistor noise coupling in 3D-ICs. A case study is presented in Section 6.4.2 to better understand noise propagation paths within a 3D environment.

6.4.1 EFFICIENT ANALYSIS OF TSV-TO-TRANSISTOR NOISE COUPLING

To characterize TSV-induced noise coupling, the physical structure depicted in Figure 6.12 is used. This structure consists of a noise injector (TSV), noise transmitter (substrate), and a noise receptor (victim transistor). Substrate contacts are also included to bias the substrate. Note that the number and placement of substrate contacts between the TSV and victim transistor play an important role in the noise coupling analysis. To analyze this physical structure, several approaches have been adopted, such as using an electromagnetic field solver, device simulator, and a highly distributed model using 3D transmission line matrix (TLM) method [82–84], as described in the following section.

6.4.1.1 Distributed Model

In the distributed model, the physical structure is discretized into unit cells (for both TSV and substrate) and each unit cell is modeled with lumped parasitic impedances. Both the measurement and 3D field solver results demonstrate that the 3D-TLM approach can accurately model the 3D physical structure including a TSV, substrate contact, and victim transistor [82,85]. Despite the reasonable accuracy achieved by the distributed model, the computational complexity is significantly high, particularly when the dimensions of the unit cells are small. This issue is exacerbated as the distance between the TSV and victim transistor increases. Furthermore, the number and location of the substrate contacts play an important role in characterizing the TSV safe zone. Re-analysis of the distributed structure when these characteristics change is computationally prohibitive. Therefore, a compact model is needed to alleviate these limitations, as described in the following section. This highly distributed model is used as a reference to validate the compact model and closed-form expressions.

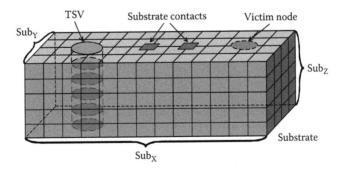

FIGURE 6.12 Physical structure used to analyze TSV-induced noise coupling.

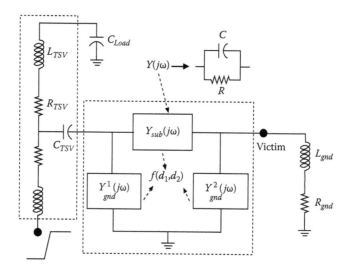

FIGURE 6.13   Compact π model to efficiently estimate the noise at the victim node in the presence of a TSV and substrate contacts.

6.4.1.2   Compact Model

A two-port, linear time-invariant network can be generally characterized with four admittances: $Y_{11}(j\omega)$, $Y_{12}(j\omega)$, $Y_{21}(j\omega)$, and $Y_{22}(j\omega)$. Utilizing this characteristic, the proposed compact model consists of a single TSV cell and an equivalent two-port π network to model noise propagation, as depicted in Figure 6.13 [73]. Each electrical element within the π network consists of a parallel RC circuit, producing an admittance $(1/R) + j\omega C$. These admittances can be obtained from the distributed mesh (based on 3D-TLM method), as described in the previous section. Specifically, the four $Y(j\omega)$ parameters of the distributed mesh are obtained through an AC analysis. The resistances and capacitances within the π network are determined such that the four $Y(j\omega)$ parameters of the compact π network are equal to the respective $Y(j\omega)$ parameters of the distributed mesh. According to this procedure, the admittances within the π network $Y_{sub}(j\omega)$, $Y_{gnd}^1(j\omega)$, and $Y_{gnd}^2(j\omega)$ are determined as follows:

- $Y_{sub}(j\omega) = (1/R_{sub}) + j\omega C_{sub} = Y_{21}(j\omega)$: represents the equivalent substrate admittance between the TSV and victim transistor.
- $Y_{gnd}^1(j\omega) = (1/R_{gnd}^1) + j\omega C_{gnd}^1 = Y_{11}(j\omega) - Y_{21}(j\omega)$: represents the equivalent substrate admittance between the TSV and ground node.
- $Y_{gnd}^2(j\omega) = (1/R_{gnd}^2) + j\omega C_{gnd}^2 = Y_{22}(j\omega) - Y_{21}(j\omega)$: represents the equivalent substrate admittance between the victim node and ground node.

In the next step, each resistive and capacitive element within each $Y(j\omega)$ of the π network are characterized as a function of d_1 and d_2, where d_1 and d_2 are depicted in Figure 6.14. Via-first and via-last TSVs are separately analyzed. To evaluate the dependencies of R and C on d_1 and d_2, AC analyses of the distributed mesh (based on 3D-TLM) described in the previous section are performed with different values of d_1 and d_2. The data obtained in this step are used to generate a 3D surface for each resistance and capacitance within $Y_{sub}(j\omega)$, $Y_{gnd}^1(j\omega)$, and $Y_{gnd}^2(j\omega)$. This surface is approximated with a logarithmic function using a 3D least square regression analysis. The logarithmic function $F(d_1, d_2)$ used to approximate the admittances of the π network as a function of the physical distances d_1 and d_2 is

$$F(d_1, d_2) = A + Bd_1 + Cd_2 + D\ln d_2 + E\ln d_1, \qquad (6.1)$$

where A, B, C, D, and E are fitting coefficients. These fitting coefficients are determined such that the resistor/capacitor value obtained from this expression reasonably approximates the actual

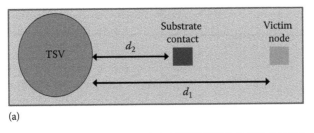

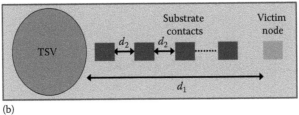

FIGURE 6.14 Substrate contact between TSV and victim node: (a) a single contact exists between TSV and victim node and (b) multiple contacts are regularly placed between TSV and victim node.

resistor/capacitor value (in the compact π model) that is obtained from the highly distributed 3D TLM model. Note that both the resistance (in kΩ s) and capacitance (in attoFarads) of each $Y(j\omega)$ within the π network are represented by the function F. Also note that the distances d_1 and d_2 are in μm. Since the π network has three admittances each consisting of a parallel RC circuit, six logarithmic functions are developed for via-first and via-last TSVs, producing a total of 12 functions. The fitting coefficients for each function are listed in Table 6.5.

TABLE 6.5

Fitting Coefficients for the Function F That Approximates the Admittances within the Compact Model (see Figure 6.13) for Via-First and Via-Last TSVs

Cases	Admittances	Fitting Coefficients					Error (%)
		A	B	C	D	E	
	$R_{sub} = 1000/F$	27.09	0	0	−0.98	−5.11	6.6
	$C_{sub} = F$	326.7	0.41	0.48	−17.26	−67.55	2.8
Via-first	$R_{gnd}^1 = 1000/F$	28.31	0	0	−8.14	1.98	1.9
	$C_{gnd}^1 = F$	301.3	−0.28	0.66	−96.94	30.09	1.6
	$R_{gnd}^2 = F$	45.91	2.30	3.08	−218.3	154.9	9.6
	$C_{gnd}^2 = 1000/F$	8.05	0.28	0.29	−20.39	12.49	10.4
	$R_{sub} = 1000/F$	27.35	−0.082	0.036	−2.11	−1.12	2.4
	$C_{sub} = F$	296.6	−0.89	0.83	−20.78	−13.12	1.9
Via-last	$R_{gnd}^1 = 1000/F$	36.16	0.028	0.39	−10.16	1.18	2.4
	$C_{gnd}^1 = F$	235.6	−1.18	−2.16	−61.86	51.91	6.9
	$R_{gnd}^2 = 1000/F$	11.57	0.11	−0.19	4.43	−5.86	11.6
	$C_{gnd}^2 = F$	129.1	1.18	−2.22	49.1	−65.23	10.9

Note: The function F is given by Equation 6.1. All resistances are in kΩ s and all capacitances are in attoFarads.

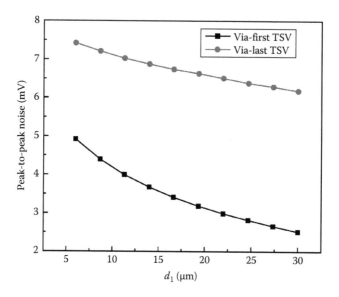

FIGURE 6.15   TSV-induced switching noise at the victim node as a function of d_1 at constant d_2 for via-first and via-last TSVs.

6.4.1.3   Results and Design Guidelines

The compact π model contains only 11 number of elements for both via-first and via-last TSVs where each element is represented by Equation 6.1 with five fitting coefficients. Alternatively, the number of circuit elements in the distributed model (based on 3D-TLM) can exceed a million when the unit cell dimension is 1 µm.

According to Table 6.5, maximum average error is slightly over 10% for certain resistances and capacitances. This error, however, does not significantly affect the electrical characteristics (and noise estimation at the victim node) since the maximum error occurs at the extreme cases when the resistance is sufficiently large and capacitance is sufficiently small. Also note that the average error over via-first and via-last TSVs is approximately 6%.

The compact model illustrated in Figure 6.13, Equation 6.1, and fitting parameters listed in Table 6.5 can be used to investigate the effect of various design and fabrication parameters such as placement of substrate contacts and TSV type. As an example, the effect of increasing d_1 (placing the victim farther from the TSV) on peak noise is investigated for both via-first and via-last TSVs. The results are shown in Figure 6.15. According to this figure, placing the victim transistor farther from the switching TSV is an effective method for via-first TSVs. Alternatively, for via-last TSVs, the noise exhibits low sensitivity to the distance between TSV and victim transistor. This phenomenon can be explained by longer height (therefore smaller substrate resistances) and larger diameter (therefore larger capacitances) of via-last TSVs.

6.4.2   CASE STUDY: SI ANALYSIS OF A 3D-INTEGRATED SoC

The primary objective of this study is to evaluate the signal integrity characteristics of a 3D-integrated mixed-signal circuit and compare the results with a 2D implementation. It is typically assumed that the noise performance of a 3D circuit is superior than a 2D system since the aggressor and victim blocks can be placed on separate planes, thereby enhancing signal isolation [5,7]. The validity of this assumption is evaluated by developing a comprehensive electrical model for both 2D and 3D versions of the same circuit. The circuit functionality, architecture, and signal integrity modeling approach are described in the following sections.

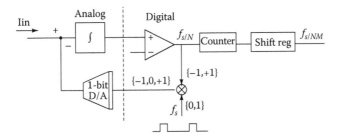

FIGURE 6.16 System level diagram of a single channel of the potentiostat. (From Stanacevic, M., Murari, K., Rege, A., Cauwenberghs, G., and Thakor, N., VLSI potenstiostat array with oversampling gain modulation for wide-range neurotransmitter sensing, *IEEE Transactions on Biomedical Circuits and Systems*, 1(1), 63–72, © 2007 IEEE.)

6.4.2.1 Potentiostat Architecture

A 16-channel VLSI potentiostat chip, designed and fabricated in 0.5 μm CMOS technology, simultaneously records neurotransmitter levels from multiple neurons [86,87]. The detection of the neurotransmitters is critical for neural pathways and the etiology of neurological diseases like epilepsy and stroke [88]. High sensitivity and wide dynamic range are critical design objectives since the measured currents are in the picoampere range. Thus, an accurate evaluation of signal integrity is critical.

To manage computational complexity, the proposed signal integrity analysis focuses on a single channel that consists of a first-order delta-sigma modulator, counter for decimation, and shift register, as depicted in Figure 6.16 [86,87]. Delta-sigma modulator consists of a current integrator, comparator, and switched-current 1-bit D/A converter in the feedback loop. The counter is the primary aggressor, whereas the sense amplifier within the current integrator is identified as one of the primary victim blocks. The switching noise that couples from the counter to the sense amplifier is analyzed for different scenarios, as described in the following sections.

6.4.2.2 3D Signal Integrity Modeling

Highly distributed 3D electrical models based on TLM method (as described in Section 6.4.1) are utilized to analyze signal integrity. Both the substrate and TSVs are discretized using unit cells consisting of *RLC* impedances, and an entire electrical model for a single channel of a neurotransmitter chip is generated in both 2D and 3D technologies.

In the 3D-integrated potentiostat, the aggressor (counter) and victim (delta-sigma modulator) are placed on separate planes. Specifically, the top plane (closer to the I/O pads) is dedicated to the victim block, whereas the bottom plane (closer to the heat sink) is dedicated to the aggressor block. A conceptual representation of the electrical model is depicted in Figure 6.17. The bulk nodes within the schematic are connected to the corresponding nodes on the substrate. Two separate substrates exist: (1) substrate of the bottom plane where the bulks of the counter are connected, and (2) substrate of the upper plane where the bulks of the delta-sigma modulator are connected. As depicted in Figure 6.17, a face-to-face bonding technology is assumed with via-first fabrication technique. In this technique, the TSVs go through the upper (analog) substrate and reach the metal layers of the analog plane. The topmost metal layer of the analog plane is connected to the topmost metal layer of the digital plane using bumps. Since only a single channel of the device is modeled, eight TSVs are required: five for the clock signals (each at 1 MHz), two for power supply voltage (3 V), and one for data signal. Signal integrity results are described in the following section.

6.4.2.3 Results

A transient analysis is performed and the noise at the bulk of the victim device is observed. Peak noise is illustrated in Figure 6.18 for two different cases: (1) nonideal TSVs with practical capacitance values, (2) ideal TSVs with zero capacitance value. Note that the result of the 2D analysis is

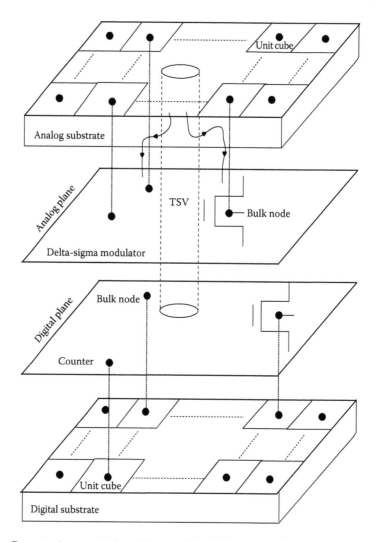

FIGURE 6.17 Conceptual representation of the overall model to analyze signal integrity in a 3D potentiostat.

also included for comparison. As depicted in this figure, the 3D system with practical TSVs, despite having separate substrates for aggressor and victim, exhibit significantly higher noise (positive peak noise exceeds 6 mV) than the 2D system.

To determine the primary noise source, the TSV capacitance C_{TSV} of the clock TSVs is removed, producing an ideal clock TSV with no coupling into the substrate. In this case, the peak noise is significantly reduced (to approximately 2 mV) and is smaller than the 2D system. Thus, in a 3D system, significant noise couples into the substrate through clock TSVs. The same conclusion is validated by observing the RMS noise over 10 μs, as listed in Table 6.6 for all of the cases. The RMS noise is reduced by more than 50% only if the TSVs are ideal, that is, TSV-to-substrate noise coupling is prevented.

For applications such as the neurotransmitter sensing as described in this section, the top plane of a 3D-IC should be dedicated to the highly sensitive frontend circuits due to physical proximity to the pads. In this case, however, clock TSVs with short rise times inject significant noise into the substrate of the analog plane since these signals need to reach the bottom plane where the digital circuit is placed. This coupling mechanism is due to the TSV-to-substrate capacitance C_{TSV} of the TSVs. Thus, the distance between an aggressor TSV and a victim device should be carefully considered. Efficient and 3D-specific signal isolation strategies should also be developed.

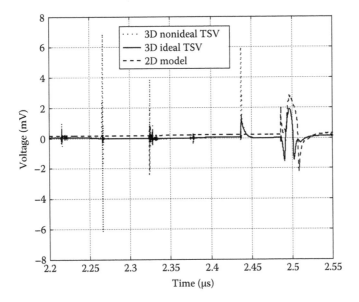

FIGURE 6.18 Transient analysis results illustrating peak noise at the bulk node of a victim device for two different cases in a 3D integrated potentiostat.

TABLE 6.6
RMS Value of the Noise at the Bulk Node over 10 μs for Different Cases

Case	RMS Noise at the Bulk Node of the Victim Block (μV)
2D potentiostat	232
3D potentiostat with nonideal TSV	242
3D potentiostat with ideal TSV	107

6.5   SUMMARY AND CONCLUSIONS

TSV-based 3D technologies facilitate higher and heterogeneous integration, and enable enhanced performance while reducing the overall power consumption. Despite various challenges at the fabrication level (such as developing low cost and high aspect ratio vertical vias and reliable bonding technologies at lower temperatures), no fundamental limitations exist. A similar argument applies to the design process where significant research has been conducted during the past decade. TSV-based 3D-ICs enable exciting opportunities not only for high-performance computing systems, but also for low-power SoCs with application to mobile computing, life sciences, and environment control.

This chapter has focused on two important issues encountered during the physical design process of 3D SoCs: power and signal integrity. An overview of existing TSV technologies was first provided to better understand the implications of each TSV technology on power delivery and signal integrity. A case study was described to illustrate power distribution in a nine-plane 3D processor-DRAM stack. According to the International Technology Roadmap for Semiconductors, processor-memory stacks are expected to be one of the early applications of TSV-based 3D-ICs where the issue of power integrity is exacerbated [89].

Next, the importance of power gating has been emphasized for 3D-ICs. It was shown that conventional decoupling capacitor topologies can degrade power integrity when one or more planes are power-gated. A reconfigurable decoupling capacitor topology was discussed to alleviate this issue and significantly enhance power integrity with relatively small overhead in physical area and power consumption.

An important concern, particularly in heterogeneous 3D integration with sensitive analog/RF blocks, is TSV-to-transistor noise coupling. This signal integrity concern was described and two approaches (that exhibit distinct accuracy versus complexity tradeoffs) were presented to analyze noise coupling in 3D-ICs. Finally, another case study of a 3D-IC with application to bioelectronics (neurotransmitter sensing) was discussed with emphasis on 3D signal integrity. It was demonstrated that the expected signal isolation advantages of 3D-ICs are possible if TSV-to-substrate noise coupling is significantly reduced.

REFERENCES

1. R. J. Gutmann et al. Three-dimensional (3D) ICs: A technology platform for integrated systems and opportunities for new polymeric adhesives. In *Proceedings of the IEEE International Conference on Polymers and Adhesives in Microelectronics and Photonics*, Potsdam, Germany, pp. 173–180, October 2001.
2. A. W. Topol et al. Three-dimensional integrated circuits. *IBM Journal of Research and Development*, 50(4/5):491–506, July/September 2006.
3. E. Salman and E. G. Friedman. *High Performance Integrated Circuit Design*, 1st edn. McGraw-Hill, New York, September 2012.
4. J. U. Knickerbocker et al. Three dimensional silicon integration. *IBM Journal of Research and Development*, 52(6):553–569, November 2008.
5. K. Banerjee, S. J. Souri, P. Kapur, and K. C. Saraswat. 3-D ICs: A novel chip design for improving deep-submicrometer interconnect performance and systems-on-chip integration. *Proceedings of the IEEE*, 89(5):602–633, May 2001.
6. V. F. Pavlidis and E. G. Friedman. Interconnect-based design methodologies for three-dimensional integrated circuits. *Proceedings of the IEEE*, 97(1):123–140, January 2009.
7. V. F. Pavlidis and E. G. Friedman. *Three-Dimensional Integrated Circuit Design*. Morgan Kaufmann, Burlington, MA, 2009.
8. C. Liu, I. Ganusov, M. Burtscher, and S. Tiwari. Bridging the processor-memory performance gap with 3D IC technology. *IEEE Design and Test of Computers*, 22(6):556–564, November–December 2005.
9. G. Loh. 3D-stacked memory architectures for multi-core processors. In *Proceedings of the International Symposium on Computer Architecture*, Beijing, China, pp. 453–464, June 2008.
10. Q. Wu, K. Rose, J.-Q. Lu, and T. Zhang. Impacts of through-dram vias in 3D processor-dram integrated systems. In *Proceedings of IEEE International Conference on 3D System Integration*, San Francisco, CA, pp. 1–6, September 2009.
11. U. Kang et al. 8 Gb 3-d ddr3 dram using through-silicon-via technology. *IEEE Journal of Solid-State Circuits*, 45(1):111–119, January 2010.
12. A. B. Kahng. Scaling: More than Moore's Law. *IEEE Design and Test of Computers*, 27(3):86–87, May–June 2010.
13. G. Q. Zhang, M. Graef, and F. van Roosmalen. The rationale and paradigm of More than Moore. In *Proceedings of the IEEE International Conference on Electronic Components and Technology*, San Diego, CA, pp. 151–157, May–June 2006.
14. G. M. Walker, J. M. Ramsey, R. K. Cavin, D. J. C. Herr, C. I. Merzbacher, and V. Zhirnov. A framework for bioelectronics discovery and innovation. Technical report, The Semiconductors Electronics Division, National Institute of Standards and Technology, Research Triangle Park, NC, February 2009.
15. S. Wong, A. El-Gamal, P. Griffin, Y. Nishi, F. Pease, and J. Plummer. Monolithic 3D integrated circuits. In *Proceedings of the IEEE International Symposium on VLSI Technology, Systems and Applications*, Hsinchu, Taiwan, pp. 23–25, April 2007.
16. P. Batude et al. GeOI and SOI 3D monolithic cell integrations for high density applications. In *Proceedings of the IEEE International Symposium on VLSI Technology*, Honolulu, HI, June 2009.
17. M. Vinet et al. 3D monolithic integration: Technological challenges and electrical results. *Microelectronic Engineering*, 88(4):331–335, April 2011.
18. E. Beyne. The rise of the 3rd dimension for system integration. In *Proceedings of the IEEE International Conference on Interconnect Technology*, Burlingame, CA, pp. 1–5, June 2006.

19. J. Miettinen, M. Mantysalo, K. Kaija, and E. O. Ristolainen. System design issues for 3D system-in-package (sip). In *Proceedings of the IEEE Electronic Components and Technology Conference*, Las Vegas, NV, pp. 610–615, June 2004.

20. W. J. Howell et al. Area array solder interconnection technology for the three-dimensional silicon cube. In *Proceedings of the IEEE Electronic Components and Technology Conference*, Las Vegas, NV, pp. 1174–1178, May 1995.

21. C. L. Chen et al. Wafer-scale 3D integration of silicon-on-insulator RF amplifiers. In *Proceedings of the IEEE Topical Meeting on Silicon Monolithic Integrated Circuits in RF Systems*, San Diego, CA, pp. 1–4, January 2009.

22. S. Koester et al. Wafer-level 3D integration technology. *IBM Journal of Research and Development*, 52(6):583–597, November 2008.

23. A. Zeng, J. Lu, K. Rose, and R. J. Gutmann. First-order performance prediction of cache memory with wafer-level 3D integration. *IEEE Design and Test of Computers*, 22(6):548–555, November-December 2005.

24. R. S. Patti. Three-dimensional integrated circuits and the future of system-on-chip designs. *Proceedings of the IEEE*, 94(6):1214–1224, June 2006.

25. J. Burns et al. Three-dimensional integrated circuits for low-power, high-bandwidth systems on a chip. In *Proceedings of the IEEE International Solid-State Circuits Conference*, San Francisco, CA, pp. 268–269, February 2001.

26. J.-Q. Lu. 3-D hyperintegration and packaging technologies for micro–nano systems. *Proceedings of the IEEE*, 97(1):18–30, January 2009.

27. T. Zhang et al. 3-D data storage, power delivery, and RF/optical transceiver—Case studies of 3-D integration from system design perspectives. *Proceedings of the IEEE*, 97(1):161–174, January 2009.

28. F. Niklaus, G. Stemme, J.-Q. Lu, and R. Gutmann. Adhesive wafer bonding. *Journal of Applied Physics*, 99(3):101.1–101.28, February 2006.

29. J.-Q. Lu et al. Stacked chip-to-chip interconnections using wafer bonding technology with dielectric bonding glues. In *Proceedings of the IEEE International Conference on Interconnect Technology*, Burlingame, CA, pp. 219–221, June 2001.

30. K. Warner et al. Low-temperature oxide-bonded three-dimensional integrated circuits. In *Proceedings of the IEEE International SOI Conference*, Williamsburg, VA, pp. 123–124, October 2002.

31. P. R. Morrow, C.-M. Park, S. Ramanathan, M. J. Kobrinsky, and M. Harmes. Three-dimensional wafer stacking via Cu–Cu bonding integrated with 65-nm strained-Si/low-k CMOS technology. *IEEE Electron Device Letters*, 27(5):335–337, May 2006.

32. S. Halder et al. Metrology and inspection for process control during bonding and thinning of stacked wafers for manufacturing 3D SIC's. In *Proceedings of the IEEE Electronic Components and Technology Conference*, Lake Buena Vista, FL, pp. 999–1002, May–June 2011.

33. Y. S. Kim et al. Ultra thinning 300-mm wafer down to 7-μm for 3D wafer integration on 45-nm node CMOS using strained silicon and Cu/low-K interconnects. In *Proceedings of the IEEE International Electron Devices Meeting*, Baltimore, MD, pp. 1–4, December 2009.

34. S. M. Satheesh and E. Salman. Power distribution in TSV-based 3-D processor-memory stacks. *IEEE Journal on Emerging and Selected Topics in Circuits and Systems*, 2(4):692–703, December 2012.

35. C. Laviron et al. Via first approach optimization for through silicon via applications. In *Proceedings of the IEEE Electronic Components and Technology Conference*, San Diego, CA, pp. 14–19, May 2009.

36. G. Pares et al. Mid-process through silicon vias technology using tungsten metallization: Process optimization and electrical results. In *Proceedings of the IEEE Electronics Packaging Technology Conference*, Singapore, pp. 772–777, December 2009.

37. G. Pares et al. Through silicon via technology using tungsten metallization. In *Proceedings of the IEEE International Conference on IC Design and Technology*, Kaohsiung, Taiwan, pp. 1–4, May 2011.

38. S. Ramaswami et al. Process integration considerations for 300 mm TSV manufacturing. *IEEE Transactions on Device and Materials Reliability*, 9(4):524–528, December 2009.

39. A. Agarwal, R. B. Murthy, V. Lee, and G. Viswanadam. Polysilicon interconnections (FEOL): Fabrication and characterization. In *Proceedings of the IEEE Electronics Packaging Technology Conference*, Singapore, pp. 317–320, December 2009.

40. M. Kawano et al. A 3D packaging technology for 4 Gbit stacked DRAM with 3 Gbps data transfer. In *Proceedings of the IEEE International Electron Devices Meeting*, San Francisco, CA, pp. 1–4, December 2006.

41. M. Puech, J. M. Thevenoud, J. M. Gruffat, N. Launay, N. Arnal, and P. Godinat. DRIE achievements for TSV covering via first and via last strategies. Technical report, ALCATEL Micro Machining Systems, Annecy, France.

42. M. Kawano et al. Three-dimensional packaging technology for stacked DRAM with 3-Gb/s data transfer. *IEEE Transactions on Electron Devices*, 55(7):1614–1620, July 2008.

43. D. H. Kim, S. Mukhopadhyay, and S. K. Lim. Through-silicon-via aware interconnect prediction and optimization for 3D stacked ICs. In *Proceedings of the International Workshop on System Level Interconnect Prediction*, New York, pp. 85–92, September 2009.

44. S. M. Satheesh and E. Salman. Effect of TSV fabrication technology on power distribution in 3D ICs. In *Proceedings of the ACM/IEEE Great Lakes Symposium on VLSI*, Paris, France, pp. 287–292, May 2013.

45. T. Dao et al. Thermo-mechanical stress characterization of tungsten-fill through-silicon-via. In *Proceedings of the IEEE International Symposium on VLSI Design Automation and Test*, Hsinchu, Taiwan, pp. 7–10, April 2010.

46. N. Kumar et al. Robust TSV via-middle and via-reveal process integration accomplished through characterization and management of sources of variation. In *Proceedings of the IEEE Electronic Components and Technology Conference*, San Diego, CA, pp. 787–793, May–June 2012.

47. J. W. Elam et al. Surface chemistry and film growth during TiN atomic layer deposition using TDMAT and NH_3. *Thin Solid Films*, 436(2):145–156, July 2003.

48. H. Kikuchi et al. Tungsten through-silicon via technology for three-dimensional LSIs. *Japanese Journal of Applied Physics*, 47(4):2801–2806, April 2008.

49. S. Arakalgud. 3D TSV interconnect program—An overview. In *Presented at SEMATECH Symposium Korea*, South Korea, October 2010.

50. V. F. Pavlidis and G. De Micheli. Power distribution paths in 3-D ICs. In *Proceedings of the IEEE/ACM Great Lakes Symposium on VLSI*, Boston, MA, pp. 263–268, May 2009.

51. S. M. Satheesh and E. Salman. Design space exploration for robust power delivery in TSV based 3-D systems-on-chip. In *Proceedings of the IEEE International System-on-Chip Conference*, Niagara, NY, pp. 307–311, September 2012.

52. H. Wang and E. Salman. Resource allocation methodology for through silicon vias and sleep transistors in 3D ICs. In *Proceedings of the IEEE International Symposium on Quality Electronic Design*, Santa Clara, CA, March 2015.

53. H. Wang and E. Salman. Decoupling capacitor topologies for TSV-based 3D ICs with power gating. *IEEE Transactions on Very Large Scale Integration (VLSI) Systems*, in press.

54. H. Wang and E. Salman. Enhancing system-wide power integrity in 3D ICs with power gating. In *Proceedings of the IEEE International Symposium on Quality Electronic Design*, Santa Clara, CA, March 2015.

55. S. Natarajan et al. A 32 nm logic technology featuring 2nd-generation high-k + metal-gate transistors, enhanced channel strain and 0.171 μm^2 SRAM cell size in a 291 Mb array. In *Proceedings of the IEEE International Electron Devices Meeting*, San Francisco, CA, pp. 1–3, December 2008.

56. R. Tatikola, M. Chowdhury, R. Chen, and J. Zhao. Simulation study of power delivery performance on flip-chip substrate technologies. In *Proceedings of the IEEE Electronic Components and Technology Conference*, Las Vegas, NV, pp. 1862–1865, June 2004.

57. Q. Wu and T. Zhang. Design techniques to facilitate processor power delivery in 3-D processor-DRAM integrated systems. *IEEE Transactions on Very Large Scale Integration (VLSI) Systems*, 19(9):1655–1666, September 2011.

58. B. Dang et al. 3D chip stack with integrated decoupling capacitors. In *Proceedings of the IEEE International Electronic Components and Technology Conference*, San Diego, CA, pp. 1–5, June 2009.

59. P. Zhou, K. Sridharan, and S. S. Sapatnekar. Congestion-aware power grid optimization for 3D circuits using MIM and CMOS decoupling capacitors. In *Proceedings of the ACM/IEEE Asia and South Pacific Design Automation Conference*, Yokohama, Japan, pp. 179–184, 2009.

60. P. Zhou, K. Sridharan, and S. S. Sapatnekar. Optimizing decoupling capacitors in 3D circuits for power grid integrity. *IEEE Design and Test of Computers*, 26(5):15–25, September 2009.

61. M. Popovich, M. Sotman, A. Kolodny, and E. G. Friedman. Effective radii of on-chip decoupling capacitors. *IEEE Transactions on Very Large Scale Integration (VLSI) Systems*, 16(7):894–907, July 2008.

62. E. Salman, E. G. Friedman, R. M. Secareanu, and O. L. Hartin. Worst case power/ground noise estimation using an equivalent transition time for resonance. *IEEE Transactions on Circuits and Systems I: Regular Papers*, 56(5):997–1004, May 2009.

63. A. Todri et al. A study of tapered 3-D TSVs for power and thermal integrity. *IEEE Transactions on Very Large Scale Integration (VLSI) Systems*, 21(2):306–319, February 2013.

64. T. Xu, P. Li, and B. Yan. Decoupling for power gating: Sources of power noise and design strategies. In *Proceedings of the ACM/IEEE Design Automation Conference*, San Diego, CA, pp. 1002–1007, June 2011.

65. North Carolina State University. FreePDK45.

66. P. Ji and E. Salman. Quantifying the effect of local interconnects on on-chip power distribution. *Microelectronics Journal*, 46(3):258–264, March 2015.

67. H. Wang and E. Salman. Power gating methodologies in TSV based 3D integrated circuits. In *Proceedings of the ACM/IEEE Great Lakes Symposium on VLSI*, pp. 327–328, May 2013.

68. X. Zhang et al. Characterizing and evaluating voltage noise in multi-core near-threshold processors. In *Proceedings of International Symposium on Low Power Electronics and Design*, pp. 82–87, September 2013.

69. Q. K. Zhu. *Power Distribution Network Design for VLSI*. Wiley, Hoboken, NJ, 2004.

70. H. Wei et al. Cooling three-dimensional integrated circuits using power delivery networks. In *Proceedings of the IEEE International Electron Devices Meeting*, San Francisco, CA, pp. 14.2.1–14.2.4, December 2012.

71. E. Salman. Noise coupling due to through silicon vias (TSVs) in 3-D integrated circuits. In *Proceedings of the IEEE International Symposium on Circuits and Systems*, Rio De Janeiro, Brazil, pp. 1411–1414, May 2011.

72. E. Salman, M. H. Asgari, and M. Stanacevic. Signal integrity analysis of a 2-D and 3-D integrated potentiostat for neurotransmitter sensing. In *Proceedings of the IEEE Biomedical Circuits and Systems Conference*, San Diego, CA, pp. 17–20, November 2011.

73. H. Wang, M. H. Asgari, and E. Salman. Compact model to efficiently characterize TSV-to-transistor noise coupling in 3D ICs. *Integration, the VLSI Journal*, 47(3):296–306, June 2014.

74. H. Wang, M. H. Asgari, and E. Salman. Efficient characterization of TSV-to-transistor noise coupling in 3D ICs. In *Proceedings of the ACM/IEEE Great Lakes Symposium on VLSI*, Paris, France, pp. 71–76, May 2013.

75. E. Salman, E. G. Friedman, and R. M. Secareanu. Substrate and ground noise interactions in mixed-signal circuits. In *Proceedings of the IEEE International SoC Conference*, Austin, TX, pp. 293–296, September 2006.

76. E. Salman, E. G. Friedman, R. M. Secareanu, and O. L. Hartin. Identification of dominant noise source and parameter sensitivity for substrate coupling. *IEEE Transactions on Very Large Scale Integration (VLSI) Systems*, 17(10):1559–1564, October 2009.

77. E. Salman, R. Jakushokas, E. G. Friedman, R. M. Secareanu, and O. L. Hartin. Methodology for efficient substrate noise analysis in large scale mixed-signal circuits. *IEEE Transactions on Very Large Scale Integration (VLSI) Systems*, 17(10):1405–1418, October 2009.

78. Z. Gan, E. Salman, and M. Stanacevic. Figures-of-merit to evaluate the significance of switching noise in analog circuits. *IEEE Transactions on Very Large Scale Integration (VLSI) Systems*, in press.

79. M. S. Peng and H.-S. Lee. Study of substrate noise and techniques for minimization. *IEEE Journal of Solid-State Circuits*, 39(11):2080–2086, 2004.

80. P. LeDuc et al. Enabling technologies for 3D chip stacking. In *Proceedings of the IEEE International Symposium on VLSI Technology, Systems and Applications*, Hsinchu, Taiwan, pp. 76–78, April 2008.

81. H. Chaabouni et al. Investigation on TSV impact on 65 nm CMOS devices and circuits. In *Proceedings of the IEEE International Electron Devices Meeting*, San Francisco, CA, pp. 6–8, December 2010.

82. J. Cho et al. Guard ring effect for through silicon via (TSV) noise coupling reduction. In *Proceedings of the IEEE CPMT Symposium*, Tokyo, Japan, pp. 1–4, August 2010.

83. M. Brocard et al. Characterization and modeling of Si-substrate noise induced by RF signal propagating in TSV of 3D-IC stack. In *Proceedings of the IEEE Electronic Components and Technology Conference*, pp. 665–672, 2012.

84. P. Saguet. The 3D transmission-line matrix method: Theory and comparison of the processes. *International Journal of Numerical Modeling: Electronic Networks, Devices and Fields*, 2(4):191–201, 1989.

85. J. Cho et al. Modeling and analysis of through-silicon via (TSV) noise coupling and suppression using a guard ring. *IEEE Transactions on Components, Packaging and Manufacturing Technology*, 1(2):220–233, February 2011.

86. M. Stanacevic, K. Murari, A. Rege, G. Cauwenberghs, and N. Thakor. VLSI potenstiostat array with oversampling gain modulation for wide-range neurotransmitter sensing. *IEEE Transactions on Biomedical Circuits and Systems*, 1(1):63–72, March 2007.

87. R. Genov, M. Stanacevic, M. Naware, G. Cauwenberghs, and N. Thakor. 16-Channel integrated potentiostat for distributed neurochemical sensing. *IEEE Transactions on Circuits and Systems I: Regular Papers*, 53(11):2371–2376, November 2006.

88. T. Wichmann and M. R. DeLong. Pathophysiology of Parkinsonian motor abnormalities. *Advances in Neurology*, 60:53–61, 1993.

89. The International Technology Roadmap for Semiconductors. ITRS Report, 2011. http://www.itrs.net. Acessed March 2014.

90. G. Katti, M. Stucchi, K. De Meyer, and W. Dehaene. Electrical modeling and characterization of through silicon via for three-dimensional ICs. *IEEE Transactions on Electron Devices*, 57(1):256–262, January 2010.

7 Design Methodology for TSV-Based 3D Clock Networks

Taewhan Kim and Heechun Park

CONTENTS

ABSTRACT

This chapter reviews the state-of-the-art design methodologies for TSV-based 3D clock networks and covers four ingredients: (1) the synthesis flows of basic 3D clock trees in comparison with the synthesis flow of 2D clock trees; (2) the importance of pre-bond testing and the support of pre-bond testability in 3D clock trees; (3) 3D clock tree design and synthesis techniques, which can tolerate clock TSV interconnect fault; (4) 3D clock tree synthesis algorithms, which can cope with various design variation parameters such as thermal variation, TSV-stress, on-package variation, and blockage awareness.

7.1 INTRODUCTION

In synchronous circuits and systems, all activities or events are driven by clock signals to synchronize them. Precisely, the synchronization is accomplished by delivering a periodic signal from a clock source to every storage element (e.g., flip-flops and latches) in the circuit or system. We refer the storage elements driven by clock signal to *clock sinks* in the clock network context. Designing a clock network of system requires a set of timing constraints to be met as well as a set of objectives to be minimized or maximized. The constraints and objectives are listed as follows:

- *Clock skew*: it is the difference of clock arrival times at two sinks.
- *Global clock skew*: it is the maximum value among all clock skews. If no confusion occurs, clock skew implicitly refers to the global clock skew in this presentation.
- *Bounded clock skew*: it is the constraint such that the (global) clock skew should be within a certain bound. That is, the clock skew is constrained not to be greater than the specified clock bound. *Bounded clock skew scheduling* is to determine the clock arrival times of all sinks while meeting the bounded clock skew constraint.
- *Useful clock skew*: it is the constraint such that for every pair of sinks, one driving the other, the clock skew between the sinks should satisfy the setup and hold time constraints between them. *Useful clock skew scheduling* is to determine the clock arrival times of all sinks while meeting the useful clock skew constraint.
- *Clock slew*: it is the time interval for clock signal transition. The clock slew should not exceed a certain threshold (usually 10% of the clock period), which is called *clock slew constraint*.
- *Clock network*: it is the structure that is used to deliver clock signals to sinks. The structure consists of wires and buffers. Tree is the most commonly used structure. Other structures are mesh, spine, and tree with links. The two roles of clock buffers are (1) to adjust the delay of clock signal to meet the clock skew constraint and (2) to strengthen the signal driving to meet the clock slew constraint.
- *Clock power*: it is the power consumed by the clock network in the course of the transition of clock signal. It is said that the clock power takes 20%–50% of total power of system [9].
- *Clock latency*: it is the delay for a clock signal to arrive at a sink from clock source. The longer the latency is, it is very likely that the more the amount of resource on clock network is as well as the more the consumption of clock power is.

In comparison with the clock networks in 2D-ICs, the clock networks in 3D-IC require an additional (new) type of resource, which is *clock TSV*. Clock TSV (through-silicon-vias) is a vertical via that connects two logic components in different layers (i.e., dies). The diameter of clock TSV is around 5–10 μm, which is comparable to or greater than the size of standard gates and the height is typically 20–50 μm [46].

Designing 3D-ICs demands many design challenges for 3D clock networks. This chapter discusses the state-of-the-art design methodologies of clock tree synthesis (CTS) with various design optimization objectives: the number or total length of TSVs, 2D wirelength (WL), and clock power consumption. In the presentation, we focus on discussing the structure of clock tree since tree is the most common and widely used one in industry. Precisely, we survey the works of the following topics.

1. **3D clock synthesis flow**: *How can the existing 2D CTS techniques be exploited/extended to implement 3D clock tree?* The typical design objective of 2D CTS is to minimize the total clock wirelength and total buffer size while constraining the clock skew in a certain limit. In 3D CST, one more design objective or constraint is included, which is the number (or total length) of clock TSVs. Two factors are added to the flow of the conventional 2D CTS. One is the flow extension to *z*-direction and the other is the inclusion of layer assignment step in the flow. The details will be discussed in the next section.
2. **Pre-bond testability**: Testing 3D-ICs can be divided into two stages: *pre-bond testing* and *post-bond testing*. Pre-bond testing is testing dies before they are bonded together, while post-bond testing is testing dies after they are bonded. Thus, the question is: *how should*

3D clock tree be structured to enable pre-bond testing as well as post-bond testing of 3D chips? The selection of "known-good-dies" (KGDs) to be stacked is highly important to make the whole 3D-IC working correctly. Thus, testing individual dies before stacking is inevitable, which requires a testing facility in 3D clock tree that offers pre-bond testability as well as post-bond testability.

3. **Clock TSV fault tolerance**: *How can 3D clock tree be designed to tolerate clock TSV faults?* The TSV implantation technology is not as stable as the other 2D process technologies. Consequently, it is necessary to design a robust 3D clock tree which is able to tolerate the clock TSV faults. It is necessary to trade-off between the hardware overhead and the degree of fault tolerance in designing fault-tolerant 3D clock trees.

4. **3D variation**: *How can the impact of variation induced by the 3D integration and clock TSV insertion on the increase of clock skew be reduced in the design stage, post-silicon stage, and bonding stage?* Three dominant variation factors are thermal-induced variation, TSV stress-induced, and on-package-induced variation.

The subtopics mentioned earlier are surveyed in detail in the subsequent sections.

7.2 3D CLOCK TREE SYNTHESIS

7.2.1 DESIGN FLOW

Traditionally, 3D CTS flow consists of a sequence of tasks: (1) abstract tree topology generation; (2) layer assignment of tree nodes; (3) clock tree routing (embedding); and (4) clock buffer insertion, as shown in Figure 7.1a, where the clock tree routing and buffer insertion are usually performed simultaneously to meet the skew and slew constraints.

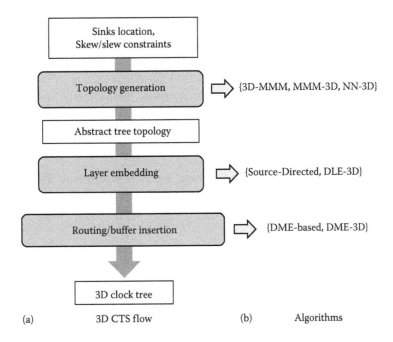

(a)   3D CTS flow   (b)   Algorithms

FIGURE 7.1 (a) The basic flow of 3D CTS. (b) The representative synthesis algorithms corresponding to the stages in the flow. (3D-MMM, Source-Directed, DME-based [31], MMM-3D, DLE-3D, DME-3D [18], and NN-3D [17]).

Tree topology generation is constructing a form of binary (sometimes 4-branch) tree with the sinks as terminals and the clock source as the root. The topology structure to be formed should be the one which leads to a least cost of total WL and clock TSVs. The layer embedding step following the topology generation step is determining the layer in 3D on which each internal node of clock tree is placed. Then, routing between the nodes in clock tree is performed together with the insertion of buffers with the objective of minimizing the total WL and total buffer size while meeting the clock skew and slew constraints. The topology generation step is the most important step since tree topology greatly affects the final quality (skew, TSVs, WL) of the design. The following sections introduce the representative algorithms for performing the stages in the flow of 3D CTS in Figure 7.1b.

7.2.2   3D-MMM-Based CTS

As mentioned previously, the abstract tree topology generation is, given the location information of the sinks, to construct a form of tree and as yet not determine locations of the internal nodes of the tree except the clock source (i.e., root). The algorithm MMM (Method of Means and Medians) [12] is the most famous topology generation algorithm used in 2D CTS. Since many works on 3D CTS are based on MMM, we briefly explain the procedure of MMM: it is a recursive top-down partitioning algorithm. At each iteration, for the sink set S in a plane to be partitioned, it finds two subsets S_L and S_R of almost equal size, which divides S around the median x-coordinate of the sink points, producing two points sets S_L and S_R. The same partitioning process is applied horizontally (i.e., in y-coordinate) to each of the sinks in S_L and S_R. The iteration of top-down plane partition stops when exactly one sink remains at every subplane. Then, from the sinks in the bottom the tree construction is performed, recursively creating internal nodes by merging the sink sets of two subplanes that were produced by the top-down partition process. Note that in addition to the topology generation, MMM determines the location of the internal nodes by computing the means of x and y values of the locations of the sinks in the merged plane. However, when the MMM's topology generation is talked, it usually refers to the MMM without determining the location of internal nodes.

The 3D-MMM [31] is one of the well-known 3D topology generation algorithms, which extends the idea of 2D-MMM [12] and takes into account not only the clock WL in 2D planes but also the length and number of clock TSVs. 3D-MMM works as follows: at each recursive partitioning process, the given set of sinks placed in layers is partitioned into two subsets S_a and S_b. The following two cases are considered, based on the TSV upper-bound constraint δ for the current sink set. (Note that 3D-MMM is applicable to 3D-ICs with *two layers only*.)

- If $\delta = 1$, the sink set is partitioned into A and B such that the sinks from the same layer belong to the same subset. The connection between A and B needs one TSV.
- If $\delta > 1$, the sink set is projected to 2D (i.e., ignoring z-coordinate) and partitioned geometrically by a horizontal or vertical line, as does in 2D MMM [12]. Since each subset contains sinks from both layers, it is expected potentially several TSVs to link them.

Here, the critical factor that influences the quality of results by 3D-MMM is how the δ value is determined in each partitioning step. 3D-MMM said "this is performed by (i) estimating the number of clock TSVs required by each set and (ii) dividing the given bound according to the ratio of the estimated TSVs. Then, the cut direction is determined such that the TSV bound is balanced in both subsets." However, detail information about how TSV bound is estimated and TSVs are balanced is not available in the description of 3D-MMM in [31]. Figure 7.2 shows the illustration of 3D-MMM under different TSV bound constraints when partitioning eight sinks in two layers.

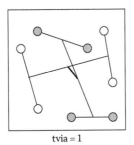
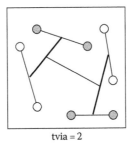
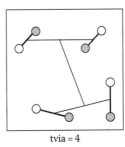

tvia = 1	tvia = 2	tvia = 4

FIGURE 7.2 3D abstract trees constructed with 3D-MMM [31] under various TSV bound constraints tvia = 1, 2, and 4. White and gray nodes are, respectively, in the top and bottom layers. Thick lines indicate connections by TSVs. The total wirelength decreases as more TSVs are used in the trees. (Taken from Minz, J., Zhao, X., and Lim, S.K., Buffered clock tree synthesis for 3D ICS under thermal variations, In *IEEE Asia and South Pacific Design Automation Conference*, Anaheim, CA, pp. 504–509, © 2008 IEEE.)

7.2.3 MMM-3D and NN-3D Based CTS

MMM-3D [18] is a topology generation algorithm which is also based on 2D-MMM [12] like 3D-MMM [31]. Unlike the use of TSV bound constraint in 3D-MMM, MMM-3D uses a designer specified parameter ρ ($0 \leq \rho \leq 1$) to control the partitioning direction between sinks in the same layers and sinks in the different layers. $\rho = 0$ means exactly 2D-MMM, which totally ignores partitioning of sinks in multiple layers, while $\rho = 1$ means that sinks are partitioned based on layers and then partitioned based on (x, y)-coordinates. MMM-3D recursively divides subsets of sinks based on the geometric median value in 2D Manhattan plane, and then the cut direction is chosen by the half perimeter wirelength of subsets. During partitioning, if the half perimeter wirelength of a subset sink is smaller than the value of $\rho \times L$, where L is the half perimeter wirelength of set of all sinks, the sinks is bi-partitioned according to the placement layers rather than (x, y)-coordinates. The advantage of MMM-3D over the 3D-MMM [31] is that since 3D-MMM assumes 3D-ICs with two layers only, all TSVs will be the same length and thus the TSV bound is a reasonable constraint. However, for 3D-ICs with arbitrary number of layers, TSVs can be of different lengths and thus exploring diverse tree topologies by MMM-3D could be useful to find globally minimal TSV allocation and wirelength.

The inherent limitation of 3D-MMM and MMM-3D is that the top-down partitioning process roughly estimates the expected number of TSVs, in many cases it fails to make an even distribution of real TSV number. For illustration, consider the placement of 16 sinks with TSV bound $\delta = 4$, as shown in Figure 7.3a. 3D-MMM equally divides TSV bound for two subsets A and B because it finds that both subsets have equal numbers of sinks in each layers. However, the resulting TSV number is 1 for subset A and 2 for subset B as shown in the clock tree of Figure 7.3a, and this is not an even distribution of real TSV number across the whole die area. For another example, consider the placement of 16 sinks shown in Figure 7.3b. Since all sinks in subset A are in layer 2, 3D-MMM propagates the whole TSV bound to subset B, which in fact restricts the maximum number of TSVs on a region rather than an even distribution of TSVs. In other words, 3D-MMM actually distributes TSV bound in a given budget rather than making an even distribution of the resulting TSV number.

Like 3D-MMM, MMM-3D also exhibits similar outcomes. This is because even though MMM-3D tries to take into account manufacturability by using TSV density, it does not consider TSVs on internal tree edges. For example, if there are 16 sinks with $\rho = 0.5$ as shown in Figure 7.3c, there exists a layer with TSV density violation, that is, one more TSV can be used on gray region. In addition, note that 3D-MMM and MMM-3D do not sufficiently take into account different

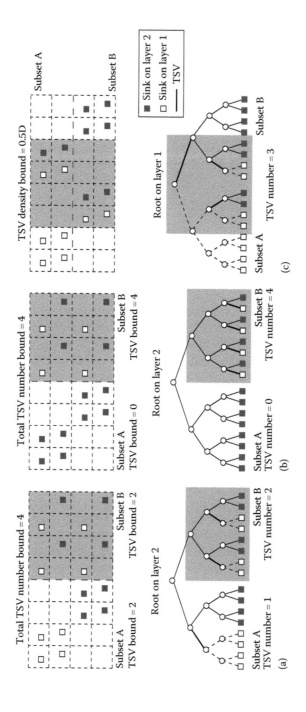

FIGURE 7.3 Examples illustrating limitations of top-down partitioning approaches 3D-MMM [31] and MMM-3D [18] of clock tree topology generation. (a) 3D-MMM for two-layered 3D-IC, (b) 3D-MMM for two-layered 3D-IC, and (c) MMM-3D for two-layered 3D-IC. (Taken from Kim, T.-Y. and Kim, T., Bounded skew clock routing for 3D stacked IC designs: Enabling trade-offs between power and clock skew, In *International Green Computing Conference*, Chicago, IL, pp. 525–532, 2010.)

costs between 2D wire and TSVs. These limitations have been overcome by NN-3D in [20], which employs a bottom-up clustering approach.

NN-3D [17] is a 3D topology generation algorithm which basically follows the idea of 2D bottom-up (nearest neighbor clustering) approach in [6]. NN-3D recursively selects a pair of sub-trees or sinks and merges the selected pair. The selection of pairs is based on a *merging* cost. First, the nearest neighbors of all subtrees and sinks are calculated according to the merging costs, and the edges in a nearest neighbor graph are sorted in nondecreasing order of their costs. Subtrees and sinks are merged in the order of the sorted edges. If one or both nodes of an edge have already been used in earlier merging, the corresponding edge is skipped. The process of nearest neighbor selection and merging is repeated until one tree is left.

The merging cost in NN-3D takes into account both wire and TSVs, in which downstream capacitances of subtree is used for the cost calculation. Performance comparison of the three topology generation algorithms 3D-MMM [31], MMM-3D [18], NN-3D [17] are given in Table 7.1, where the same algorithms for the layer embedding and wire routing were applied to the results of the tree

TABLE 7.1

Performance Comparison of the Topology Generation Algorithms 3D-MMM [31] (TSV Bound = 100%), MMM-3D [18] (TSV Density Constraint $\rho = 0.0$), and NN-3D [17]

	3D-MMM	MMM-3D	NN-3D	3D-MMM	MMM-3D	NN-3D	3D-MMM	MMM-3D	NN-3D
ispd09		#TSVs			Wirelength (µm)			#BUFs	
f11	42	45	39	134,429	132,710	128,500	163	164	149
f12	43	42	36	126,649	118,413	116,355	157	145	142
f21	40	38	37	139,343	136,955	135,843	162	164	156
f22	28	28	30	79,458	79,458	74,732	100	100	91
f31	88	85	79	305,783	279,793	276,008	357	334	318
f32	60	61	54	245,118	222,160	211,259	288	267	246
f33	67	66	61	231,311	222,634	206,845	275	266	247
f34	52	51	48	203,634	183,504	175,622	236	223	202
f35	63	63	62	213,738	198,277	193,452	254	229	226
fnb1	109	112	38	36,553	32,586	38,499	89	78	73
fnb2	138	136	92	90,375	86,145	82,693	164	158	130
Ratio	1.00	1.00	0.85	1.00	0.94	0.93	1.00	0.95	0.88
		Delay (ns)			Skew (ps)			Power (mW)	
f11	0.36	0.35	0.36	27	21	23	57.50	57.10	54.80
f12	0.35	0.33	0.34	24	20	23	54.50	51.40	50.40
f21	0.39	0.37	0.37	32	29	25	57.40	57.10	56.30
f22	0.27	0.27	0.27	17	17	22	35.60	35.60	33.80
f31	0.56	0.54	0.54	32	27	29	125.90	117.00	114.90
f32	0.51	0.55	0.51	31	37	24	99.40	91.30	86.90
f33	0.49	0.50	0.47	25	29	27	95.70	92.70	87.10
f34	0.48	0.52	0.48	25	23	26	82.60	75.60	72.30
f35	0.49	0.47	0.49	35	17	25	88.60	82.80	81.10
fnb1	0.13	0.10	0.16	17	15	24	27.60	25.80	25.80
fnb2	0.22	0.20	0.20	23	19	26	51.60	49.80	46.30
Ratio	1.00	0.97	1.00	1.00	0.89	0.99	1.00	0.95	0.92

Source: Taken from Kim, T.-Y. and Kim, T., Bounded skew clock routing for 3D stacked IC designs: Enabling trade-offs between power and clock skew, in: *International Green Computing Conference*, Chicago, IL, pp. 525–532, 2010.

Note: Skew and power values were measured with SPICE simulation.

topologies. It is shown that NN-3D outperforms 3D-MMM and MMM-3D, reducing the number of TSVs by 15%, WL by 7%, the number of buffers by 12%, and clock power by 8% on average while the clock skew is well managed under 30 ps (3% of 1 ns clock period).

7.2.4 LAYER EMBEDDING

Layer embedding is assigning layers to the internal nodes of the clock tree. Since the result of layer embedding directly determines the number of clock (unit-length) TSVs to be inserted, a careful layer embedding is required. Figure 7.4 shows possible layer embedding in two-layered 3D-IC where sinks s_1 and s_3 are located on layer 1 and sinks s_2 and s_4 are on layer 2. There are eight possible ways of embedding layer assignment for internal nodes x_1, x_2, and root node s_0, as shown in the middle columns of the table in Figure 7.4. The number in each entry of the last column of the table shows the number of TSVs needed for the corresponding embedding layer assignment. For example, optimal embedding layer assignments are $\langle x_1, x_2, s_0 \rangle = \langle 1, 1, 1 \rangle$ and $\langle 2, 2, 2 \rangle$.

DLE-3D [18] is a layer embedding algorithm which guarantees producing a minimal number of TSVs in two-layered 3D-ICs and a minimal total length of TSVs in more than two-layered 3D-ICs. In DLE-3D, the number of TSVs in an embedded clock topology T_x can be recursively defined as

$$nv(T_x) = \begin{cases} 0 & \text{if } x \text{ is a sink node,} \\ nv(T_{left(x)}) + nv(T_{right(x)}) + nv(x, left(x)) + nv(x, right(x)) & \text{if } x \text{ is an internal node,} \end{cases} \quad (7.1)$$

which is the sum of the numbers of TSVs allocated in the subtrees rooted at the two children of x and the number of TSVs needed in routing from x to the children.

If the two children of x are sinks, then the layers of the children are given. Then, from the embedding layers of the children, one will find the set of embedding layers of x, denoted as $el(x)$, that leads to a minimum value of $nv(T_x)$. On the other hand, if the two children are internal nodes, by recursive process, their embedding layers are sets of embedding layers. Then, from the sets of embedding layers of the two children, one will find the set of embedding layers of x (i.e., $el(x)$) that leads to a minimum value of $nv(T_x)$.

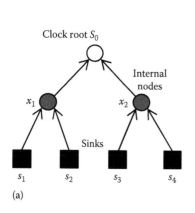

Sink location				Embedding layer			TSVs
s_1	s_2	s_3	s_4	x_1	x_2	s_0	
1	2	1	2	1	1	1	2
						2	3
				1	2	1	3
						2	3
				2	1	1	3
						2	3
				2	2	1	3
						2	2

(a)   (b)

FIGURE 7.4 TSV allocation examples for two-layered 3D-IC. (a) Rooted binary clock tree and (b) possible embedding layer assignments. (Taken from Kim, T.-Y. and Kim, T., Clock tree embedding for 3D ICS. In *IEEE Asia, and South Pacific Design Automation Conference*, Taipei, Taiwan, pp. 486–491, © 2010 IEEE.)

7.2.5 ROUTING

Routing clock wire and inserting clock buffers are performed in the last step in 3D CTS. Minimizing the planar total WL and total buffer size is the main objective in this step. Unlike the objective in 2D counterpart, the determination of TSV location is added in 3D CTS. The most well-known routing algorithm in 2D is DME (called deferred merging embedding algorithm in [1]) and its variants (e.g., [3,4,10]). Consequently, most 3D routing algorithms in 3D CTS extend the DME algorithm by adding the vertical directions, that is, through z-coordinate. Since the TSV cost has already been fixed, the determination of TSV location is exploited to minimize the total planar WL. DME-3D [18,25] and DME-based [31] are some of the representative 3D clock routing algorithms that achieve this goal.

7.3 CLOCK TREE SYNTHESIS WITH PRE-BOND TESTABILITY

7.3.1 BASICS

Testing is an important issue in 3D-ICs as well as in 2D-ICs. Contrary to 2D-ICs, testing 3D-ICs consists of two sub-testings: pre-bond testing and post-bond testing. Before testing full 3D chip constructed by bonding all 2D dies (i.e., *post-bond testing*), there is a step of testing 2D dies to be stacked separately to filter KGDs (known-good-dies), (i.e., *pre-bond testing*). If pre-bond test is skipped, there is a possibility that bad (faulty) dies are bonded with good dies, which results in a yield loss since those good 2D dies stacked are not usable along with bad dies. Therefore, pre-bond testing is a new and essential step for 3D-ICs to filter out bad dies before stacking dies.

7.3.2 MAIN ISSUES IN PRE-BOND TESTABILITY

Figure 7.5 shows the 3D-IC test flow. In the pre-bond testing, every die is tested by using conventional 2D-IC testing techniques before bonding them, and only KGDs are passed to the next stage of die stacking.

Since there is a single clock source for a complete 3D clock tree, some dies without the clock source (e.g., die1 in Figure 7.6) have incomplete forms of clock trees, as shown in Figure 7.6. For this

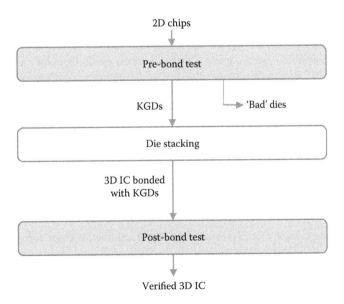

FIGURE 7.5   Testing flow of 3D-ICs.

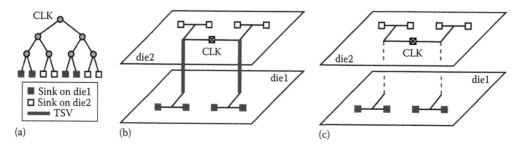

FIGURE 7.6 Example of two-layered 3D clock tree without pre-bond testability [22]. (a) Abstract tree topology. (b) Routed 3D clock tree. (c) Before die stacking. (Taken from Kim, T.-Y. and Kim, T., Resource allocation and design techniques of prebond testable 3-D clock tree, *IEEE Transactions on Computer-Aided Design of Integrated Circuits and Systems*, 32(1), 138–151, © 2013 IEEE.)

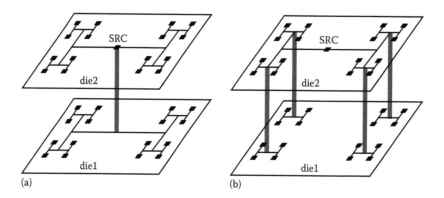

FIGURE 7.7 Example of 3D clock trees with different number of clock TSVs. (a) Using one TSV and (b) using four TSVs.

case, although the die with the clock source (e.g., die2 in Figure 7.6) can perform pre-bond testing, the rest of dies are not able to deliver clock signal to the clock sinks, disabling the pre-bond testing for the dies. Consequently, to enable pre-bond testability on *n*-layered 3D-IC, the *n* − 1 dies which have no clock source should have complete (2D) clock trees, which meet the clock skew constraint.

Pre-bond testing of dies can be simply solved by constructing a complete 2D clock tree for each die without clock source and connecting them to the clock source by using one TSV as shown in Figure 7.7a. However, this form of clock tree would consume a lot of wire resource compared to the form in Figure 7.7b, which connects distributed clock subtrees with one TSV for each.

Moreover, if we may include new logic or wires for the purpose of pre-bond testing of dies, they should be exclusively used at the stage of pre-bond testing. Thus, minimizing the amount of this new resources is important.

7.3.3 Pre-Bond Testable 3D Clock Tree

Figure 7.8a shows an example of a two-layered pre-bond testable 3D clock tree, in which additional tree is added to the die without clock source (die2). Such additional tree is used only in pre-bond stage and will do nothing after the pre-bond test. The added tree is called a 2D *redundant tree*. *Transmission gates* (TGs) together with a control signal line are required between redundant tree and clock subtrees in die2. TGs are ON in the pre-bond stage to connect the redundant tree with clock subtrees in die2 and allow the signal of test clock (TCLK) to flow into those subtrees (see Figure 7.8b). While TGs are OFF after pre-bond test to disconnect the redundant tree from the

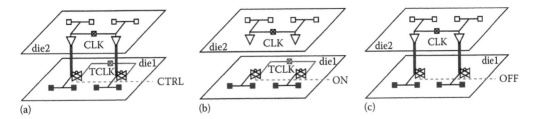

FIGURE 7.8 Example of two-layered 3D clock tree with pre-bond testability, extended from Figure 7.6. (a) Pre-bond testable 3D clock tree. Added (TCLK to transmission gates in die 1) wire represents redundant tree. (b) Pre-bond test stage. (c) Post-bond operation stage. (Taken from Kim, T.-Y. and Kim, T., Resource allocation and design techniques of prebond testable 3-D clock tree, *IEEE Transactions on Computer-Aided Design of Integrated Circuits and Systems*, 32(1), 138–151, © 2013 IEEE.)

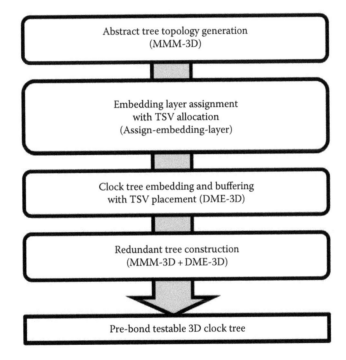

FIGURE 7.9 Flow of pre-bond testable 3D CTS [19]. (Taken from Kim, T.-Y. and Kim, T., Clock tree synthesis with pre-bond testability for 3D stacked IC designs, In *ACM/IEEE Design Automation Conference*, pp. 723–728, © 2010 IEEE.)

clock tree (see Figure 7.8c). One thing to be observed is that since no TSVs exist at the time of pre-bond testing, the nodes which have not yet been connected to TSVs will have different downstream capacitance values from those in the post-bond stage, which results in a wrong pre-bond test of die1. To resolve this, two buffers, the so-called *TSV-buffer*, are inserted at the top of two TSVs in Figure 7.8a to make constant downstream capacitances regardless of testing stages, and permit accurate pre-bond testing of the die with clock source (die1).

Figure 7.9 shows the pre-bond 3D CTS flow. The first three steps are identical to that in the previous section. The only difference is that a TSV-buffer is added at the top of each TSV for accurate pre-bond testing. In the last step, redundant trees are routed to all 2D dies to construct complete 2D clock trees for the dies. Detail process of this step is summarized in Figure 7.10. Firstly, clock subtrees in the 2D die are extracted and are named as clock "sinks," as shown in Figure 7.10a, and their

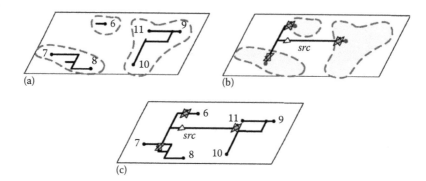

FIGURE 7.10 Example of redundant tree construction [45]. (a) Extract clock sinks. (b) Construct redundant tree and insert TGs. (c) Final pre-bond testable clock tree. Control signal wires for TGs are omitted for simplicity. (Taken from Zhao, X., Lewis, D.L., Lee, H.-H.S., and Lim, S.K., Low-power clock tree design for pre-bond testing of 3-D stacked ICS, *IEEE Transactions on Computer-Aided Design of Integrated Circuits and Systems*, 30(5), 732–745, © 2011 IEEE.)

timing and capacitance information are calculated. Then, existing zero-skew 2D CTS algorithm is applied to the 2D die to construct a redundant tree, followed by inserting TGs between the generated redundant tree and clock subtrees, as shown in Figure 7.10b. Finally, pre-bond testable clock tree for 2D die is constructed as in Figure 7.10c.

With this pre-bond testable 3D CTS flow, total cost of WL, power consumption, clock skew, etc., of overall clock tree (3D clock tree and 2D redundant trees) can be minimized by applying conventional 2D and 3D CTS algorithms.

7.3.4 ADDITIONAL ISSUES

As mentioned previously, redundant tree is added for pre-bond testing. Thus, the resources to make the tree such as wires, buffers, and TGs are not reusable in the post-bond operation stage, which means that reducing the amount of resources is necessary.

An intuitive method of reducing the cost of redundant tree is to enlarge clock subtrees in 2D dies, which results in reduced portion of redundant tree in 2D pre-bond testing tree. This method is implemented by changing the embedded layers of internal nodes in the clock tree topology. Figure 7.11 shows the effect of layer embedding of internal clock nodes. Although the layer embedding of clock sinks are identical, it can be found that the layer embedding result in Figure 7.11b is more regular than that in Figure 7.11a. If both results are implemented in two-layered 3D-ICs with pre-bond

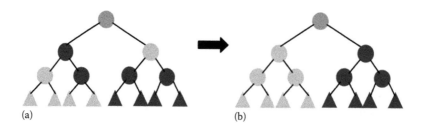

FIGURE 7.11 Effect of layer embedding of internal clock nodes [42]. Black and light gray color represents for two different layer embedding. Gray colored root node (source) represents that layer embedding of this node is undetermined. (a) Irregular embedding. (b) Regular embedding. (Taken from Wang, S.-J., Lin, C.-H., and Li, K.S.-M., Synthesis of 3D clock tree with pre-bond testability. In *IEEE International Symposium on Circuits and Systems*, Beijing, China, pp. 2654–2657, © 2013 IEEE.)

testable 3D CTS algorithm, the result in Figure 7.11a will have clock subtrees on both two dies, and redundant tree construction is needed for pre-bond testability, regardless of the layer embedding of the clock source. On the other hand, the result in Figure 7.11b will have complete 2D clock trees for both dies, which means redundant tree construction is not needed. As complete clock trees for all dies may cause a large wire overhead, this process should be applied right after the topology generation to be effective in finding a best trade-off between pre-bond testability and overhead reduction.

Another issue of reducing redundant tree cost is the control signal line to TGs. TGs are placed at the locations that connect the redundant tree and clock subtrees to control the connection and disconnection of redundant tree according to current state (pre-/post-bond). To control TGs simultaneously, tree-structured wires for TG control signal (e.g., green dashed lines in Figure 7.8) are inserted in all dies with TGs. Overhead of these wires is considerably large, about 17% of all WL of pre-bond testable 3D clock tree on average [22]. To reduce this overhead, a new TG structure called *self-controlled clock transmission gate* (SCCTG) is proposed. The structure of SCCTG is given in Figure 7.12, which consists of an original TG (mp5 + mn5), three inverters (u1, u3, u4), and a PMOS(mp2)/NMOS(mn2) transistor. SCCTG autonomously detects clock-like (periodically oscillating) signal without other control signal, as in right waveforms of Figure 7.12. When TCLK generates test clock signal during pre-bond test and the signal arrives at the input of the SCCTG, control signal is self-generated and the gate becomes ON. When the pre-bond test is finished, control signal from the inside of SCCTG becomes low and the SCCTG automatically becomes OFF. By replacing the original TGs with SCCTGs, the TG control signal tree can be completely eliminated and a large amount of WL will be reduced for pre-bond testable 3D clock tree.

There are several techniques which address the redundant tree cost reduction for pre-bond testable 3D clock tree, such as *TSV-buffer aware clock topology generation* [22,45] to reduce the number of *bad* TSV-buffers that give rise to additional buffer/wire insertion for clock skew balancing.

7.4 CLOCK TSV FAULT-TOLERANT TECHNIQUES

7.4.1 BASICS

Since TSV implant technology is not as stable as 2D process technology, TSV-based 3D-IC fabrication process faces serious chip reliability problem that attenuates its advantages in high density and low power dissipation. TSV may have random open defect during fabrication as in Figure 7.13a or induce crack to the chip as in Figure 7.13b. Moreover, different process, voltage, and temperature (PVT) variations across different dies in 3D-IC makes 3D process technology challenging [7].

Fault of a clock TSV in 3D-IC affects more serious than that of a logic signal TSV. As shown in Figure 7.14, if a clock TSV is faulty, the entire clock sinks under the subtree rooted as the faulty TSV cannot receive clock signal, and all signals and logic gates from those clock sinks are incorrect, which results in fatal malfunction of whole chip. Therefore, clock TSV faults should be minimized for the robustness of 3D-ICs. Since TSV faults cannot be completely removed in fabrication, clock TSV fault-tolerant CTS problem, which delivers clock signal to all clock sinks even when clock TSV fault(s) occurs, have been addressed in many works.

7.4.2 REDUNDANT TSV METHODS

A naive solution to the clock TSV fault tolerance problem is to insert spare TSVs and use them when some of the original clock TSVs are faulty. Since spare TSVs are redundant resources only for clock TSV fault tolerance, this technique is called *redundant TSV technique*.

Figure 7.15 shows two examples of redundant TSV techniques. The redundant TSV structure in Figure 7.15a, called *Double-TSV structure* [23], adds a spare TSV (STSV) for each of the original clock TSVs. By applying this structure to all clock TSVs, there are two TSVs for each inter-die clock signal path, and whole 3D-IC may have twice more TSV fault tolerance than before. However,

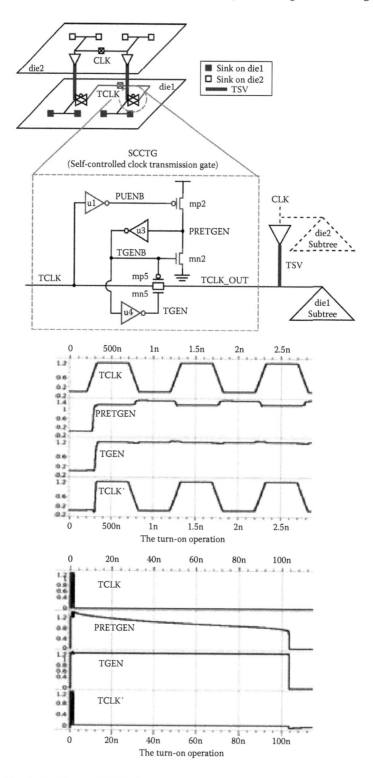

FIGURE 7.12 Circuit of self-controlled clock transmission gate (SCCTG) [22] and its operations [19]. (Taken from Kim, T.-Y. and Kim, T., Resource allocation and design techniques of prebond testable 3-D clock tree, *IEEE Transactions on Computer-Aided Design of Integrated Circuits and Systems*, 32(1), 138–151, © 2013 IEEE.)

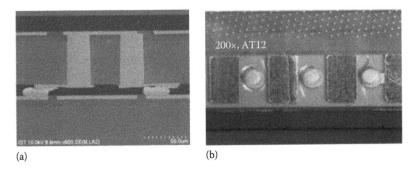

(a) (b)

FIGURE 7.13 Example of chip defects by TSV fault [28]. (a) Open defect of TSV by undercut. (b) TSV-induced crack. (Taken from Lung, C.-L., Chien, J.-H., Shi, Y., and Chang, S.-C., TSV fault-tolerant mechanisms with application to 3D clock networks, *IEEE International SoC Design Conference*, Jeju, Korea, pp. 127–130, © 2011 IEEE.)

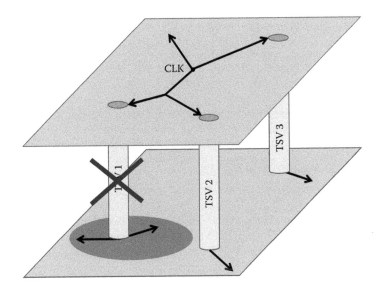

FIGURE 7.14 Impact of a clock TSV fault on the function of clock tree. Entire subtree rooted at the failed TSV cannot receive correct clock signal.

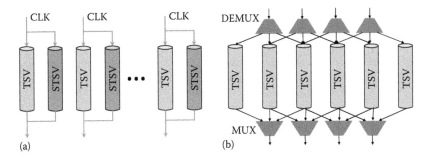

FIGURE 7.15 Examples of redundant TSV techniques. (a) Double-TSV structure [23]. (b) 4:2 shared-spare TSV structure [14]. (Taken from Park, H. and Kim, T., Synthesis of TSV fault-tolerant 3D clock trees, *IEEE Transactions on Computer-Aided Design of Integrated Circuits and Systems*, 34(2), 266–279, © 2015 IEEE.)

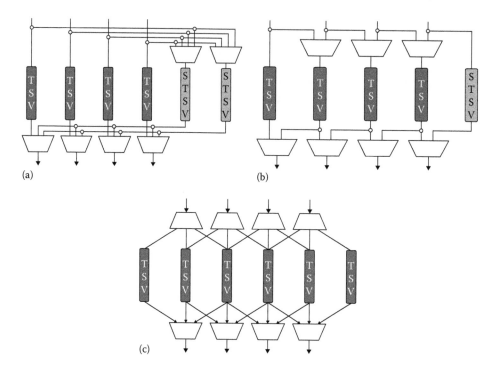

FIGURE 7.16 Different shared-spare TSV structure types introduced in [29]. (a) 4:2 shared-spare TSV with dedicated spare TSVs [16]. (b) 4:1 shared-spare TSV with dedicated spare TSV [26]. (c) 4:2 shared-spare TSV without dedicated spare TSVs [14] (same figure as Figure 7.15b). (Taken from Lung, C.-L., Su, Y.-S., Huang, H.-H., Shi, Y., and Chang, S.-C., Through-silicon via fault-tolerant clock networks for 3-D ICS, *IEEE Transactions on Computer-Aided Design of Integrated Circuits and Systems*, 32(7), 1100–1109, © 2013 IEEE.)

despite of its robustness and simplicity, double-TSV structure is not popularly used because of the considerable size of redundant TSVs added to the structures, since the size of a TSV is relatively large (up to 10 μm) and large number of TSVs may result in more TSV-induced stresses to other logics near those TSVs, other unpredictable variations occur.

Figure 7.15b shows another redundant TSV structure, called *shared-spare TSV structure* [14]. In this structure, one clock TSV will not necessarily have one dedicated spare TSV and for m clock TSVs, n ($<m$) redundant TSVs are added to work as spare TSVs. This structure can slightly reduce the number of redundant TSVs by using $m + n$ ($<2m$) TSVs for m inter-die clock signal paths. In addition, as indicated in the examples in Figure 7.16, the number of spare TSVs or the number of TSVs for given inter-die clock signal paths as well as the structure of shared-spare TSV can vary.

On the other hand, to use shared-spare TSV structure, additional multiplexers (MUXs) and demultiplexers (DEMUXs) are necessary to route m clock signals to m fault-free TSVs, and the routing wires to connect m distributed TSVs also become a large cost. Moreover, since clock path changes dynamically according to fault state of TSVs in shared-spare TSV structures, clock skew analysis of overall 3D clock tree will be complicated.

7.4.3 Advanced TSV Fault-Tolerant Techniques

As mentioned earlier, the redundant TSV methods impose high TSV cost. To reduce this cost, several TSV fault-tolerant techniques specialized for clock TSVs have been proposed. These techniques utilize unique properties of clock signal, which are (1) periodical oscillation; (2) spreading over entire chip through clock wire; (3) arrival at all clock sinks within limited timing difference.

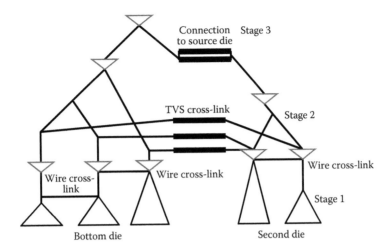

FIGURE 7.17 Utilization of TSV cross-links and wire cross-links in 3D clock tree to take into account TSV fault. Thick horizontal lines represent TSVs. (Taken from Ewetz, R., Udupa, A., Subbarayan, G., and Koh, C.-K A TSV-cross-link-based approach to 3D-clock network synthesis for improved robustness. In *Great Lakes Symposium on VLSI*, Houston, TX, pp. 15–20, 2014.)

Figure 7.17 shows a robust CTS technique which uses the concept of *cross-link* [7]. In this technique, during the bottom-up merging step of 3D CTS, wire cross-links are inserted between clock subtrees on the same die, while TSV cross-links are inserted between subtrees that reside in different dies. With TSV cross-links in Figure 7.17, clock sinks in the second die can receive clock signal even when the original clock TSV (double-lined TSVs between Stage 3 and Stage 2) is faulty. Clock skew variation caused by the cross-link can be minimized by regulating the start point of the cross-link. The robustness of 3D clock tree can be further improved by applying double-TSV structure to clock TSVs at higher stage, as does double-lined TSV in Figure 7.17. With this technique, redundant TSVs are added only for a small number of clock TSVs near the clock source and for several TSV cross-links, and the number of redundant TSVs is far less than that of the redundant TSV techniques mentioned earlier. Moreover, clock skew bound constraint is considered as well.

Another technique to aggressively reduce the number of redundant resources is to apply a new structure with more than two TSVs, called *TSV fault-tolerant unit* (TFU) [29,30]. Figure 7.18 shows the structure of TFU. The main idea of TFU is to eliminate redundant TSVs by allowing two TSVs in a TFU to work as the spare TSV for each other. This approach is feasible in that all clock TSVs deliver the same signal. If one of two TSVs in a TFU (TSV1 and TSV2 in Figure 7.18) is faulty, clock signal from the other TSV can alternate the original clock signal through *redundant path* between both TSVs. (Note that "redundant path" in TFU and "redundant tree" in pre-bond testable clock tree should be differentiated.) To dynamically change the clock signal flow in the TFU, two TGs (TG1 and TG2) are inserted in the redundant path, and self-control units (SCU [19,22], Figure 7.18c) control those TGs. Redundant resources are further reduced by reusing redundant path that has been constructed for pre-bond testable clock tree. During redundant tree construction of pre-bond testable 3D-CTS, one TG (TG3) is inserted between the generated redundant tree and redundant path of TFU, and the final pre-bond test clock tree can reuse the redundant path, as shown in Figure 7.18b.

Figure 7.19 shows the clock signal flow in a TFU for all possible states (pre-bond or post-bond, faulty or fault-free). In post-bond stage, TG3 automatically becomes OFF since TCLK does not generate any signal to it. When both TSVs in a TFU are fault-free, TG1 and TG2 become OFF and clock signal from TSVs directly flows into the subtree under each TSV, as shown in Figure 7.19a. When one of two TSVs, say TSV1, is faulty, TG1 and TG2 become ON and the subtree under TSV1 receives clock signal from redundant path, as shown in Figure 7.19b. In pre-bond stage, all TGs

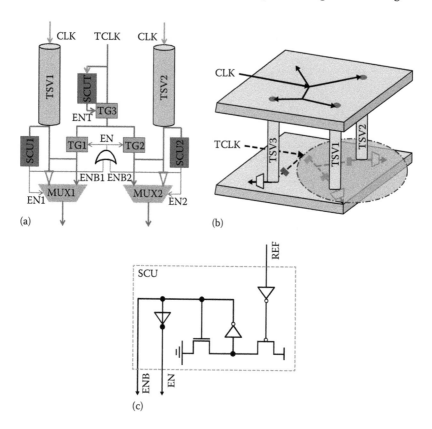

FIGURE 7.18 (a) Full structure of a TSV fault-tolerant unit (TFU) [29,30]. (b) Example of pre-bond testable 3D clock tree with TFU applied. (Control units are omitted for simplicity) (c) Structure of self-control unit (SCU) [19,22]. (Taken from Park, H. and Kim, T., Comprehensive technique for designing and synthesizing TSV fault-tolerant 3D clock trees, In *ACM/IEEE International Conference on Computer-Aided Design*, San Jose, CA, pp. 691–696, © 2013 IEEE.)

become ON and TCLK flows to both subtrees through TGs. Changes of ON/OFF state of TGs and select signal of MUXs are generated from SCUs (SCU1, SCU2, and SCUT) in the TFU.

When this TFU structure is applied to all clock TSVs in a 3D-IC, the number of redundant resources, such as TSVs, logic gates, and wires, is much less than the redundant TSV methods. Moreover, local clock skew problem is also solved in this structure by inserting two delay at one input of MUXs, as shown in Figure 7.18a, while ensuring pre-bond testability.

7.4.4 Additional Issues

In spite of those advantages of 3D clock tree with TFUs, it is still hard to apply TFU structure into 3D clock trees because of other serious issues. The most critical issue is the global clock skew violation problem by the clock signal coming from redundant path in TFU.

Figure 7.20 shows an example of global clock skew violation caused by adding a TFU. Although the two delay buffers can tune the local clock skew between s2 and s3, delayed clock signal by those buffers may incur global clock skew between clock sinks in the subtrees rooted at TSVs in the TFU and the sinks outside of TFU, which results in overall clock skew of 5 as shown in Figure 7.20.

Another problem is the bad slew rate problem of clock signal received from redundant path. Since an additional clock path is connected to the TSV, TSV may suffer from heavy load capacitance problem and sometimes the downstream capacitance of the TSV in a TFU becomes larger than the maximum load capacitance constraint (C_{max}), which results in clock signal distortion. Figure 7.21

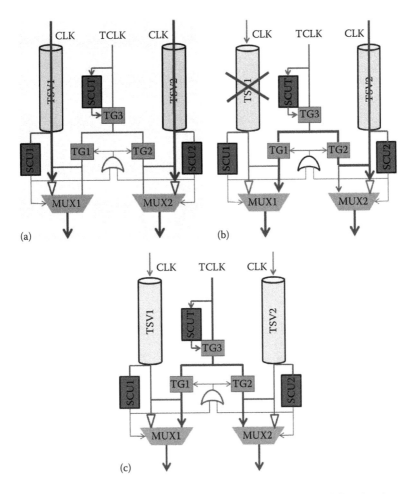

FIGURE 7.19 Clock signal flow (thick arrows) in a TFU for various stages. (a) Post-bond stage when both TSVs are fault-free. (b) Post-bond stage when TSV1 is faulty. (c) Pre-bond test stage.

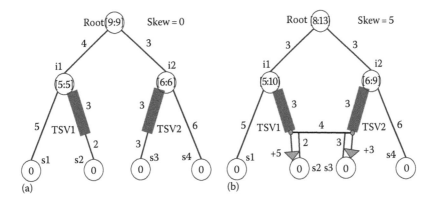

FIGURE 7.20 Global clock skew violation caused by TFU insertion. Interval numbers in clock internal node represent minimum to maximum clock latency from the node to a clock sink of the subtree rooted at the internal node. Each number along the wire represents the latency of the wire. (a) Clock skew before TFU insertion. (b) Clock skew after TFU insertion. (Taken from Park, H. and Kim, T., Synthesis of TSV fault-tolerant 3D clock trees, *IEEE Transactions on Computer-Aided Design of Integrated Circuits and Systems*, 34(2), 266–279, © 2015 IEEE.)

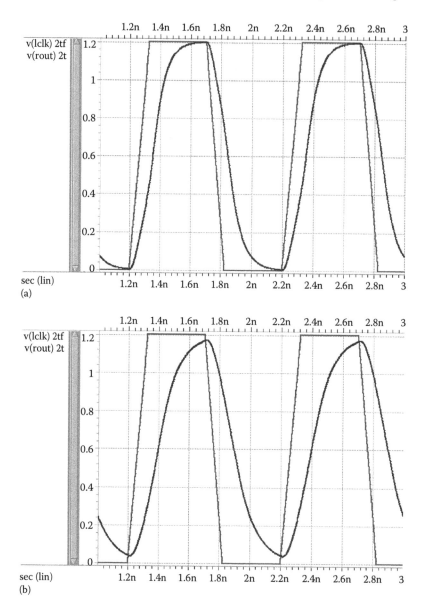

FIGURE 7.21 Distortion problem of clock signal coming from redundant path. Trapezoid waveform is original clock signal and corn waveform represents clock signal detected at the output of redundant path. (a) Short redundant path. (b) Long redundant path. (Taken from Park, H. and Kim, T., Synthesis of TSV fault-tolerant 3D clock trees, *IEEE Transactions on Computer-Aided Design of Integrated Circuits and Systems*, 34(2), 266–279, © 2015 IEEE.)

shows that clock signal from redundant path in TFU is distorted. In Figure 7.21a, when the redundant path of the TFU is short, clock signal from redundant path is distorted, but still usable as clock signal since it reaches the highest and lowest voltage values. However, when the redundant path is long as indicated in Figure 7.21b, clock signal from redundant path is even more distorted than its highest and lowest peak voltages of the original clock, which means the functionality of clock signal cannot be guaranteed. This slew problem can be simply solved by adding load buffers along the redundant path, but buffer is not allowed in redundant path, since redundant path is bidirectional and adding a buffer blocks the signal flow of one direction.

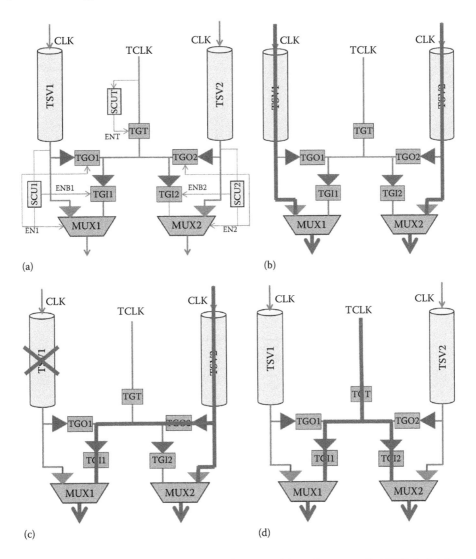

FIGURE 7.22 Structure and clock signal flow (thick arrows) in a slew-controlled TSV fault-tolerant unit (SC-TFU) [34,35] for various stages. (a) Structure of SC-TFU. (b) Post-bond stage when both TSVs are fault-free. (c) Post-bond stage when TSV1 is faulty. (d) Pre-bond test stage. SCUs for SC-TFU are omitted for (b), (c), and (d) for simplicity. (Taken from Park, H. and Kim, T., Synthesis of TSV fault-tolerant 3D clock trees, *IEEE Transactions on Computer-Aided Design of Integrated Circuits and Systems*, 34(2), 266–279, © 2015 IEEE.)

In addition to the problems described earlier, there are other problems that hinder the use of TFU to 3D-IC. More information about this can be found in [35].

To cope with bad slew rate problems, a modified TFU structure, called slew-controlled TSV fault-tolerant unit (SC-TFU) [34,35], is proposed. Full structure and clock signal flows for various stages (pre-bond and post-bond) and fault states (fault-free and faulty) of a SC-TFU are shown in Figure 7.22. Although clock signal flows of SC-TFU are similar to those of TFU, it was observed from Figure 7.22b through d that all TGs in SC-TFU (TGIs, TGOs, and TGT) work unidirectionally, and therefore load buffer insertion for redundant path is enabled to solve slew rate problem. Figure 7.23 compares slew rate of clock signal from redundant path. It is concluded that slew rate of SC-TFU is far better than that of TFU, without regard to redundant path length. In addition, those load buffers apply constant downstream capacitance value regardless of current stage and fault

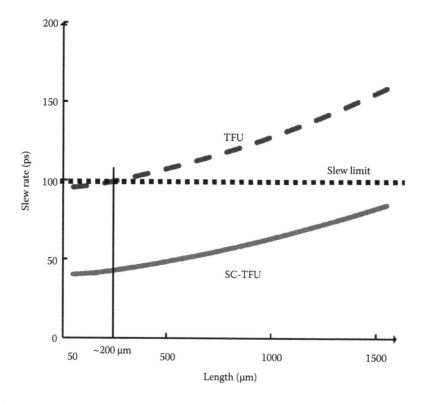

FIGURE 7.23 Graphical comparison of clock slew rate between TFU and SC-TFU, according to redundant path length. (Taken from Park, H. and Kim, T., Synthesis of TSV fault-tolerant 3D clock trees. *IEEE Transactions on Computer-Aided Design of Integrated Circuits and Systems*, 34(2), 266–279, © 2015 IEEE.)

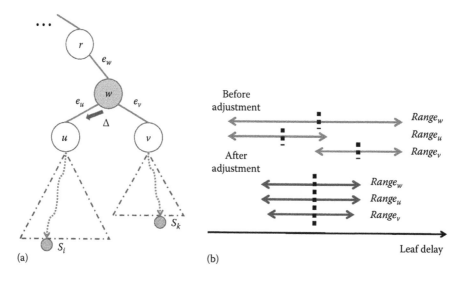

FIGURE 7.24 Working view of adaptive wire adjustment (AWA) algorithm [39]. (a) Relocation of internal clock node. (b) Resulting adjustment of clock latency. (Taken from Park, H. and Kim, T., Comprehensive technique for designing and synthesizing TSV fault-tolerant 3D clock trees. *ACM/IEEE International Conference on Computer-Aided Design*, San Jose, CA, pp. 691–696, © 2013 IEEE.)

state of SC-TFU. Therefore, more accurate clock skew analysis is possible for 3D clock tree with SC-TFUs than with TFUs. (This function of load buffers is similar to that of TSV buffers mentioned in the previous section, which enables accurate pre-bond test for 2D die with clock source.)

To further reduce global clock skew of 3D clock tree with SC-TFUs, the concept of adaptive wire adjustment (AWA) algorithm [39] is applied after SC-TFU insertion. Figure 7.24 shows the working view of AWA algorithm. When clock skew violation between two clock sinks is detected, the nearest common parent, node w of Figure 7.24a, is found by back traversing, and adjust the location of this node toward its child that has the larger delay sink in its subtree. After one such adjustment, length of clock latency range of the subtree starts from w in Figure 7.24b, which means the clock skew of the subtree rooted at w can be reduced as shown in Figure 7.24b. Therefore, global clock skew is reduced for each adjustment, and it is possible to iteratively adjust clock internal nodes until global clock skew becomes lower than the given skew constraint.

7.5 VARIATION TOLERANT TECHNIQUES AND OTHERS

Designing 3D-ICs requires more attention on the effect of variations on the final design quality than designing 2D-ICs. This is because a few unique features due to TSV implantation should be taken into account. Those are *thermal-induced variation*, *TSV stress-induced variation*, *on-package-induced variation*, and *TSV-related blockage awareness*.

7.5.1 THERMAL VARIATION AWARE CTS

While conventional 2D chips generate heats through a single layer, 3D chips generate heat through all layers. Thus, spreading the heat from the layers quickly is more challenging due to the use of (lower thermally conductive) dielectric materials between the layers. The difference in temperature in a single layer and across layers increases because of the structural difference and nonsymmetrical execution activities. Increasing the temperature changes the resistance of interconnect wires according to the relation $R = R_0[1 + \alpha(T - T_0)]$, where R_0 is the interconnect resistance at the nominal temperature T_0, and α is the temperature coefficient of resistance for the interconnect material. Consequently, as opposed to 2D CTS, meeting the clock skew in 3D CST is much harder due to the more dynamic delay variation induced by the thermal difference between layers.

BURITO in [31] is one of the most notable works of CTS which have considered thermal variation. BURITO proposed the so-called "Balanced Skew Theorem," which claims that the construction of a clock tree (either buffered or nonbuffered) with balanced skew under thermal profiles T_A and T_B is equivalent to constructing a zero-skew tree under profile $T_m = (T_A + T_B)/2$, in which the temperature of each grid in T_m is the average between T_A and T_B of the corresponding location. The significance of the theorem is that it simplifies the clock tree construction under thermal variations considerably because the clock tree now has to be optimized for one thermal profile instead of two. However, if more than two thermal profiles are considered, the theorem is not likely to be applicable.

Contrary to the static control of clock skew like BURITO, dynamic control of clock signal delay in accordance with the thermal variation does make more sense in practice. The design method in [32] is one of the dynamic controls of clock latency. The idea is shown in Figure 7.25. The idea is that temperature sensors are placed close to the buffers and sense the ambient temperatures and convert the temperatures to voltages that are processed by a wave-shaping circuitry and finally used for dynamically changing the driving strengths of the clock buffers to reduce the overall skew. In addition to the adaptive current method, a dynamic body biasing technique is applied to dynamically tune the threshold voltage of the devices in the buffer. The wave-shaping circuitry will provide the controlled voltages required for body biasing.

On the other side, an accurate thermal-induced delay modeling is critical in thermal aware CTS. The work in [40] pointed out this observation. The observation by [40] said that for a 3D clock

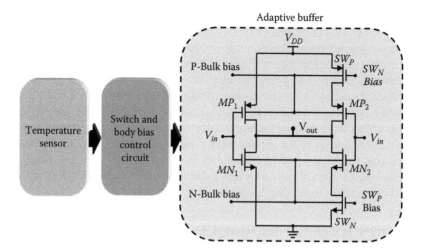

FIGURE 7.25 Thermally adaptive clock buffer schematic. (Taken from Mondal, M. et al., Thermally robust clocking schemes for 3D integrated circuits, in: *Design, Automation and Test in Europe*, Acropolis, France, pp. 1–6, 2007.)

tree distribution network with TSVs, the delay becomes a nonlinear function with temperature $D_{TSV} = k_0 + k_1 T + k_2 T^2 + k_3 T^3$, which is significantly different from the 2D case in which the delay is only linearly dependent on temperature $\tau = d_0' + k_0' T$. This is because in 3D-IC, the TSV is mainly modeled as nonlinear temperature-dependent capacitor while the wire is mainly modeled as linear temperature-dependent resistor. Thus, Reference 40 showed that the electrical–thermal nonlinear coupling from signal TSV may significantly increase the clock skew due to the large temperature gradient in 3D-IC. The proper design by applying thermal TSVs for heat removal and further balancing the clock skew thereby becomes an important approach to be explored for 3D clock tree.

7.5.2 TSV-Stress Aware CTS

Strained silicon has been used to enhance I_{on} of a transistor. However, in 3D-IC manufacturing, unwanted stress is caused by CTE mismatch between copper TSV and silicon. Investigations showed that at 200°C, an anneal time of 30–60 min is required to achieve reasonable copper properties. Since CTE of copper is larger than that of silicon, after annealing, copper has less volume compared with silicon. Several papers have been published to simulate TSV-induced stress [5,27] using finite element analysis (FEA) simulation. They showed that TSV can cause tensile stress of more than 200 MPa. Thus, it is important to systematically deal with the clock buffer variation due to TSV stress in CTS of 3D-ICs.

Reference 40 is one noticeable CTS work which takes into account the nonuniform distribution of wire and buffer delays caused by the stress of TSV implantation. The CTS used in [40] has consisted of slightly different schemes for layer embedding and clock routing.

Figure 7.26 shows a clock path spreading along two dies. The clock buffers connected to F1 and F2 for path inputs are called type-A and type-B buffers, respectively. Let F1 and B be placed in die0 and L1 and F2 be placed in die1. If z-location of clock buffer A is flexible, clock buffer A can be assigned to either die0 or die1. Intuitively, it is desirable to assign A to die0 to avoid TSV insertion between A and F1. However, it is necessary to consider covariance between A and other cells during z-location determination for the buffer to mitigate the induced delay variation.

In [40] (in the bottom-up process of 3D-MMM [31]) the determination of z-location of buffer is done by using an ILP (integer linear programming) formulation to minimize covariance. ILP formulation uses the clock tree information which has been constructed so far, and logical path

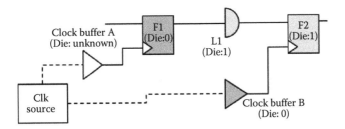

FIGURE 7.26 Illustrating how placing clock buffer A to mitigate the variation between the buffer and other cells. (Taken from Shang, Y., Zhang, C., Yu, H., Tan, C.S., Zhao, X., and Lim, S.K., Thermal-reliable 3D clock-tree synthesis considering nonlinear electrical–thermal-coupled TSV model, *IEEE Asia and South Pacific Design Automation Conference*, Yokohama, Japan, pp. 693–698, © 2013 IEEE.)

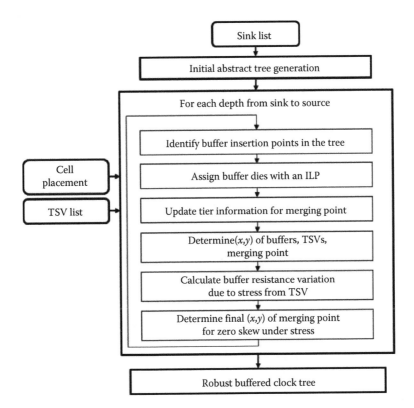

FIGURE 7.27 CTS flow considering TSV stress-induced delay variation in all steps. (Taken from Shang, Y., Zhang, C., Yu, H., Tan, C.S., Zhao, X., and Lim, S.K., Thermal-reliable 3D clock-tree synthesis considering nonlinear electrical–thermal-coupled TSV model, *IEEE Asia and South Pacific Design Automation Conference*, Yokohama, Japan, pp. 693–698, © 2013 IEEE.)

information to make the optimal z-location of newly inserted buffer. If the z-location of buffer determined by the ILP formulation is different from the z-location of child node, a TSV is inserted between child node and buffer. If buffers on two edges are assigned to the same layer and merging point is not, the merging point layer is substituted to buffer layer to reduce the number of TSVs. The stress-aware CTS is performed in all steps of CTS, including the determination of (x, y) location for clock buffers and the generation of wire length of each edge. Figure 7.27 shows the steps performed in [40] where all steps except the first one considers the TSV stress-induced delay variation.

7.5.3   ON-PACKAGE AWARE CTS

A 3D-stacked IC is made by stacking independently manufactured dies. This means that each die assembled in a 3D package can be manufactured at different process corner compared to the other dies. In other words, die-to-die process variation in 2D-ICs acts as the on-package variation in a 3D-IC. Figure 7.28a shows an example of a two-die stacked 3D-IC. Conventionally, during timing closure activity, it is assumed that all the circuit elements are manufactured at the same process corner as shown in Figure 7.28b. However, there are additional process corner combinations to consider on-package variation derived from the die-to-die variation as shown in Figure 7.28c.

If we do not consider the on-package variation, we cannot guarantee the correct operation of a final stacked 3D-IC due to the possibility of speed degradation and (especially) system function failure. Figure 7.29 shows examples of the on-package variation effect on design timing. Logic gates on the launch clock and data paths (CP2, FF2, and DP) are manufactured on die2, and gates on the capture clock path (CP1 and FF1) on die1. If die2 is a slow corner sample and die1 is a fast corner sample as shown in Figure 7.29a, the conventional all slow (or all fast) corner-based timing sign-off shown in Figure 7.28b does not work. Die-to-die variation between 2D dies becomes on-package variation in a 3D system and chip performance can be degraded due to the unintended fast clock propagation on the capture clock path (CP2). Similarly, if die2 is a fast corner sample and die1 is a slow corner sample as shown in Figure 7.29b, system function failure due to the unintended hold time violation can occur for the fast data paths with small logic gates.

As shown in Figure 7.29, the conventional all slow (or all fast) corner-based timing sign-off cannot guarantee the correct operation of 3D-stacked ICs due to the unintended delay mismatches caused by

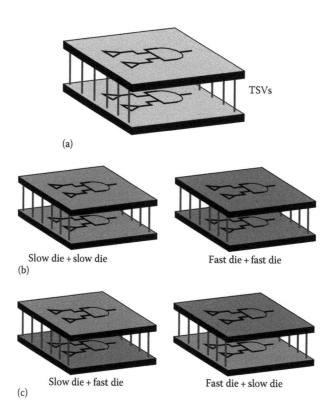

(a)

Slow die + slow die
(b)

Fast die + fast die

Slow die + fast die
(c)

Fast die + slow die

FIGURE 7.28   On-package variation in a 3D-IC. (a) An example of two-die stacked 3D-IC. (b) Conventional process corners. (c) New process corners. (Taken from Kim, T.-Y. and Kim, T., *J. Semicond. Technol. Sci.*, 12(2), 139, 2012.)

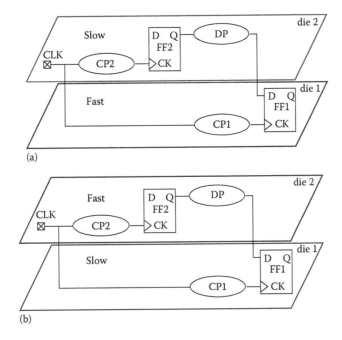

FIGURE 7.29 On-package variation effect on design timing. (a) Speed degradation due to the unintended setup time violation. (b) System function failure due to the unintended hold time violation (CP#: clock path, DP: data path). (Taken from Kim, T.-Y. and Kim, T., *J. Semicond. Technol. Sci.*, 12(2), 139, 2012.)

the on-package variation. Therefore, the on-package variation effect should be taken into account for 3D-IC designs, more importantly for the commercial ASIC designs that demand high yield even at corner cases. This observation concludes that the selection of dies to be stacked to form a 3D-IC (i.e., *die matching* problem) and routing clock wires and TSVs (i.e., *3D routing* problem) are carefully carried out, so that the global clock skew of the 3D clock network should be satisfied. If the clock skew is still not met, *post-silicon tuning* techniques need to be applied in a similar way as does in 2D-ICs.

- *Post-silicon tuning*: Body biasing [2] is one of the most effective control techniques of device characteristics at post-silicon stage. Body biasing technique controls device characteristics by changing the body voltage of NMOS and PMOS transistors. If the NMOS (PMOS) body bias voltage is higher (lower) than the source voltage, it is called forward body biased (FBB) with increased speed and leakage. If the NMOS (PMOS) body bias voltage is lower (higher) than the source voltage, it is called reverse body biased (RBB) with reduced speed and leakage. To mitigate the on-package variation, global body biasing voltages can be applied to individual dies.

  As an illustration, Figure 7.30 in [21] shows buffer delay trends at various process corners with varying body bias voltages under 1.2 V power supply voltage in 45 nm process technology. If it is allowed to use body bias voltages in 0.2 V as shown in Figure 7.30, it is not feasible to tune all the devices to have similar delay. For example, devices at points A and C can be tuned to points B and D by applying 0.2 V FBB and RBB voltages. However, devices at regions E and F cannot be tuned to have similar delay with each other because the allowable body bias voltage is limited. So, if we want to use body biasing as a post-silicon 3D clock skew management technique, it needs a smart method to maximize the parametric yield with regard to 3D clock skew under a limited body biasing voltage range. The die matching problem to be explained in the next section considers the enhancement of parametric yield of 3D-ICs.

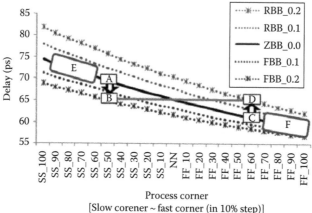

FIGURE 7.30 Buffer delay trends at various process corners, in 45 nm technology, with varying body bias voltages. Allowed from 0.2 V RBB to 0.2 V FBB in which devices at region E and F cannot be tuned to have similar delay characteristic. (SS_100: slow corner, NN: nominal corner, FF_100: fast corner, SS/FF_##: interpolated corners between slow/fast corner and nominal corner in 10% step). (Taken from Kim, T.-Y. and Kim, T., *J. Semicond. Technol. Sci.*, 12(2), 139, 2012.)

- *Die matching*: Let us consider the problem of matching dies in two die sets where one set contains the upper dies when stacked and the other set contains the lower dies when stacked. It assumes that each die in the sets has been characterized (in terms of timing in this example) and classified into one of the three process corners: slow, fast, and normal. In addition, it assumes that a slow (fast) die in a set can be stacked with a slow (fast) or a normal die in the other set without timing violation by applying a proper body biasing to the dies, but stacking a fast (slow) die to a slow (fast) die causes timing violation whatever body biasing technique is applied. It is called a matching (i.e., stacking) of two dies *infeasible* if the matching incurs timing violation that can never be resolved with body biasing technique. Otherwise, the matching is called *feasible*. (Note that the feasibility and infeasibility of die matching will be determined according to the classification of process corners and body biasing voltage range. In addition, the definition of feasibility and infeasibility can be similarly extended to the case of more than two dies to be stacked.) Figure 7.31 shows examples of feasible and infeasible die-to-die matchings when body biasing is used as post-silicon tuning technique, in which it is assumed that no body biasing is able to fix the timing violation when a slow die is matched with a fast die.
  The die-to-die feasible matching problem is
  Die-to-die matching (DDM) problem: *Given K die sets $\{D_1, ..., D_K\}$ where each die set D_i, $i = 1, ..., K$, consists of N dies $d_1^i,...,d_N^i$ such that their process corners have been identified by wafer-level testing and the body biasing voltage range $(V_{BB.MIN} \sim V_{BB.MAX})$, find a maximal K-dimensional feasible matching from the sets $D_1, ..., D_K$.*
  Some of representative works addressing the die-to-die matching problem can be found in [8,36,38,41].
- *Clock Routing*: Die-to-die variation can be mitigated by carefully embedding clock edges (i.e., wires) to the layers. On-package variation aware clock edge embedding problem can be stated as
  Clock edge embedding (CEE) problem: *For an input 3D clock tree in which the clock node embedding process has been done, assign the clock edges to layers so that the clock skew variation is minimized while maintaining the TSVs defined by the node embedding.*

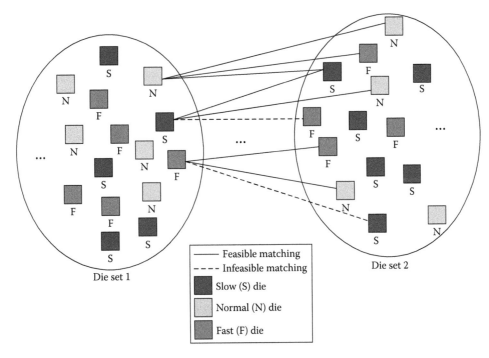

FIGURE 7.31 Examples of feasible (solid line) and infeasible (dashed line) die-to-die matchings when body biasing is used as a post-silicon tuning technique. It is assumed that no body biasing can resolve the timing violation when a slow die is matched with a fast die. (Taken from Park, S. and Kim, T., Die matching algorithm for enhancing parametric yield of 3D ICS, *IEEE International SoC Design Conference*, Jeju, Korea, pp. 143–146, © 2013 IEEE.)

First, we illustrate how the clock skew variation in a 3D clock tree can be changed as the clock edges are embedded on the layers.

Let us consider two sinks (e.g., flip-flops) s_i and s_j that are driven by the clock signal on a clock tree. Let $D(s_i)$ and $D(s_j)$ denote the arrival times of clock signal at s_i and s_j, respectively, and assume that the arrival time difference, $\Delta_{i,j}$, between $D(s_i)$ and $D(s_j)$ follows the Gaussian distribution:

$$\Delta_{i,j} = D(s_i) - D(s_j) \sim N(\mu_{\Delta_{i,j}}, \sigma_{\Delta_{i,j}}),$$

$$\mu_{\Delta_{i,j}} = \mu_{D(s_i)} - \mu_{D(s_j)}, \tag{7.2}$$

$$\sigma_{\Delta_{i,j}}^2 = VAR(D(s_i)) + VAR(D(s_j)) - 2COV(D(s_i), D(s_j)).$$

Then, according to the expression above, we can derive the distribution of clock skew $\Delta_{1,2}$ for two different clock edge embeddings: embedding clock edge e_1 to layer 1 and embedding e_1 to layer 2. It can be observed that since the correlation between random variables $D(s_1)$ (defined by e_1) and $D(s_2)$ (defined by e_2) come from the variation in technology parameters, which will be uncorrelated, if e_1 and e_2 are placed on different layers as shown in Figure 7.32a, $D(s_1)$ and $D(s_2)$ are independent, that is, $COV(D(s_1), D(s_2)) = 0$ [43]. On the other hand, the random variables $D(s_1)$ and $D(s_2)$ are correlated if edges e_1 and e_2 are both placed on the same layer as shown in Figure 7.32b. Precisely, since the within-die variation is spatially correlated, the variance of skew distribution $\sigma'_{\Delta_{1,2}}$ in Figure 7.32b would be much smaller than $\sigma_{\Delta_{1,2}}$ in Figure 7.32a. (Note that $COV(D'(s_1), D'(s_2)) > 0$.) It can be

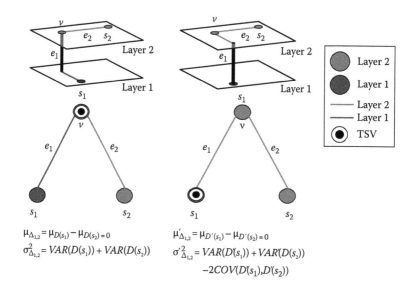

FIGURE 7.32 Clock skew comparison between two different edge embeddings. Left figure: an edge embedding without layer sharing between edges e_1 and e_2 under TSV usage constraint of 1. Right figure: an edge embedding with layer sharing between edges e_1 and e_2 under TSV usage constraint of 1. (Taken from Park, S. and Kim, T., *Integr. VLSI J.*, 47(4), 476, 2014.)

observed from the analysis that by embedding the edges on two clock paths into the common layers as many as possible, the resulting clock skew variation between the two paths can be reduced.

The work in [43] demonstrated empirically under the use of 3D H-tree structure that the skew variation between a pair of clock sinks can be reduced if the sinks and their connecting edges are all together placed in the same layer. The work in [47] further elaborated the idea in that they aggressively attempted to route the most part of clock paths to the same layer at the expense of the increase of TSV usage. In addition, the work in [44] proposed an on-package variation aware 3D CTS methodology to reduce the clock signal variation. The idea is to place the clock buffers in clock tree on as many uncorrelated dies as possible during the process of CTS. Even though this method is effective in mitigating the effect of on-package variation on the critical path, it demands much more TSVs than necessary. The work in [37] solved the clock edge embedding problem of mitigating on-package variation in two steps. In the first step, it maximizes the total amount of layer sharing among the edges on all clock signal paths from the clock source to sinks. This leads to minimize the impact of on-package variation on the total sum of the variations of delay difference between all clock paths. In the second step, the edge embedding result obtained in the first step is refined locally in order to reduce the worst variation of the clock skew.

7.5.4 BLOCKAGE AWARE CTS

TSVs are placed in the whitespace among macro blocks or cells and an improper arrangement of TSVs may cause longer WL since the available TSV might be far away from its connected cells. Nowadays, intellectual property (IP) and standard cell-based design has been extensively used to reduce design cost, but after floorplan and placement, only few whitespace blocks are reserved for clock TSVs. Unfortunately, there are not many 3D clock routing works (e.g., [46,?]) which have considered blockage awareness. For some of the existing 2D works on obstacle-aware clock routing, Kahng and Tsao [13] proposed DME-based obstacle expansion rules to determine feasible embedding locations for the internal nodes. Kim and Zhou [15] presented a planar obstacle-aware routing scheme to clean up overlaps between clock nets and obstacles. Huang et al. [11] proposed another DME-based clock routing method to avoid obstacles with the help of a track graph.

The work in [46] solved a practical 3D clock routing problem which stems from all kinds of TSV-induced obstacles, such as P/G, signal and clock TSV. The work analyzed the behavior of TSV obstacles as

1. TSVs for logic signals: they occupy silicon area only, and work as placement obstacles for clock buffers and clock TSVs. Thus, clock TSVs and clock buffers are not allowed to overlap with existing signal TSVs. Furthermore, clock nets are allowed to be routed over the signal TSVs because their landing pads are in M1 and free up the metal spaces above. Figure 7.33a shows an illustration.
2. TSVs for power/ground: they occupy both silicon area and metal layers, and thus function as both placement and routing obstacles. This means that clock TSVs and clock buffers should avoid overlap with existing power/ground TSVs and the clock net is not allowed to route over the power/ground TSV. Figure 7.33 shows an illustration.
3. TSVs for clock: besides power/ground TSVs and signal TSVs, 3D CTS itself also inserts clock buffers and clock TSVs. They become the same kind of clock routing obstacles as signal TSVs if added in an iterative manner like DME-based clock tree embedding.

The work in [46] developed a TSV-induced obstacle-aware deferred-merge embedding (DME) method to construct a buffered clock tree which can avoid those obstacles with the help of newly defined merging segments.

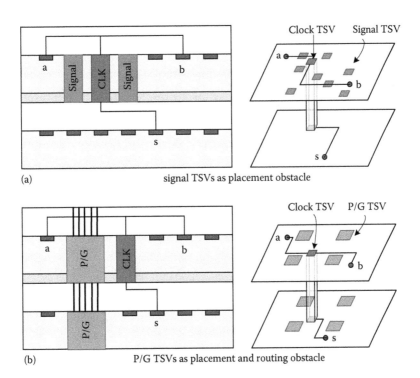

FIGURE 7.33 Side and top-down view of via-first power/ground (P/G) TSVs, clock TSVs, and signal TSVs. (a) P/G TSVs use many local vias in between vertically. (b) Size of the TSV cells (= TSV + keep-out-zone) in terms of the standard cell row height (45 nm technology). (Taken from Zhao, X. and Lim, S.K., Through-silicon-via-induced obstacle-aware clock tree synthesis for 3D ICS, in: *IEEE Asia and South Pacific Design Automation Conference*, Sydney, Australia, pp. 347–352, © 2012 IEEE.)

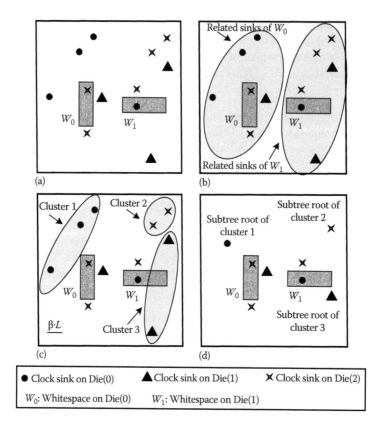

FIGURE 7.34 Sink pre-clustering. (a) Before pre-clustering. (b) Arranging sinks to their related whitespace blocks according to distance. (c) For each whitespace block, generate clusters for its related sinks. (d) Subtree roots of the clusters are reserved as new sinks while other nodes in the cluster are neglected. (Taken from Li, X., Liu, W., Du, H., Wang, Y., Ma, Y., and Yang, H., Whitespace-aware TSV arrangement in 3D clock tree synthesis, *IEEE Computer Society Annual Symposium on VLSI*, Natal, Brazil, pp. 115–120, © 2013 IEEE.)

Contrary to [46], the work in [24] pointed out that too many obstacles exist and only few whitespace blocks are reserved for clock TSVs after floorplan and placement in IP and Standard cell-based designs. Long wire snaking is inevitable in such scenarios and a new whitespace-aware algorithm is necessary. To this end, the work proposed a whitespace-aware TSV arrangement algorithm in 3D CTS, which consists of three steps: (Step 1) sink pre-clustering: sinks far away from their related whitespace are clustered to merge subtrees, only the root node of the subtree is reserved and will play as a new sink after that. Figure 7.34 shows an illustration. (Step 2) clock tree topology generation: the 3D-MMM is extended by judging whether the current x/y-cut between multiple dies is appropriate, considering whitespace to make sure that each sink set contains whitespace. (Step 3) merging segment reconstruction: the work modifies the merging segment of the internal nodes having TSVs with the consideration of TSV geometries and whitespace occupation, which would bring convenience to detail routing and TSV arrangement.

7.6   CONCLUSION

In this chapter, we surveyed the topics which were highly important in designing 3D clock networks. Besides the basic 3D clock network synthesis flow, considerable attention should be paid to designing with pre-bond testability, TSV fault aware design, and many practical issues such as thermal variation, TSV stress, on-package variation, and blockage awareness.

REFERENCES

1. T.-H. Chao, Y.-C. Hsu, J.-M. Ho, K. D. Boses, and A. B. Kahng. Zero-skew clock routing with minimum. *IEEE Transactions on Circuits and Systems*, 39(11):799–814, 1992.
2. B. Choi and Y. Shin. Lookup table-based adaptive body biasing of multiple macros for process variation compensation and low leakage. *Journal of Circuits, Systems, and Computers*, 19(7):1449–1464, 2010.
3. J. Cong, A. B. Kahng, C.-K. Koh, and C.-W. A. Tsao. Bounded-skew clock and steiner routing under elmore delay. In *ACM/IEEE International Conference on Computer-Aided Design*, San Jose, CA, pp. 66–71, 1995.
4. J. Cong, A. B. Kahng, C.-K. Koh, and C.-W. A. Tsao. Bounded-skew clock and steiner routing. *ACM Transactions on Design Automation of Electronic Systems*, 3(3):341–388, 1998.
5. T. Dao, D. H. Triyoso, M. Petras, and M. Canonico. Through silicon via stress characterization. In *IEEE International Conference on IC Design and Technology*, Austin, TX, pp. 1–3, 2009.
6. M. Edahiro. A clustering-based optimization algorithm in zero-skew routings. In *ACM/IEEE Design Automation Conference*, Dallas, TX, pp. 612–616, 1993.
7. R. Ewetz, A. Udupa, G. Subbarayan, and C.-K. Koh. A TSV-cross-link-based approach to 3D-clock network synthesis for improved robustness. In *Great Lakes Symposium on VLSI*, Houston, TX, pp. 15–20, 2014.
8. C. Ferri, S. Reda, G. Smith, and R. I. Bahar. Parametric yield management of 3D ICS: Models and strategies for improvement. *ACM Journal on Emerging Technologies in Computing Systems*, 4(4):19:1–19:22, 2008.
9. P. E. Gronowski, W. J. Bowhill, R. P. Preston, M. K. Gowan, and R. L. Allmon. High-performance microprocessor design. *IEEE Journal of Solid-State Circuits*, 33(5):676–686, 1998.
10. D. J.-H. Huang, A. B. Kahng, and C.-W. A. Tsao. On the bounded-skew clock and steiner routing problems. In *ACM/IEEE Design Automation Conference*, San Francisco, CA, pp. 508–513, 1995.
11. H. Huang, W.-S. Luk, W. Zhao, and X. Zeng. DME-based clock routing in the presence of obstacles. In *IEEE International Conference on ASIC*, Quilin, China, pp. 429–434, 2008.
12. M. A. B. Jackson, A. Srinivasan, and E. S. Kuh. Clock routing for high performance ICS. In *ACM/IEEE Design Automation Conference*, Orlando, FL, pp. 573–179, 1990.
13. A. B. Kahng and C.-W. A. Tsao. More practical bounded-skew clock routing. In *ACM/IEEE Design Automation Conference*, Anaheim, CA, pp. 594–599, 1997.
14. U. Kang, H.-J. Chung, S. Heo, D.-H. Park, H. Lee, J.-H. Kim, S.-H. Ahn et al. 8 Gb 3D DDR3 dram using through-silicon-via technology. In *IEEE International Solid-State Circuits Conference*, pp. 130–131, 2009.
15. H. Kim and D. Zhou. Efficient implementation of a planar clock routing with the treatment of obstacles. *IEEE Transactions on Computer-Aided Design of Integrated Circuits and Systems*, 19(10):1220–1225, 2000.
16. J. Kim, F. Wang, and M. Nowak. Method and apparatus for providing through silicon via (TSV) redundancy. United State Patent Application Publication, number U.S. 20,100,295,600, 2010.
17. T.-Y. Kim and T. Kim. Bounded skew clock routing for 3D stacked IC designs: Enabling tradeoffs between power and clock skew. In *International Green Computing Conference*, Chicago, IL, pp. 525–532, 2010.
18. T.-Y. Kim and T. Kim. Clock tree embedding for 3D ICS. In *IEEE Asia, and South Pacific Design Automation Conference*, Taipei, Taiwan, pp. 486–491, 2010.
19. T.-Y. Kim and T. Kim. Clock tree synthesis with pre-bond testability for 3D stacked IC designs. In *ACM/IEEE Design Automation Conference*, pp. 723–728, 2010.
20. T.-Y. Kim and T. Kim. Clock tree synthesis for TSV-based 3D IC designs. *ACM Transactions on Design Automation of Electronic Systems*, 16(4):48:1–48:21, 2011.
21. T.-Y. Kim and T. Kim. Post silicon management of on-package variation induced 3D clock skew. *Journal of Semiconductor Technology and Science*, 12(2):139–149, 2012.
22. T.-Y. Kim and T. Kim. Resource allocation and design techniques of prebond testable 3-D clock tree. *IEEE Transactions on Computer-Aided Design of Integrated Circuits and Systems*, 32(1):138–151, 2013.
23. M. Laisne, K. Arabi, and T. Petrov. Systems and methods utilizing redundancy in semiconductor chip interconnects. United State Patent Application Publication, number U.S. 20,100,060,310, 2010.
24. X. Li, W. Liu, H. Du, Y. Wang, Y. Ma, and H. Yang. Whitespace-aware TSV arrangement in 3D clock tree synthesis. In *IEEE Computer Society Annual Symposium on VLSI*, Natal, Brazil, pp. 115–120, 2013.
25. C. Liu and S. K. Lim. Ultra-high density 3D SRAM cell designs for monolithic 3D integration. In *Proceedings of IEEE International Interconnect Technology Conference*, San Jose, CA, pp. 1–3, 2012.
26. I. Loi, S. Mitra, T.H. Lee, S. Fujita, and L. Benini. A low-overhead fault tolerance scheme for TSV-based 3D network on chip links. In *ACM/IEEE International Conference on Computer-Aided Design*, San Jose, CA, pp. 598–602, 2008.

27. K. H. Lu, X. Zhang, S.-K. Ryu, R. Huang, J. Im, and P. S. Ho. Thermomechanical reliability of 3-D ICS containing through silicon vias. In *IEEE Electronic Components and Technology Conference*, San Diego, CA, pp. 630–633, 2009.

28. C.-L. Lung, J.-H. Chien, Y. Shi, and S.-C. Chang. TSV fault-tolerant mechanisms with application to 3D clock networks. In *IEEE International SoC Design Conference*, Jeju, Korea, pp. 127–130, 2011.

29. C.-L. Lung, Y.-S. Su, H.-H. Huang, Y. Shi, and S.-C. Chang. Through-silicon via fault-tolerant clock networks for 3-D ICS. *IEEE Transactions on Computer-Aided Design of Integrated Circuits and Systems*, 32(7):1100–1109, 2013.

30. C.-L. Lung, Y.-S. Su, S.-H. Huang, Y. Shi, and S.-C. Chang. Fault-tolerant 3D clock network. In *ACM/IEEE Design Automation Conference*, New York, pp. 645–651, 2011.

31. J. Minz, X. Zhao, and S. K. Lim. Buffered clock tree synthesis for 3D ICS under thermal variations. In *IEEE Asia and South Pacific Design Automation Conference*, Anaheim, CA, pp. 504–509, 2008.

32. M. Mondal, A.J. Ricketts, S. Kirolos, T. Ragheb, G. Link, N. Vijaykrishnan, and Y. Massoud. Thermally robust clocking schemes for 3D integrated circuits. In *Design, Automation and Test in Europe*, Nice Acropolis, France, pp. 1–6, 2007.

33. S. Panth, K. Samadi, Y. Du, and S. K. Lim. Power-Performance study of block-level monolithic 3D-ICs considering InterTier performance variations. In *Proceedings of ACM Design Automation Conference*, San Francisco, CA, pp. 1–6, 2014.

34. H. Park and T. Kim. Comprehensive technique for designing and synthesizing TSV fault-tolerant 3D clock trees. In *ACM/IEEE International Conference on Computer-Aided Design*, San Jose, CA, pp. 691–696, 2013.

35. H. Park and T. Kim. Synthesis of TSV fault-tolerant 3D clock trees. *IEEE Transactions on Computer-Aided Design of Integrated Circuits and Systems*, 34(2):266–279, 2015.

36. S. Park and T. Kim. Die matching algorithm for enhancing parametric yield of 3D ICS. In *IEEE International SoC Design Conference*, Jeju, Korea, pp. 143–146, 2012.

37. S. Park and T. Kim. Edge layer embedding algorithm for mitigating on-package variation in 3D clock tree synthesis. *Integration, the VLSI Journal*, 47(4):476–486, 2014.

38. S. Reda, G. Smith, and L. Smith. Maximizing the functional yield of wafer-to-wafer 3D integration. *IEEE Transactions on Very Large Scale Integration Systems*, 17(9):1358–1362, 2009.

39. H. Saaied, D. Al-Khalili, A. Al-Khalili, and M. Nekili. Adaptive wire adjustment for bounded skew clock distribution network. In *IEEE Asia and South Pacific Design Automation Conference*, Kitakyushu, Japan, pp. 243–248, 2003.

40. Y. Shang, C. Zhang, H. Yu, C. S. Tan, X. Zhao, and S. K. Lim. Thermal-reliable 3D clock-tree synthesis considering nonlinear electrical–thermal-coupled TSV model. In *IEEE Asia and South Pacific Design Automation Conference*, Yokohama, Japan, pp. 693–698, 2013.

41. G. Smith, L. Smith, S. Hosali, and S. Arkalgud. Yield consideration in the choice of 3D technology. In *International Symposium on Semiconductor Manufacturing*, Santa Clara, CA, pp. 1–3, 2007.

42. S.-J. Wang, C.-H. Lin, and K.S.-M. Li. Synthesis of 3D clock tree with pre-bond testability. In *IEEE International Symposium on Circuits and Systems*, Beijing, China, pp. 2654–2657, 2013.

43. H. Xu, V. F. Pavlidis, and G. D. Micheli. Effect of process variations in 3D global clock distribution networks. *ACM Journal on Emerging Technologies in Computing Systems*, 8(3):20:1–20:25, 2012.

44. J. S. Yang, J. Pak, X. Zhao, S. K. Lim, and D. Z. Pan. Robust clock tree synthesis with timing yield optimization for 3D-ICS. In *IEEE Asia and South Pacific Design Automation Conference*, Yokohama, Japan, pp. 621–626, 2011.

45. X. Zhao, D. L. Lewis, H.-H.S. Lee, and S. K. Lim. Low-power clock tree design for pre-bond testing of 3-D stacked ICS. *IEEE Transactions on Computer-Aided Design of Integrated Circuits and Systems*, 30(5):732–745, 2011.

46. X. Zhao and S. K. Lim. Through-silicon-via-induced obstacle-aware clock tree synthesis for 3D ICS. In *IEEE Asia and South Pacific Design Automation Conference*, Sydney, Australia, pp. 347–352, 2012.

47. X. Zhao, S. Muhopadhyay, and S. K. Lim. Variation-tolerant and low-power clock network design for 3D ICS. In *IEEE Electronic Components and Technology Conference*, Lake Buena Vista, FL, pp. 2007–2014, 2011.

48. C. Liu and S. K. Lim, A design tradeoff study with monolithic 3D integration. In *Proceedings of International Symposium on Quality Electronic Design*, Santa Clara, CA, pp. 529–536, 2012.

49. H. Geng, L. Marecsa, B. Cronquist, Z. Or-Bach, and Y. Shi. Monolithic three-dimensional integrated circuits: Process and design implications. In *ECS Transactions on Electrochemical Society Meeting*, Orlando, FL, 2014.

8 Design Methodology for 3D Power Delivery Networks*

Aida Todri-Sanial

CONTENTS

ABSTRACT

Design of power delivery network (PDN) is a constrained optimization problem. An ideal PDN must limit voltage drop that results from switching circuits transients, satisfy current density constraints that arise from electromigration limits, yet use only minimal metal resources so that design density targets can be met. It should also provide an efficient thermal conduit to address heat flux. Further, ideal PDN should be a regular structure to facilitate design productivity and manufacturability, yet be resilient to address varying power demands across its distribution area. In 3D-ICs, these problems are further constrained by the need to minimize TSV area and bridge power lines of different dimensions across tiers, while addressing varying power demands in lateral and vertical directions.

* From Todri-Sanial, A. et al., *IEEE Trans. Very Large Scale Integr. Syst.*, 22(10), 2131, October 2014.

In this paper, we propose an unconventional power-grid optimization solution that allows us to resize each tier individually by applying tier-specific constraints and yet be optimal in a multitier network, where each tier is locally resized while globally constrained. Tier-specific constraints are derived from electrical and thermal targets of 3D PDNs. Two resizing algorithms are presented that optimize 3D PDNs stand-alone or 3D PDNs together with TSVs. We demonstrate these solutions on a three-tier setup where significant area savings can be achieved.

8.1 INTRODUCTION

Three-dimensional (3D) integration presents a path toward higher performance, denser circuits, and heterogeneous implementation while delivering a smaller footprint [17]. Ironically, these advantages are also at the root of power and thermal integrity problems in 3D-ICs. Even more so for power delivery networks (PDNs).

Power delivery networks deliver power and ground voltage from package to the devices. Variations on the supplied voltage can lead to the degradation of circuit's performance and reliability. These effects become more critical in 3D-ICs due to longer current paths and exacerbated thermal dissipation from stacking multiple tiers.

From early on in the design cycle, designers are concerned to know the worst-case voltage drop on each tier and possibly find ways to alleviate it. While much work has been done to understand the resiliency of 2D PDNs, designers are faced with new challenges for ensuring robustness of 3D PDNs. Some of the unique questions related to 3D PDNs addressed in this paper are as follows:

1. What is the impact of physical design constraints (i.e., uniform or nonuniform dies with various power densities) on 3D PDNs?
2. What is the impact of manufacturing constraints (i.e., thinned die or wafer) on the resiliency of 3D PDNs?
3. What is the impact of package on voltage drop and thermal dissipation in 3D PDNs?
4. What is the impact of through-silicon-vias (TSVs)? How does TSV sizing affect power and thermal integrity of 3D PDNs?
5. How does nonuniform temperature distribution inter- and intra-tier affect voltage drop on PDNs?

We will take up these intricate issues, since not only are they pivotal to the design of 3D PDNs, but they provide an important framework for evaluating the benefits and costs of various manufacturing and design constraints unique to 3D integration. To do so, we develop a platform to perform 3D PDN electrothermal analysis and optimization.

8.1.1 POWER AND THERMAL INTEGRITY ISSUES ON 3D PDNs

To address these issues, we take a deep look at the source of power and thermal integrity problems in 3D-ICs. 3D PDNs are structured as multitier networks that are bridged together using TSVs. PDN of each tier is also structured as a multilayer mesh grid with power lines spanning the entire tier, where top- and low-level metal layers devise the global and local PDN, respectively. TSVs facilitate the continuity between PDNs of each tier and depending on the stacking technology, they connect: (1) global to global, (2) local to local, or (3) global to local PDNs of two adjacent tiers. Figure 8.1 depicts a three-tier system with face-to-back TSVs connecting global to local PDNs.

Wire sizing is a widely effective applied method on 2D PDNs to meet voltage drop and thermal constraints; however, for 3D PDNs, this problem appears more intricate due to the current and heat flow attributes in multitier networks.

In 3D-ICs, current flows hierarchically from package to the PDN of the first tier, then through TSVs to the next tiers PDN and so on, till the very last tier. Such conduct results in large amount of

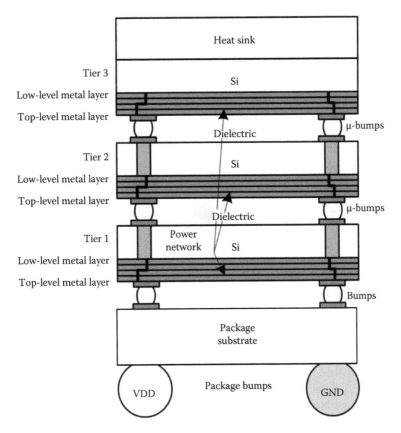

FIGURE 8.1   Three-tier network using face-to-back TSVs.

current flow in the PDN of the bottom tier (next to package) that reduces gradually as it reaches PDNs of top tiers. Hence, each tier depending on their stack location would require individual PDN sizing to accommodate the hierarchical current flow. Furthermore, large amounts of current flowing through parasitic branches of PDNs along with fast switching frequencies induce excessive and nonuniform voltage drops on power and ground distributions of each tier. Inconveniently, TSVs also introduce additional parasitics to power networks and contribute to voltage drop. Moreover, TSVs act as a medium for power supply noise propagation from one tier to the next. In the worst case, power supply noise of one tier can reach the adjacent tier when it is already experiencing excessive self-induced voltage drop.

Thermal effects are another important phenomena that strain 3D PDNs. Frequent switching circuits generate heat and lead to elevated and nonuniform temperatures on each tier. Heat sink generally located on the top of the stack provides immediate cooling to the top tier, which causes additional temperature nonuniformity among tiers. Consequently, this leads to exacerbated voltage drop, as electrical resistance is temperature-dependent. Besides, TSVs also propagate heat from one tier to the next, which in conjunction with voltage drop can cause detrimental effects, that is, timing failures and reliability issues.

8.1.2   Prior Work

In open literature, there are several works that look into electrical modeling and IR drop analysis of 3D PDNs. In [9], authors provide detailed analytical modeling of 3D PDNs. Authors in [24] analyze 3D PDNs while considering TSVs inserted in array shape. In [8,15], authors investigate different TSVs technologies and explore novel TSV topologies. In [36], authors investigate the impact of TSVs on IR drop of 3D PDNs. In [7], floorplanning along with power network cosynthesis is investigated. In [29], authors investigate TSV tapering for power and thermal integrity of 3D PDNs. In [38], decoupling capacitors insertion is explored as an alternative for congestion-aware power-grid optimization.

In comparison to literature on IR drop assessment of 3D PDNs, there is significantly fewer works on thermal analysis of 3D PDNs. Early works from [6,16] identified the criticality of high temperatures in 3D-ICs and presented simple thermal models for predicting temperature. More recently, authors in [11,20,23,35] developed heat transfer models for multitier designs. While these analytical models predict the temperature rise in each tier, they are not directly applicable to 3D PDNs. Although, the close coupling effects between electrical and thermal behavior is widely acknowledged, only few papers have looked into electrothermal 3D PDN analysis [21,34].

In contrast to these previous works, we present an electrothermal optimization methodology for 3D PDNs. To the best of our knowledge, this is the first work that investigates power and thermal integrity driven 3D PDN optimization.

The contributions of our work are threefold.

First, we present an efficient thermal model to compute temperature on 3D PDNs by exploiting the electrical–thermal duality. Instead of splitting the metal tracks in small segments, temperature is computed on grid branches based on the temperature of vias. The resulting 3D PDN thermal model has the same number of nodes and topology as the electrical network.

Second, we present a methodology for electrothermal optimization of 3D PDNs. As each tier can have different topology and power density distributions, we develop a tier-based optimization method that satisfies multitier network constraints—meaning that each tier is locally optimized while globally constrained. We define tier-specific constraints that are derived from multitier networks' electrical and thermal targets, TSVs sizing, and power density distribution.

Third, we study power and thermal integrity of 3D PDNs while introducing manufacturing (stacking) and design constraints. We investigate the impact of fully or partially identical dies, wafer thinning, and package choice on electro-thermal optimization of 3D PDNs.

The rest of the chapter is organized as follows. Preliminaries on the 3D technology, TSV dimensions and power tracks are provided in Section 8.2. Section 8.3 contains two motivational experiments to highlight that re-use of already optimized 2D PDNs is not viable and that nonuniform temperature distribution requires circuit area increase. We describe our 3D PDNs electrothermal analysis method in Section 8.4. In Section 8.5, we describe our two proposed algorithms for performing electro-thermal optimization of 3D PDNs. Experiments are presented in Section 8.6. Section 8.7 concludes the paper.

8.2 PRELIMINARIES

In high-performance designs, PDNs are commonly structured as multilayered grids from top to bottom metal layers resulting in global and local power meshes. Furthermore, PDN topology can be designed as uniform or nonuniform meshes, that is, grids can be globally irregular and locally regular, or power-grid branches are intentionally removed to ease signal routing [22,27]. Regardless of their structure, PDNs are designed to handle worst-case switching currents while keeping voltage drop within margins.

In 3D-ICs, supply voltage is brought in by package bumps and distributed to each tier by using TSVs. On each tier, supply voltage is distributed from global to local grids and finally reaching to the devices.

Figure 8.1 depicts a 3D power network for a three-tier structure with Si substrate of $\rho = 10\ \Omega$ cm and connected using face-to-back TSVs. Current flows from C4 package bumps with attributes as: 200 W/cm², 100 µm diameter on 200 µm pitch [33]. In this work, we consider high-density

TABLE 8.1

Parameters Used in This Work

PDN Parameters

Silicon substrate	$rho = 10\ \Omega$ cm
Power mesh	300 μm × 300 μm
Mesh granularity	120 × 120 mesh
TSVs size	2 μm diameter, 20 μm thickness, 60 μm pitch
TSVs density	10,000 TSV/mm²
C4 bumbs	100 μm diameter, 200 μm pitch
Power density	1.5 μW/μm²
Decap. density	5 fF/μm²

PDN Parasitics

	R (mΩ)	L (pH)	C (fF)
C4 package	10	6.2	9.5
Cu TSVs	44.5	34.2	35.8
Cu metal tracks	45.38	3.5	0
Decaps	0	0	450,000
Current (μA)	$I_{peak} = 100$ μA		$T_{rise} = 50$ ps
	$I_{leakage} = 20$ μA		$T_{fall} = 100$ ps
Supply voltage	$V_{DD} = 1$	Voltage drop	Margin 10% V_{DD}
Frequency	3.2 GHz		
Temperature	$T_{ambient} = 27°C$		

face-to-back Cu TSVs of 2 μm diameter with thickness of 20 μm for densities up to 10,000 TSV/mm². As reported by [2], high-density TSVs enable to reduce parasitic effects associated with capacitance and inductance, while increasing static resistance compared to medium-density TSVs.

PDN consists of copper metal layers and we consider an area of 300 μm × 300 μm. Power-grid model includes parasitic resistance of the metal tracks [9]. TSVs include parasitic resistance, capacitance, and inductance [2]. Package is represented by its inductive and resistive parasitics [13].

From a power-grid analysis perspective, circuits draw current from the grid and are usually modeled as current sources. In this work, switching circuits are represented as time-varying current sources in triangular-like current waveform model to present their rise time, tr, fall time, tf, peak time, tp, current peak, I_{peak} and leakage current, $I_{leakage}$. To further set up our analysis, we use parameters of 45 nm Intel Xeon [1] X5482 quad-core processor with die size of 214 mm², which consumes 150 W (1.5 μA/μm² power density) with 1 V supply voltage.

Decoupling capacitance (decaps) is represented by intentionally inserted capacitance and non-switching circuits decoupling capacitance. Current design practices apply on-chip decaps with range of 1–30 fF/μm². We apply 5 fF/μm² per tier resulting to 450 pF per each tier. Please note that in this work we only consider on-die power and ground network while ignoring PCB parasitics. Table 8.1 summarizes the applied parameters.

8.3 MOTIVATIONAL EXPERIMENT

Due to 3D integration attributes, the top tier is located next to the heat sink, while the bottom tier is next to the package. Such arrangement imposes that the current flows through longer parasitic paths from package to reach the top tier in comparison to the current paths to reach the bottom tier. Hence, a voltage drop increase in the upward direction can be expected. Similarly, the top tier located next to the heat sink provides immediate cooling and relief for thermal dissipation. While the bottom tier can experience higher temperature levels, hence a progressive temperature increase in downward direction can be expected. Furthermore, circuits power density and switching activity on each tier impact voltage drop and thermal distribution, and may further aggravate their directional tendencies.

Electrical and thermal codependencies also contribute to these directional tendencies. Since, electrical resistivity is a function of temperature, voltage drop is worsened by nonuniform thermal distribution. In addition, heat generated by switching circuits causes even more nonuniform temperature distribution. This coupled behavior further exacerbates voltage drop and temperature variations among tiers.

Because of the aforementioned issues, we show that already optimized 2D PDNs cannot be directly applied to 3D PDNs due to excessive voltage drop and elevated temperatures. Secondly, we show that to cope with the nonuniform voltage drop and temperature distribution would require circuit area increase in order to meet performance constraints, which is also unsuitable solution.

8.3.1 RE-UTILIZATION OF OPTIMIZED 2D PDNs ON 3D-ICs

To illustrate these effects, we perform HSPICE transient voltage drop and thermal analysis on a three-tier network. For the sake of simplicity, the underlying circuit is represented as an identical core at 45-nm technology with parameters as in Table 8.1. The mesh grid is initially designed and optimized for delivering supply voltage to the core while meeting electrical and thermal constraints. The same PDN is used for each tier and connected using TSVs. TSVs are inserted in array shape with pitch of 60 μm and dimensions as in Table 8.1.

We measure voltage drop and temperature levels on each tier for various setups and results are shown in Table 8.2.

As shown in Table 8.2, we note that for each scenario, the worst-case voltage drop is measured on the tier that is the farthest from package (tier three, T3), while the maximum temperature is measured on the tier that is the farthest from heat sink (tier one, T1), corresponding to two distinct tiers. This indicates that to satisfy both voltage drop and thermal constraints, it requires investigating all tiers simultaneously. Furthermore, we note that for individual active tiers voltage drop is within 10% of V_{DD} (96.8 mV at T1, 97.5 mV at T2, 98.9 mV at T3) and progressively increases as more tiers become active (up to 135.6 mV when T1, T2, and T3 active). A similar behavior is observed with temperature distribution. This suggests that PDNs that were designed and optimized stand-alone for 2D implementation might not necessarily be optimal for a multitier network.

One way of mitigating excessive voltage drop and temperature is by increasing metal track widths. We apply uniform upsizing to all tiers by increasing their metal track widths. Pitches of track widths are based on ITRS predictions of global wire width [10]. Voltage drop constraint is

TABLE 8.2

Worst-Case Voltage Drop and Temperature

Active Tiers	Max Voltage Drop	Max Temperature
T1 only	96.8 mV at T1	36.8°C at T1
T2 only	97.5 mV at T2	36.38°C at T1
T3 only	98.9 mV at T3	28.63°C at T1
T1 and T2	113.2 mV at T2	38.13°C at T1
T1 and T3	113.5 mV at T3	36.94°C at T1
T2 and T3	117.5 mV at T3	36.65°C at T1
T1, T2, T3	135.6 mV at T3	38.33°C at T1

TABLE 8.3
Maximum Area Increase due to Voltage Drop and Temperature

Active Tiers		Max Voltage Drop (mV)	Max Temperature (°C)	Delay Ratio	Area Ratio
All	T1	110	44	1.29	1
uniformly	T2	135	39.8	1.32	1
sized buffers	T3	143	29.4	1.34	1
After resizing	T1	114	46.27	1	1.02
buffers	T2	136	41.5	1	1.04
	T3	145	29.7	1	1.05

set to 10% of V_{DD} drop and temperatures up to 32°C such that timing constraints are still met. Please note that these constraints are not based on an actual design, but simply used to illustrate the increase of PDN area for satisfying these constraints. By applying uniform upsizing, we report that a PDN area increase in 52% is required to meet both voltage drop and thermal constraints. This highlights the severity of the problem and special attention should be given to 3D PDN design since reuse of 2D PDNs is not as straightforward.

8.3.2 Circuit Area Increase due to Nonuniform Voltage Drop and Temperature Rise

Another way to look at this problem is to resize the underlying circuits to account for nonuniform voltage drop and temperature distributions among tiers such that timing constraints are met.

For illustration, we insert buffer chains on each tier and measure their performance when they are uniformly sized while they experience different voltage drop and temperatures on each tier. Afterwards, we resize the buffers such that timing constraints are met. Please note that all tiers are active and PDNs are identical and remain unchanged. Results are shown in Table 8.3.

Initially, buffers have the same area and a delay up to 34% increase from nominal delay is measured at T3 due large voltage drop (143 mV) and temperature (29.4°C). Buffer resizing is performed for each tier and a maximum of 5% increase in buffer area for T3 is obtained. Despite being a small area increase, for a real chip a 5% of overall circuit area increase can impose additional resources (i.e., routing, TSVs, extension of power/clock networks, power overhead), which may be simply impractical for current designs with already tight area and power budgets. It can also impose higher manufacturing costs due to additional processing area.

These experiments clearly show that either reuse of already optimized 2D PDNs or resizing of circuits to cope with nonuniform voltage drop and temperatures are unfeasible design choices for 3D-ICs. These observations further motivate us to investigate the power-grid sizing problem for multitier networks and develop an electro-thermal optimization strategy for 3D PDNs.

8.4 3D PDN ANALYSIS

Here, we first discuss electrical analysis followed by thermal analysis method based on electrical–thermal duality principle. We then discuss our proposed optimization method.

8.4.1 Electrical Modeling and Analysis

8.4.1.1 Electrical Modeling

As mentioned in Section 8.2, 3D PDN electrical model includes package, power meshes, TSVs, decaps, and switching circuits. Topology and granularity of PDNs can vary on size and power density applied on them. In this work, we adopt electrical models that have been widely applied

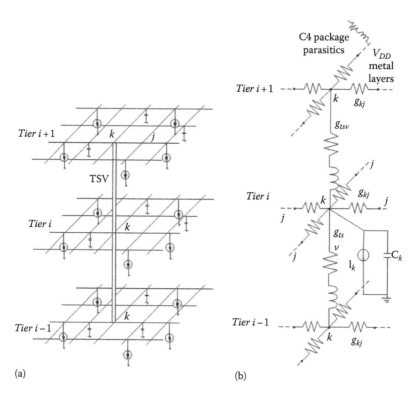

FIGURE 8.2   (a) Illustration of 3D PDN electrical model and (b) node k on 3D PDN.

in literature [3,9–38] and capture well PDNs behavior. In our case, we have extended these models to include TSVs and several PDNs of stacked tiers. Figure 8.2a shows an illustration of a three-tier PDNs with TSVs, current sources, and decoupling capacitances.

8.4.1.2   Electrical Analysis

3D PDNs are modeled as *RLC* networks to represent parasitics of metal tracks for each tier, package, and TSVs where underlying circuits are modeled as time-varying current waveforms. The behavior of such network can be described by first-order differential equations formulated using modified nodal analysis (MNA) method [4].

Each voltage node k of tier i on the multitier network can be expressed as follows:

$$V_k^i = \frac{\sum_{j \in \{neighbors\,of\,k\}} V_j^i g_{kj}^i + \sum_{r \in \{i+1,i-1\}} V_k^r g_{tsv}^{i,r} + I(s)_k^i}{\sum_{j \in \{neighbors\,of\,k\}} g_{kj}^i + \sum_{r \in \{i+1,i-1\}} g_{tsv}^{i,r} + sC_k^i} \qquad (8.1)$$

where
   i is the current tier
   $i + 1$ and $i - 1$ represent top and bottom tiers
   j represents the neighboring nodes to node k
   g_{kj} represents the impedance admittance between nodes k and j
   g_{tsv} represents the TSV impedance admittance connecting two tiers at node k
   $I(s)$ represents the current waveform in s-domain of the underlying switching circuit at node k
        for the current tier i
   C is the decoupling capacitance at node k

Figure 8.2b shows node k on 3D PDN. In matrix form, node equations can be devised as follows:

$$\left(\begin{bmatrix} G_t^1 & G_{tsv}^{12} & 0 \\ G_{tsv}^{21} & G_t^2 & G_{tsv}^{23} \\ 0 & G_{tsv}^{32} & G_t^3 \end{bmatrix} + s \begin{bmatrix} C_t^1 & 0 & 0 \\ 0 & C_t^2 & 0 \\ 0 & 0 & C_t^3 \end{bmatrix} \right) \begin{bmatrix} V_t^1 \\ V_t^2 \\ V_t^3 \end{bmatrix} = \begin{bmatrix} B_t^1 \\ B_t^2 \\ B_t^3 \end{bmatrix} \tag{8.2}$$

where $G_{t_{n1\times n1}}^1$, $G_{t_{n2\times n2}}^2$, $G_{t_{n3\times n3}}^3$ are sparse, and symmetric conductance matrices for each tier. Their sizes $n1 \times n1$ for tier 1, $n2 \times n2$ for tier 2, and $n3 \times n3$ for tier 3 can be different based on power network delivery topology and granularity of each tier. $G_{tsv_{n1\times n2}}^{12}$ and $G_{tsv_{n2\times n3}}^{23}$ are diagonal matrices that represent the conductance of the TSVs connecting tier 1–2 and tier 2–3, respectively. $G_{tsv_{n2\times n1}}^{21}$ and $G_{tsv_{n3\times n2}}^{32}$ are their transpose. We remark that several diagonal entries of $G_{tsv}^{i,r}$ have zero values as there is not a TSV connected to each node given that TSV granularity is sparse on the power grid. $C_{t_{ni\times ni}}^i$ are diagonal matrices representing decoupling capacitances for each tier i. $V_{t_{ni\times 1}}^i$ are vectors representing node voltages and $B_{t_{ni\times 1}}^i$ are the vectors representing current and voltage sources for each tier i. Utilizing node voltage equations for each tier i, the current density $J_{branch_{kj}}^i$ of branch kj is computed as follows:

$$J_{branch_{kj}}^i = \frac{I_{branch_{kj}}^i}{h_{kj} w_{kj}} = \frac{\left(V_k^i - V_j^i \right) g_{kj}^i}{h_{kj} w_{kj}} \tag{8.3}$$

where $I_{branch_{kj}}^i$ is the branch current derived from node voltages V_k^i, V_j^i and their respective branch impedance, g_{kj}^i. h_{kj} and w_{kj} are height and width of the track where branch kj resides. In this work, we analyze 3D power grids and current flow by considering only *RL* parasitics of the power metal tracks, while ignoring the current return paths.

8.4.2 Thermal Modeling and Analysis

8.4.2.1 Thermal Modeling

Heat generated from transistors is dissipated through the silicon tiers, TSVs, heat sink, package to the air. As heat sink and package are located on opposite ends, there are two heat path flows. In this work, we consider both heat path flows and use thermal models for PDN, TSVs, package and circuits.

In this work, we utilize electrical–thermal duality principle [4] to represent a thermal network for performing fast thermal analysis. Due to this duality, temperature is analogous to voltage; heat can be modeled by current flowing through thermal resistances driven by current sources, which represent power consumption. Traditionally, thermal models are based on partitioning each power metal track into small wire segments [32]; hence, it results into an extremely large thermal network. Instead, we apply an improved lumped thermal model as initially introduced for 2D PDNs by [37], to reduce the number of thermal nodes and make thermal computation feasible.

The idea is to construct a thermal network that has the same structure and number of nodes as the electrical network. The motivation is to have a one-to-one correspondence between electrical and thermal networks which facilities our analytical formulation and analysis.

Thermal resistance is derived similarly as electrical resistance while using thermal conductivity coefficients. As the electrical branch resistance on the power grid is between two vias, thermal resistance is also represented similarly and derived by applying temperature at the vias as boundary condition for solving heat equation on a wire. Such representation is valid as long as the wire length between two vias is within the range of characteristic thermal length, L_H [5,25,37]. Within this length the Joule heating generated at a wire will flow through the vias to the next metal layer rather than through the dielectric layer. As a result, wire segment temperature can be derived based on the temperature at vias. Thus, we can build a thermal model at the vias similarly as the electrical model. The final thermal network has the same structure as the electrical network, where thermal resistance is equivalent to electrical resistance [37] as follows:

$$R_{thermal} = \frac{R_{electrical}}{\rho_0 k_m} \tag{8.4}$$

where

ρ_0 is the resistivity at reference temperature
k_m is the thermal conductivity of the metal layer

A thermal resistance represents the difference in temperature necessary to transfer a certain amount of heat and unit is °C/W or K/W. Derivation of thermal resistances represents the thermal PDN of each tier.

Thermal capacitance for each tier is derived based on the thickness and area of the die. Thermal capacitance represents the materials heat capacity that can be stored or removed to increase or decrease temperature and unit is J/°C or J/K. We utilize a distributed thermal capacitance model that is introduced on each node of thermal network. Single die thermal capacitance [25,28] can be derived as

$$C_{thermal} = c_p t A \tag{8.5}$$

where c_p is thermal capacitance of the material, while t and A are die thickness and area. From published data [26], silicon has $c_p = 10^6$ J/m^3 K and $\rho_0 = 10^{-2}$ mK/W.

Heat transfer from die to heat sink is modeled by a thermal resistance between every grid node to heat sink node. As a case study, we assume a heat sink with thermal resistance of 2 K/W. As circuit transients are in order of ps compared to temperature variations in seconds at the heat sink [25], we model heat sink at a constant temperature represented by a constant voltage source. For example, a heat sink with thermal capacitance of 30 J/K and thermal resistance of 2 K/W, we can determine the time constant to cooling/heating as $RC = 30$ J/K $\times$ 2 K/W $= 60$ s.

TSVs are represented as RC model based on their thermal characteristics as described by [14,30].

Package with C4 bumps provide a secondary heat path for dissipating the generated heat [12,19]. As a case study, a package thermal resistance of 20 K/W is assumed.

Heat generated by switching circuits is presented as current sources proportional to their power consumption obtained from electrical analysis.

Overall, we aim to represent a complete thermal 3D PDN model (i.e., including package, heat sink, circuits, TSVs, power meshes) while also compact (i.e., same size and structure as electrical network) to compute temperature distribution on 3D PDNs and feed it back to electrical analysis. Figure 8.3a illustrates thermal 3D PDN model.

8.4.2.2 Thermal Analysis

Similar to electrical network, a thermal network can be devised where temperature at node k can be expressed as

$$T_k^i = \frac{\sum_{j \in \{neighbors\ of\ k\}} T_j^i g_{th_{kj}}^i + \sum_{r \in \{i+1, i-1\}} T_k^r g_{th_{tsv}}^{i,r} + Q_k^i}{\sum_{j \in \{neighbors\ of\ k\}} g_{th_{kj}}^i + \sum_{r \in \{i+1, i-1\}} g_{th_{tsv}}^{i,r}} \tag{8.6}$$

where

T_k represents the temperature at node k
T_j represent the temperatures at neighboring nodes j
$g_{th_{kj}}$ represents the thermal impedance between nodes k and j
$g_{th_{tsv}}$ represents the thermal impedance for a TSV as characterized in [23]
Q_k represents the heat dissipation (or power consumption for electrical network) at node k

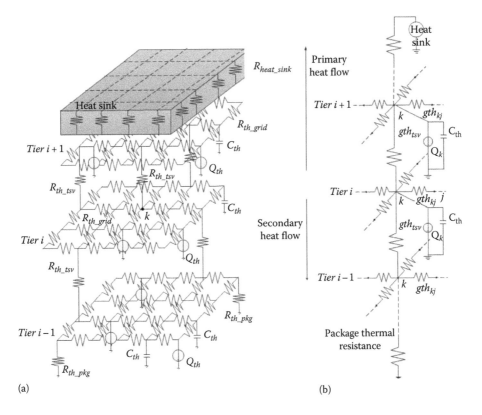

FIGURE 8.3   (a) Illustration of 3D PDN thermal model and (b) node k on 3D PDN.

Thermal networks can be devised in matrix form as follows:

$$
\begin{bmatrix}
P_t^1 & P_{tsv}^{12} & 0 \\
P_{tsv}^{21} & P_t^2 & P_{tsv}^{23} \\
0 & P_{tsv}^{32} & P_t^3
\end{bmatrix}
\begin{bmatrix}
T_t^1 \\
T_t^2 \\
T_t^3
\end{bmatrix}
=
\begin{bmatrix}
Q_t^1 \\
Q_t^2 \\
Q_t^3
\end{bmatrix}
\tag{8.7}
$$

where

$P_{t_{n_i \times n_i}}^i$ are the sparse and symmetric matrices representing thermal conductance of each tier i

$P_{tsv_{n1 \times n2}}^{12}$ and $P_{tsv_{n2 \times n3}}^{23}$ are the diagonal matrices of thermal conductances for TSVs connecting tier

1–2 and tier 2–3, respectively

$P_{tsv_{n2 \times n1}}^{12}$ and $P_{tsv_{n3 \times n2}}^{32}$ are their transposes

$T_{t_{n_i \times 1}}^i$ are the vectors of node temperatures for each tier i

$Q_{t_{n_i \times 1}}^i$ are the vectors representing the heat dissipation for each tier i

We also consider Joule heating of power metal tracks due to their current flow. Joule-heating power computed as $Q_{self} = I^2 R_{electrical}$ which is a function of thermal conductivity of wire and temperature. Self-heating power Q_{self} of power-grid wires are taken into account at Q_k of each node. For example, a power-grid wire between two nodes (i.e., nodes k and $k + 1$), Q_{self} is proportionally distributed

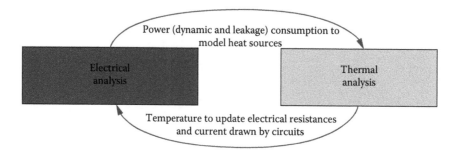

FIGURE 8.4 Electrothermal analysis iterative process.

between these two nodes. The behavior of electrical and thermal networks can be coupled together since electrical resistivity, ρ, is dependent to temperature and expressed as

$$\rho = \rho_0[1 + \beta(T - T_0)] \tag{8.8}$$

where
ρ_0 is electrical resistivity at reference T_0 (ambient temperature at 27°C)
$\beta = 0.0039/°C$ is the temperature coefficient for resistance

Thus, the resistance of the power metal tracks can be computed as

$$R_i = \rho \frac{L_i}{H_i W_i}[1 + \beta(T - T_0)] \tag{8.9}$$

where L_i, W_i, and H_i are the length, width, and height of the metal track, respectively. As temperature changes, $R_{electrical}$ changes, which consequently affects voltage drop and current flow. Naturally, node voltage equation in Equation 8.1, the conductance parameters (i.e., g_{kj}, g_{tsv}) are temperature dependent, which lead to voltage drop and current flow change on power wires. The temperature changes are fed to electrical analysis engine for updating impedances and circuit current model (i.e., $I_{leakage}$ and I_{peak}) in order to recompute voltage droop and power consumption (i.e., dynamic and leakage power). Power dissipation serves as an input to thermal analysis for deriving heat sources. Due to this closely coupled relation, electrical and thermal analyses run in several loops until final voltage drop and temperature distributions are obtained also as shown in Figure 8.4. Equations 8.2 and 8.7 can be solved fast and accurately utilizing different linear algebra solvers (i.e., direct or iterative methods). In this work, multigrid method is applied to reduce the size of matrices while exploiting sparsity of matrices and power-grid topology [18]. Multigrid method initially relaxes the power grid by reducing its granularity from finer to coarser grid. Once the coarse grid is solved, its solution is mapped back to the original size power grid. Multigrid method has been widely and effectively employed for power-grid analysis [18,31].

8.5 3D PDN OPTIMIZATION METHOD

We propose a tier-based optimization approach for resizing the 3D PDN to satisfy both voltage drop and thermal constraints. Given that each tier can have different PDN topology, circuit functionality and power density, we treat them individually while imposing 3D PDN-aware power and thermal constraints. Thus, the essence of our 3D PDN optimization method is to *locally optimize each tier while applying global constraints.* Such approach reduces computational complexity of the problem while treating each tier individually in the context of many tiers.

Each tiers constraints are derived based on the knowledge of the 3D PDN electrical and thermal targets, current demand, and power density of each tier, and TSV sizing. We define the tier-specific voltage drop constraints for a tier i as follows:

$$V_{droop}^i = \frac{\max V_{droop}^{\text{for all tiers}}}{\text{number of tiers}} - R_{tsv}^{i,i+1} \sum_{j=1}^{\text{top tiers}} I^j \tag{8.10}$$

where $\max V_{droop}^{\text{for all tier}}$ represents the total amount of voltage droop allowed for the whole system such as $10\% V_{DD}$ (i.e., 100 mV for $V_{DD} = 1$ V). The accumulated resistance of all TSVs connecting two adjacent tiers is represented as $R_{tsv}^{i,i+1}$ and I^j as the upward current flow from package to each tier through TSVs. As face-to-back TSVs are considered, global power grid of bottom tier is connected to local power grid of top tier. Thus, resistance R_{int}^i (Ω) aims to represent the cumulative parasitic impedance from local to global power grid for each tier including local, intermediate, and global grids. For a three-tier system, voltage drop constraints can be derived as in Figure 8.5. In a similar way, each tiers' thermal constraints can be derived with respect to number of tiers, heat dissipation, and TSV sizing. We define tier-specific thermal constraint for a tier i as follows:

where the constraints are

$$V_{droop}^1 = \frac{\max V_{droop}^{\text{for all tiers}}}{3} \tag{8.11}$$

$$V_{droop}^2 = \frac{\max V_{droop}^{\text{for all tiers}}}{3} - R_{tsv}^{12}(I^2 + I^3) \tag{8.12}$$

$$V_{droop}^3 = \frac{\max V_{droop}^{\text{for all tiers}}}{3} - R_{tsv}^{23}I^3 \tag{8.13}$$

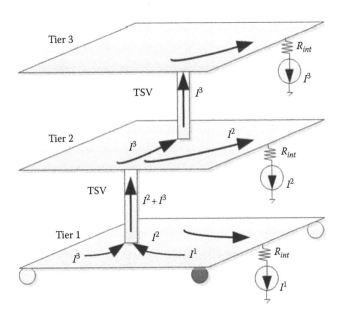

FIGURE 8.5 Current flow on a three-tier 3D PDN for computing voltage drop constraints.

In a similar way, each tier's thermal constraints can be derived with respect to number of tiers, heat dissipation, and TSV sizing. We define tier-specific thermal constraint for a tier i as

$$T_{max}^i = \frac{\max T_{difference}^{tiers}}{\text{number of tiers}} - P_{tsv}^{i,i+1} \sum_{j=i-1}^{\text{bottom tiers}} Q^j \tag{8.14}$$

where $\max T_{difference}^{tiers}$ is the maximum allowed temperature difference between top and bottom tiers, $P_{tsv}^{i,i+1}$ is the accumulated thermal resistance of all TSVs connecting two adjacent tiers, and Q^j represents the heat flow from each tier to the heat sink (located on the top). Thus, V_{droop}^i and T_{max}^i serve as tier-specific constraints, which are derived from the multitier constraints in order to account for the voltage droop and heat generated in other tiers and TSVs. Such constraints facilitate to investigate each tiers' power-grid sizing problem individually without imposing any power/thermal issues to the rest of the tiers. Additionally, the computational complexity of the problem is significantly reduced as the number of circuit nodes to be analyzed is same as in 2D PDN optimization problem, whereas a full 3D problem would have at least m times more circuit nodes (m, number of tiers).

In the next sections, we describe two proposed electro-thermal algorithms.

8.5.1 3D PDN STAND-ALONE OPTIMIZATION

Here, we describe the optimization technique applied only to the metal tracks of the 3D PDN.

Problem formulation: Derive the metal track widths for the PDN of each tier while applying 3D PDN-aware electrothermal constraints such that the minimum PDN area is obtained.

Optimization method can be applied to any tier (regardless of stacking order) together with its specific voltage drop and thermal constraints. Please note that the optimization technique can be applied on either coarse or original size grids. The problem of PDN sizing with voltage drop and thermal constraints can be formulated as a linear optimization problem. We aim to minimize the PDN area by minimizing the summation of metal track widths as

$$\min \sum w \tag{8.15}$$

such that the following constraints are met:

1. Voltage droop:

$$V_k^i \geq V_{DD} - V_{droop}^i \tag{8.16}$$

   based on node voltage equations on Equation 8.1 and 3D PDN-aware voltage drop constraints as in Equation 8.10.
2. Temperature:

$$T_k^i \leq T_{amb} + T_{max}^i \tag{8.17}$$

   based on node temperature equations on Equation 8.6 and 3D PDN-aware temperature constraint as in Equation 8.11.
3. Track width:

$$w_{min} \leq w \leq w_{max} \tag{8.18}$$

   track width within an upper and lower range.
4. Electromigration:

$$J_{branch}^i \leq J_{max} \tag{8.19}$$

   based on branch current flow as in Equation 8.3 to ensure that current density for PDN branch, *kj* does not surpass maximum allowed current density, J_{max}.

5. Optimization parameter:

$$0 \leq \alpha \leq 1 \qquad\qquad (8.20)$$

where $\alpha = 1$ enables voltage droop optimization and $\alpha = 0$ enables thermal optimization. Constraints are weighted as $\alpha V_k^1 + (1-\alpha)T_k^i$.

Please note that TSV sizes remain unchanged during this optimization flow. One way to satisfy all the constraints is to upsize the widths of all metal tracks on the grid. However, this would result in a significantly large power-grid area that is unnecessarily over-designed. Instead, we propose to upsize all tracks by maximum allowed width, w_{max}, and iteratively reduce by w_{step} the widths of those tracks that have minimal impact on power and thermal integrity while still satisfying voltage drop and thermal constraints. Some tracks due to electromigration constraints may not be reduced much as they can pose current density issues. Thus, we aim to reduce the PDN area by maximizing the track widths to be downsized. Figure 8.6 shows the flowchart and the steps of the optimization algorithm.

Variable w_{step} is a user-selected parameter, which can impact the quality of the solution and the number of iterations. Similarly, α is a user-selected parameter to vary the weight between voltage drop and thermal optimization. Note that in Step 3, we compute voltage drop ($O_{V_{drop}}$) and thermal (O_T) objective functions (also defined in Figure 8.6b) as ratios to the tier-specific constraints. This allows them to be used as unitless numbers for computing the objective function in Step 4 regardless of representing two different entities.

As the final step, a voltage drop and thermal analysis is performed where all optimized PDNs are considered simultaneously. We find that tier-based electro-thermal optimization guarantees power and thermal integrity of 3D PDNs.

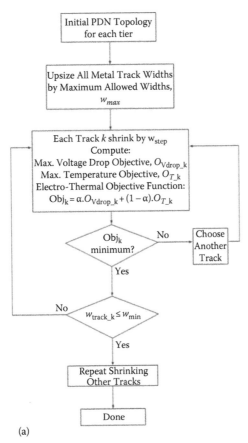

(a)

FIGURE 8.6 (a) Flowchart of the 3D PDN electro-thermal optimization method.   *(Continued)*

ALG 1: 3D PDN Only Electrothermal Optimization
STEP 1: Upsize all tracks widths by w_{max} such that both voltage drop and thermal constraints are met
STEP 2: Shrink each track k by w_{step} where n is total number of tracks on power grid.
STEP 3: Compute: Voltage drop objective: $O_{vdrop_k} = \dfrac{V_{drop\_shrunk\_track_k}}{V_{drop}^i}$ Thermal objective: $O_{T_k} = \dfrac{T_{max\_shrunk\_track_k}}{T_{max}^i}$
STEP 4: Compute: $Obj_k = \alpha \cdot O_{vdrop_k} + (1-\alpha) \cdot O_{T_k}$
STEP 5: Compute: $\min(Obj_1, Obj_1, \ldots Obj_k, \ldots, Obj_n)$ Identify track j with the minimum electro-thermal objective function, $\boldsymbol{Obj_j}$

STEP 6: Does track j have widths, $w_j \geq w_{min}$?
      – Yes, Accept width reduction of track j, Update widths $w_j = w_j - w_{step}$
      – No, Go back to Step 2.

STEP 7: Repeat shrinking widths (Steps 2–6) while electrothermal constraints are met as $\boldsymbol{Obj_j} \leq 1$
STEP 8: Done. New track width distributions are found such that electro-thermal constraints are met.

(b)

FIGURE 8.6 (*Continued*)   (b) Step-by-step algorithm description.

Initially, upsizing all power metal tracks ensure that both voltage droop and temperature constraints are met. This step also serves as the basis for convergence for our optimization method. Meaning that, the starting point of algorithm is a good (not optimal) solution, which guarantees the optimization engine to converge by further finding an optimal solution. The complexity of the algorithm is $O(nn_{step})$, where n is the number of power metal tracks and n_{step} is the number of shrinking steps. Algorithm evaluates voltage drop and temperature each time a track is shrunk and for each shrinking step (w_{step}).

8.5.2   3D PDN AND TSV-SIZING CO-OPTIMIZATION

Here, we describe that the electro-thermal optimization method is applied to both metal track widths and TSV sizes.

Problem formulation: Derive the metal track widths and TSV sizes (diameter) while applying 3D PDN-aware electro-thermal constraints such that the minimum PDN area and TSV sizes are obtained.

Only the diameter is considered as variable for sizing the TSVs, whereas their thickness and pitches are assumed constant (due to stacking approach). As mentioned in Section 8.2, we consider high-density TSVs inserted in array shape where the diameters can vary from 1 to 5 µm [2]. Please note, we consider uniformly sized TSVs where all TSVs have the same radii. Also, TSVs continue to remain uniformly sized even after our PDN-TSV co-optimization method.

An important aspect of the tier-based proposed algorithm is the treatment of TSVs such as how are TSVs optimized and how are they included to the PDN optimization. These are critical aspects and can impact 3D PDNs resiliency because TSVs can be sized optimally for one tier but may create voltage drop or thermal issues for the next tier.

Most sensitive tiers to voltage drop and thermal are middle tiers as they are remotely connected to either package pins or heat sink. Thus, our electrothermal PDN-TSV optimization method is initially applied to middle tiers in order to account for TSV impacts. The obtained TSV sizes then serve as lower bound constraints for the rest of the tiers during their PDN-TSV co-optimization. Figure 8.7 describes our PDN-TSV electrothermal co-optimization method.

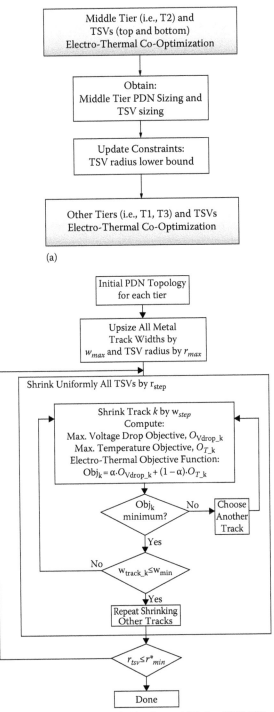

(a)

(b)

r^*_{min}: lower bound TSV radius imposed either from middle tiers PDN-TSV optimization or manufacturing contraint.

FIGURE 8.7 (a) 3D PDN-TSV co-optimization flow. Initially middle tiers and TSVs are co-optimized where the obtained TSV sizes serve as lower bound constraints for other PDN-TSV optimization. (b) Flowchart of the 3D PDN-TSV electrothermal optimization method.   *(Continued)*

ALG 2: 3D PDN and TSV Electrothermal Optimization
STEP 1: Upsize all tracks widths by w_{max} and TSV sizes by r_{max} such that both voltage drop and thermal constraints are met.
STEP 2: Shrink all TSVs by r_{step}.
STEP 3: Shrink each track k by w_{step} where n is total number of tracks on power grid.
STEP 4: Compute: Voltage drop objective: $O_{vdrop_k} = \dfrac{V_{drop\_shrunk\_track_k}}{V_{drop}^i}$ Thermal objective: $O_{T_k} = \dfrac{T_{max\_shrunk\_track_k}}{T_{max}^i}$
STEP 5: Compute: $Obj_k = \alpha \cdot O_{Vdrop_k} + (1-\alpha) \cdot O_{T_k}$
STEP 6: Compute: $min(Obj_1, Obj_1, ...Obj_k, ..., Obj_n)$ Identify track j with the minimum electro-thermal objective function, Obj_j
STEP 7: Does track j have widths, $w_j \geq w_{min}$?
– Yes, Accept width reduction of track j Update widths $w_j = w_j - w_{step}$
– No, Go back to Step 3.
STEP 8: Repeat shrinking widths (Steps 3–7) while electro-thermal constraints are met as $Obj_j \leq 1$
STEP 9: Repeat shrinking TSV radius (Step 2–8).
STEP 10: Done. New track width distributions and TSV radius are found such that electrothermal constraints are met.

(c)

FIGURE 8.7 (*Continued*)   (c) Step-by-step algorithm description.

The problem of PDN-TSV sizing with voltage drop and thermal constraints can be formulated as a linear optimization problem. We aim to minimize the PDN and TSV area by minimizing the summation of their widths as

$$\text{Objective}: \min \sum \beta w_{track} + \sum (1-\beta) r_{tsv} \tag{8.21}$$

such that the following constraints are met:

1. Voltage droop:

$$V_k^i \geq V_{DD} - V_{droop}^i \tag{8.22}$$

   based on node voltage equations on Equation 8.1 and 3D PDN aware voltage drop constraints as in Equation 8.10.
2. Temperature:

$$T_k^i \leq T_{amb} + T_{max}^i \tag{8.23}$$

   based on node temperature equations on Equation 8.6 and 3D PDN-aware temperature constraint as in Equation 8.11.
3. Metal track width:

$$w_{min} \leq w \leq w_{max} \tag{8.24}$$

track width within an upper and lower range.

4. TSV radius:

$$r_{min}^* \leq r \leq r_{max} \tag{8.25}$$

where r_{min}^* can be either the minimum radius range or the imposed lower bound radius constraint from PDN-TSV optimization of the middle tiers.

5. Electromigration:

$$J_{branch}^i \leq J_{max} \tag{8.26}$$

based on branch current flow as in Equation 8.3 to ensure that current density for PDN branch, kj, does not surpass maximum allowed current density, J_{max}.

6. Optimization parameter:

$$0 \leq \alpha \leq 1 \tag{8.27}$$

where $\alpha = 1$ enables voltage droop optimization and $\alpha = 0$ enables thermal optimization. Constraints are weighted as $\alpha V_k^i + (1 - \alpha) T_k^i$.

Parameter β is introduced to provide a weight coefficient between power track sizes and TSV radius. It can vary as $0 \leq \beta \leq 1$.

In summary, the proposed globally constrained locally optimized methodology is flexible to be applied to any PDN regardless of topology or technology. As heterogeneous technologies and functionalities can be implementable in 3D, our electrothermal optimization problem helps to study each tier stand-alone while applying constraints that take into account multitier network and TSVs.

8.6   EXPERIMENTS

We have implemented our tier-based power-grid-resizing algorithm in MATLAB® and tested it on various 3D PDN setups. We applied circuit parameters as described in Table 8.2 along with wire pitches w_{min}, w_{max}, and current density, J_{max} as in [10]. Voltage drop and temperature constraints were set to 10% of V_{DD} and $T_{max} = 10°C$ difference between topmost and bottommost tiers. Ambient temperature is set to $T_{amb} = 27°C$. Please note that both T_{max} and T_{amb} are not based on any specific design or technology parameter but simply chosen as a case study. Both parameters are user-dependent. Track width sizing parameter w_{step} and r_{step} are set to 5% of the original track widths and TSV radius.

In our experiments, we applied both ALG1 and ALG2 as described in Figures 8.6 and 8.7. There are no prior works that developed tools for 3D PDNs electrothermal optimization, which is why we utilize uniform resizing (UR) algorithm for comparison. We conducted our experiments on a three-tier network with PDNs of mesh topology. We utilize three types of power density blocks distributed on a tier and through tiers: high (h), medium (m), and low (l) power density blocks. We studied the impact of optimization parameter α, power density distribution among tiers, and design constraints on power and thermal integrity of 3D PDNs.

8.6.1   Impact of Optimization Parameter, α

We applied our globally constrained locally optimized optimization method on the three-tier power network while varying the electro-thermal optimization parameter, α. α values vary from 0 to 1 where 0 refers to thermal only, 1 refers to electrical only, and 0.5 refers to electrothermal optimization.

To capture α effect on the power-grid area, we conducted experiments on PDNs with identical power density distributions of 0.6 µW/µm². We apply algorithms ALG1 and ALG2 and the results are shown in Figures 8.8 and 8.9, respectively. Results are represented with respect to voltage drop objective, O_{Vdrop}, thermal objective, O_T, and 3D PDN area ratio. As derived in Section 8.5, voltage drop objective (O_{Vdrop}) is unitless number to indicate the correlation to the voltage drop constraint.

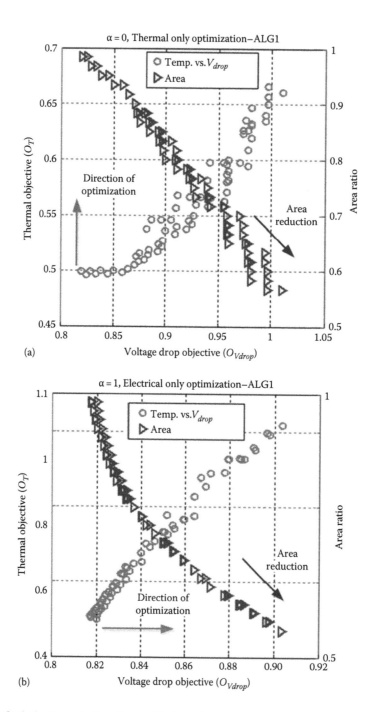

FIGURE 8.8 Optimization objectives for ALG1 when (a) α = 0 and (b) α = 1. *(Continued)*

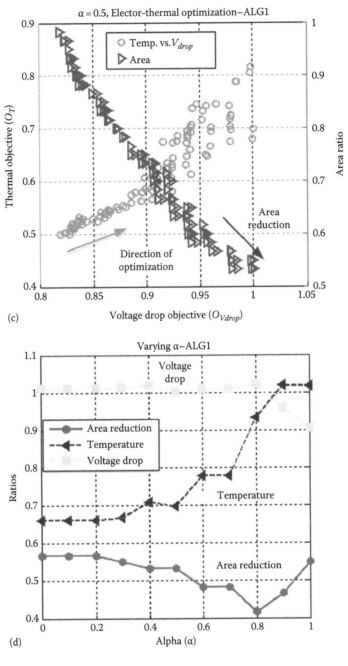

(c)

(d)

FIGURE 8.8 (*Continued*) Optimization objectives for ALG1 when (c) $\alpha = 0.5$ and (d) optimized power-grid area with varying α.

For example $O_{Vdrop} = 1$ indicates that voltage drop is equal to maximum allowed voltage drop of 100 mV. Similarly, thermal objective (O_T) is a unitless number to indicate the correlation to the maximum temperature, T_{max} constraint, where $O_T = 1$ is equal to 37°C (where $T = T_{amb} + T_{max} = 27°$ C + 10°C = 37°C derived from temperature constraints as in Section 8.5) and $O_T = 0$ corresponds to 27°C. Area ratio displayed on the second y-axis (right-hand side on figures) represent the area savings from the optimization algorithms in comparison to UR algorithm where w_{max} (maximum metal track width) and r_{max} (maximum TSV radius) were applied such that both electrical and thermal constraints were met. For example, area ratio = 1 represents the 3D PDN area when maximum upsizing for ALG1 (metal tracks only) and ALG2 (metal tracks and TSVs) are applied.

In Figure 8.8a, we observe that for $\alpha = 0$ (thermal optimization only), there is up to 42% of total area reduction from uniformly upsizing all tiers by w_{max}. Track widths are shrunk where thermal objective, O_T, varies from 0.5 to 0.68 and voltage drop objective, O_{Vdrop}, varies from 0.8 to 1, thus

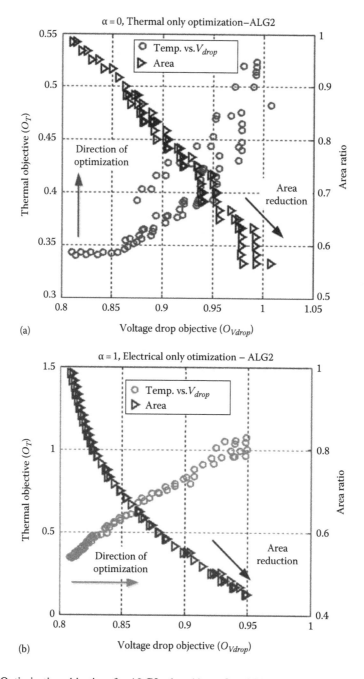

(a)

(b)

FIGURE 8.9 Optimization objectives for ALG2 when (a) $\alpha = 0$ and (b) $\alpha = 1$. *(Continued)*

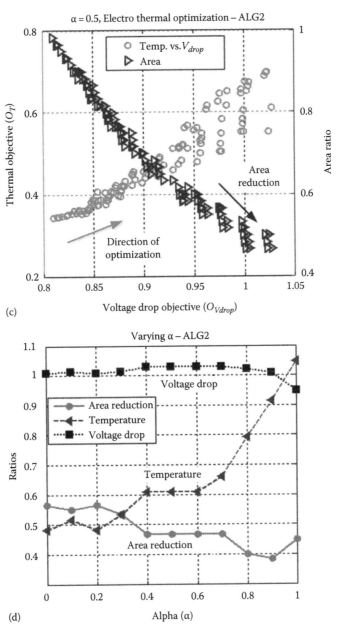

FIGURE 8.9 (*Continued*) Optimization objectives for ALG2 when (c) $\alpha = 0.5$ and (d) optimized power-grid area with varying α.

both electrical and thermal constraints remain satisfied. Additionally, we note that there are oscillations during the optimization. This is due to various track width distributions that provide different amount of voltage drop without much change in temperature. In the case of $\alpha = 1$ (voltage drop optimization only), there is a total of 45% of area reduction from uniform upsizing all tiers as shown in Figure 8.8b. We note that there is a minimal impact on O_{Vdrop} that varies from 0.82 to 0.9 while O_T varies from 0.5 to 1. Figure 8.8c shows the optimization objectives for $\alpha = 0.5$ or equally weighted electro-thermal optimization. A 48% area reduction is achieved while O_{Vdrop} varies from 0.82 to 1 and O_T varies from 0.5 to 0.82.

In Figure 8.8d, we show the reduced multitier power network area for various α values with respect to their maximum voltage drop and temperature objectives. In general, we note that voltage is less sensitive than temperature with varying α. This indicates that the changes in power track widths provide significant improvement on temperature distribution while less difference on worst-case voltage drop. This is due to wider power tracks enabling lateral heat dissipation on a tier. Whereas voltage drop modeled as *IR* drop, *L di/dt* and *RC dv/dt* effects experiences less improvement with power track widths. This suggests that *L di/dt* and *RC dv/dt* effects continue to be significant even as power track widths are varied.

We note that power-grid area savings varies from 42% (α = 0) to 58% (α = 0.8). We observe that voltage drop remains almost flat with some minimization around α = 1, while temperature decreases gradually toward α = 0. At α = 0.8, both voltage drop and temperature are at their maximum allowed objective (1), which results in the minimum power-grid area obtained. Another observation from Figure 8.8d is that power-grid area from voltage drop optimization (α = 1) is larger than area obtained from thermal optimization (α = 0). Even though power-grid area is larger for α = 1, it may not be viable for satisfying thermal constraints as track width distribution on each tier is more important than the total area. This suggests that track width distribution affects the quality of the grid and performing voltage drop only optimization is not sufficient for meeting thermal constraints.

In Figure 8.9 we show the optimization results for ALG2, which includes the optimization of TSV sizing together with metal track widths. Overall, we notice that there is more 3D PDN area savings obtained when both TSVs and metal tracks are considered. For example, up to 43% for α = 0 (Figure 8.9a), 55% for α = 1 (Figure 8.9b), and 53% for α = 0.5 (Figure 8.9c). Figure 8.9d shows the area for various values of α. We note up to 62% of area savings for α = 0.9.

These results indicate the importance of considering TSV sizing along with the design of PDN. We note that additional 3D PDN area savings can be obtained when both metal track widths and TSV radii are considered. TSVs are crucial for delivering current and removing heat from one tier to the next. Thus, resizing TSVs provides headroom for 3D PDNs to meet power and thermal constraints.

8.6.2 Impact of Identical and Nonidentical Dies

Given that each tier can have different dies inserted on them, we study the impact of fully and partially identical dies. Each die can have different network topology, functionality, and power density distribution. We investigate three kind of dies represented as single-cores of different power density distributions. We study their different stacking options on a three-tier setup.

Three power density distributions are as follows: (1) h, high power density of 1.5 µW/µm², (2) m, mid power density of 1 µW/µm², and (3) l, low power density of 0.8 µW/µm². We note that initially all PDNs were uniformly upsized such that voltage drop and thermal constraints were met. We utilize ALG1 algorithm to perform electro-thermal optimization. We consider six different cases (cases of 1–6 in Table 8.4) with nonidentical dies inserted on each tier. Additionally, we consider three cases with identical tiers (cases 7–9) as listed in Table 8.4.

For each case, report voltage drop objective, O_{Vdrop}, and thermal objective, O_T, before and after ALG1 was applied while α = 0.5. First column list the various cases. Power density distributions for each tier are provided in the second column. Before and after optimization voltage drop and thermal objectives are listed in columns 3–6. Area savings are listed in columns seven followed by runtime in column eight.

Some cases have O_T objective of order 0.9 as simultaneously O_{Vdrop} objective reached max 1. Further optimization of thermal, O_T, would lead to increase of voltage drop, O_{Vdrop}, hence optimization stops and the values of O_T and O_{Vdrop} are reported. When nonidentical dies were used, we note that case 3 has the least area savings and this is due to high voltage drop and thermal objectives before optimization. Case 2 is also similar to case 3. In case 2, high-density tier (h)

TABLE 8.4

Area Savings for Different Power Density Distribution

	T3–T2–T1	O_{Vdroop} Before	O_{Vdroop} After	O_T Before	O_T After	Area Savings (%)	Runtime (s)
1	h–m–l	0.87	1	0.46	0.56	25	1440
2	h–l–m	0.91	1	0.5	0.56	20	2160
3	m–h–l	0.91	1	0.56	0.61	18.3	1550
4	m–l–h	0.84	1	0.6	0.72	26.7	1550
5	l–m–h	0.79	1	0.63	0.82	33.3	1780
6	l–h–m	0.87	1	0.58	0.70	26.7	1200
7	h–h–h	0.98	1	0.66	0.70	8	900
8	m–m–m	0.88	1	0.45	0.90	36	780
9	l–l–l	0.86	1	0.34	0.82	41	1140

is at T3 farthest away from package (high-voltage drop) while it is located next to the heat sink (immediate cooling), thus more area saving is obtained in case 2 than 3. For cases of nonidentical dies, the most area savings is obtained in case 5 due to small O_{Vdroop} before optimization. Case 5 appears similar to case 6. However, the location of high-density tier in case 5 cause it to have less voltage drop (as is next to package) while in case 6 it is located in the middle tier which results in more voltage drop but at lower temperatures (closer to heat sink), thus more area saving is obtained in case 5.

When identical dies were used, we note that case 9 with low power density distributions have the least voltage drop and thermal objectives, thus the most area savings up to 41% is obtained. In contrary, identical dies with high power density distributions, case 7 lead to the least area savings, up to 8% from uniform upsizing. Such results indicate the sensitivity of 3D PDNs to identical dies with high power density distributions, which might limit the benefits of 3D integration. Furthermore, it highlights the importance of workload distribution among tiers while satisfying power and thermal constraints. We also report the run times for each case. We observe that the initial quality of the design impacts the run times. Additionally, the number of iterations for some cases can vary due to the sensitivity of voltage drop and temperature on track width distribution.

8.6.3 Impact of Design Constraints

8.6.3.1 Package as Secondary Heat Path

Package can have an important impact on temperature distribution across tiers in 3D-ICs. Generated heat from devices is conducted from one silicon die to another, thermal interface material, heat spreader to the heat sink then dissipated into air. This is the primary heat flow. There is also a secondary path as heat can flow from devices, interconnects, TSVs, I/O pads, package bumps reaching to the board. In this work, we perform experiment to demonstrate the impact of package as the second heat path on a three-tier network with identical power density distributions. As mentioned in Section 8.3, package thermal model is represented by thermal resistances (lateral and vertical) based on its thermal conductivity.

Figure 8.10 shows the temperature distribution on each tier for both power and ground delivery networks with no package model. We note that temperatures can vary considerably inter and intra-tier. Tier three (T3) is next to heat sink and experiences temperatures from 28.5°C to 30°C on power network or 32°C on ground network. Whereas, tier one (T1) is next to package and undergoes

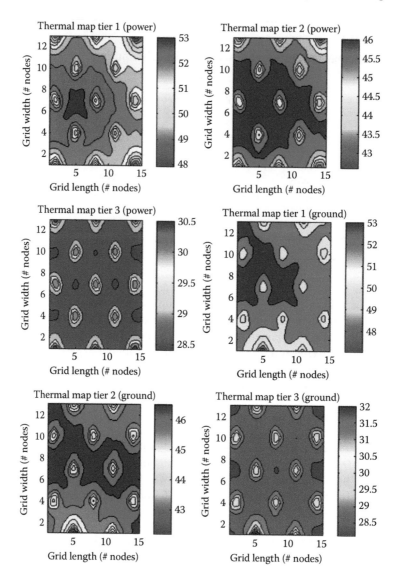

FIGURE 8.10 Thermal maps on power and ground networks for each tier when package model not included.

temperatures from 48°C to 53°C on both power and ground networks. Tier two being in the middle of the stack experiences temperatures from 43°C to 46°C. We deduce that TSVs between T1 and T2 are faced with high temperature difference (up to 13°C) between tiers. TSVs from T2 and T3 experience temperature difference up to 2°C between tiers. Such temperature distribution indicates the criticality of thermal impacts on TSVs and tiers further away from heat sink. Elevated temperatures can lead to performance, reliability, and power consumption issues.

Figure 8.11 shows the temperature distribution on each tier with the package model. Overall, the temperature levels are reduced in comparison to Figure 8.10. Temperatures reach up to 31°C for T1, 43.5°C for T2, and 48°C for T3. A temperature difference up to 5°C for T3 is obtained or 10% of temperature reduction. Similarly, TSVs undergo slightly lower temperatures such as 10°C between T1 and T2 and 1°C between T2 and T3. This experiment highlights the importance of secondary heat path through package as it helps to remove some of the generated heat. Also, neglecting the package effects can lead to inaccurate temperature estimates.

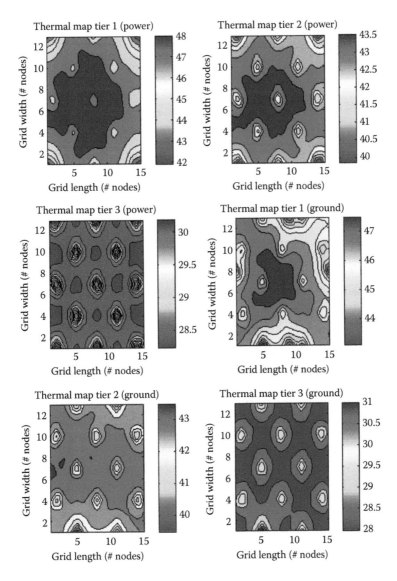

FIGURE 8.11 Thermal maps on power and ground networks for each tier when package model is included.

8.6.3.2 Wafer Thinning

Different manufacturing approaches for 3D-ICs are categorized by the TSV size (diameter, pitch, and aspect ratio) and wafer thickness. Wafer thickness can vary depending on the wafer-to-wafer or die-to-wafer stacking schemes. TSVs with diameter between 2 and 10 μm require one die to be thinned between 25 and 70 μm, which makes wafer and die handling very challenging. Moreover, wafer thinning can pose thermal dissipation problems and introducing additional complexity to 3D thermal management problem. Wafer thinning is modeled by the electrical and thermal resistance and capacitance of each tier derived based on the thickness and area of the die as in [25,28].

To examine the impact of wafer thinning on the power and thermal integrity of 3D PDNs, we study two types of wafer thickness, 50 and 775 μm. We utilize the three-tier network where identical power density distributions were applied to each tier. Initially, we study 3D PDNs with equal thickness wafer of 775 μm. Then, we study wafer thinning up to 50 μm when applied to top two tiers. We investigate temperature distribution on each tier and compare the impact of wafer thinning on 3D PDN area.

TABLE 8.5

Area Savings for Different Wafer Thicknesses

Wafer Thinning	Electrothermal Analysis		3D PDN Area		
	O_{Vdroop}	O_T	UR	ALG1	ALG2
Two top wafers thinned at 50 μm	0.82	0.62	1	0.31	0.28
Two top wafers thinned at 50 μm	0.81	0.54	1	0.25	0.22

Table 8.5 summarizes our findings. Columns two and three show the voltage drop and thermal objective for the two types of wafer thickness, respectively. We observe that wafer thickness has a considerable impact on thermal objective with $O_T = 0.54$ for thickened wafers of 775 μm to $O_T = 0.62$ for thinned wafers of 50 μm. This indicates that thinned wafers experience higher temperature levels. Whereas, there is little impact on voltage drop objective due to wafer thinning.

We investigate the impact of wafer thinning on 3D PDN area by applying greedy resizing (uniformly upsize all metal tracks by w_{max}), and our proposed electro-thermal optimization algorithms, ALG1 and ALG2, where results are shown in columns 4–6 of Table 8.5. The area obtained from ALG1 and ALG2 are shown in proportional (ratio) to the area obtained from greedy resizing. For wafer of 775 μm thickness, we obtain 0.25 for ALG1 or 75% of area savings from GR and 0.22 for ALG2 or 78% of area savings from GR. As ALG2 considers both metal track and TSV sizing provides more flexibility on saving 3D PDN area. In the case of thinned wafers, we obtain 0.31 for ALG1 (or 69% of area savings from GR) and 0.28 for ALG2 (or 72% of area savings from GR). These results demonstrate that less 3D PDN area savings are obtained for thinned wafer or larger power grids are required to satisfy power and thermal integrity.

Overall, we observe that wafer thinning increases temperature levels and poses additional thermal challenge to 3D-ICs. This is because the thermal capacity of the die to absorb heat is greatly diminished with thinning which as consequence reduces the capability of the die to spread some of the heat. Consequently, it impacts the 3D PDNs and their area for satisfying power and thermal constraints.

In summary, in this work, we perform power and thermal integrity driven 3D PDN optimization. Our approach performs tier-level optimization while applying multitier level constraints. Electrical thermal models are represented as *RLC* network representing power grid, package, switching circuits, and TSVs parasitics. Coupling between TSVs is ignored in this work. Thermal models are based on *RC* networks representing package, heat sink, switching circuits, power grid, die thickness, and TSV parasitics. We exploit electro-thermal duality to device efficient and compact models for computing voltage droop and temperature in a complex multitier power network. As future work, we aim to investigate signal routing and placement issues on 3D-ICS and their impact on 3D PDNs power and thermal integrity. The two proposed approaches based on globally constrained locally optimized algorithm resize each tier individually while applying multitier constraints.

8.7 CONCLUSIONS

In this paper, we presented a voltage drop and thermal constraints driven 3D power network delivery optimization. We have shown that already optimized 2D power networks cannot be directly applicable as 3D PDNs and moreover, circuits might require upsizing to meet timing constraints due to nonuniform voltage drop and temperature in a multitier system. We proposed a novel tier-based resizing technique that individually optimizes each tier while enforcing multitier constraints. Compact and efficient electrical and thermal models were applied for performing electrothermal analysis. Furthermore, we presented two optimization algorithms for resizing 3D PDNs stand-alone or together with TSVs for satisfying power and thermal constraints.

REFERENCES

1. Intel Corp. Santa Clara, CA, USA [Online]. Available: http://www.interl.com.
2. C. Bermond, L. Cadix, A. Farcy, T. Lacrevaz, P. Leduc, and B. Flechet. High frequency characterization and modeling of high density TSV in 3D integrated circuits. In *IEEE Workshop on Signal Propagation on Interconnects,* Strasbourg, France, *2009. SPI'09,* pp. 1–4, May 2009.
3. R.L. Boylestad. *Introduction to Circuit Analysis.* 2000.
4. H.H. Chen and D.D. Ling. Power supply noise analysis methodology for deep-submicron VLSI chip design. In *Proceedings of the 34th Design Automation Conference,* Anaheim, CA, *1997,* pp. 638–643, June 1997.
5. T.-Y. Chiang, K. Banerjee, and K.C. Saraswat. Effect of via separation and low-k dielectric materials on the thermal characteristics of cu interconnects. In *Electron Devices Meeting, 2000. IEDM '00. Technical Digest. International,* San Francisco, CA, pp. 261–264, December 2000.
6. T.-Y. Chiang, S.J. Souri, C.O. Chui, and K.C. Saraswat. Thermal analysis of heterogeneous 3D ICS with various integration scenarios. In *Electron Devices Meeting, 2001. IEDM 01. Technical Digest. International,* San Francisco, CA, pp. 31.2.1–31.2.4, December 2001.
7. P. Falkenstern, Y. Xie, Y.-W. Chang, and Y. Wang. Three-dimensional integrated circuits (3D IC) floorplan and power/ground network co-synthesis. In *15th Asia and South Pacific Design Automation Conference (ASP-DAC),* Taipei, Taiwan, *2010,* pp. 169–174, January 2010.
8. M.B. Healy and S.K. Lim. A novel TSV topology for many-tier 3D power-delivery networks. In *Design, Automation Test in Europe Conference Exhibition (DATE), 2011,* Grenoble, France, pp. 1–4, March 2011.
9. G. Huang, M. Bakir, A. Naeemi, H. Chen, and J.D. Meindl. Power delivery for 3D chip stacks: Physical modeling and design implication. In *Electrical Performance of Electronic Packaging, 2007 IEEE,* Atlanta, GA, pp. 205–208, October 2007.
10. ITRS. The International Technology Roadmap for Semiconductors (ITRS), Interconnects, 2011, http://www.itrs.net/, 2011. Accessed December 2014.
11. A. Jain, R.E. Jones, R. Chatterjee, and S. Pozder. Analytical and numerical modeling of the thermal performance of three-dimensional integrated circuits. *IEEE Transactions on Components and Packaging Technologies,* 33(1):56–63, March 2010.
12. A. Jain, R.E. Jones, R. Chatterjee, S. Pozder, and Z. Huang. Thermal modeling and design of 3d integrated circuits. In *11th Intersociety Conference on Thermal and Thermomechanical Phenomena in Electronic Systems, 2008. ITHERM 2008,* Orlando, FL, pp. 1139–1145, May 2008.
13. G. Katti, M. Stucchi, K. de Meyer, and W. Dehaene. Electrical modeling and characterization of through silicon via for three-dimensional ICS. *IEEE Transactions on Electron Devices,* 57(1):256–262, January 2010.
14. G. Katti, M. Stucchi, D. Velenis, B. Soree, K. de Meyer, and W. Dehaene. Temperature-dependent modeling and characterization of through-silicon via capacitance. *Electron Device Letters, IEEE,* 32(4):563–565, April 2011.
15. N.H. Khan, S.M. Alam, and S. Hassoun. Power delivery design for 3-D ICS using different through-silicon via (TSV) technologies. *IEEE Transactions on Very Large Scale Integration (VLSI) Systems,* 19(4):647–658, April 2011.
16. M.B. Kleiner, S.A. Kuhn, P. Ramm, and W. Weber. Thermal analysis of vertically integrated circuits. In *Electron Devices Meeting, 1995. IEDM '95, International,* San Francisco, CA, pp. 487–490, December 1995.
17. J.U. Knickerbocker, P.S. Andry, B. Dang, R.R. Horton, M.J. Interrante, C.S. Patel, R.J. Polastre et al. Three-dimensional silicon integration. *IBM Journal of Research and Development,* 52(6):553–569, November 2008.
18. J.N. Kozhaya, S.R. Nassif, and F.N. Najm. A multigrid-like technique for power grid analysis. *IEEE Transactions on Computer-Aided Design of Integrated Circuits and Systems,* 21(10):1148–1160, October 2002.
19. G.B. Kromann. Thermal modeling and experimental characterization of the c4/surface-mount-array interconnect technologies. *IEEE Transactions on Components, Packaging, and Manufacturing Technology, Part A,* 18(1):87–93, March 1995.

20. Y.-J. Lee and S.K. Lim. Co-optimization and analysis of signal, power, and thermal interconnects in 3-D ICS. *IEEE Transactions on Computer-Aided Design of Integrated Circuits and Systems*, 30(11):1635–1648, November 2011.

21. Z. Luo, J. Fan, and S.X. Tan. Localized statistical 3D thermal analysis considering electro-thermal coupling. In *IEEE International Symposium on Circuits and Systems, 2009. ISCAS 2009*, Taipei, Taiwan, pp. 1289–1292, May 2009.

22. T. Mitsuhashi and E.S. Kuh. Power and ground network topology optimization for cell based VLSIs. In *Proceedings of the 29th ACM/IEEE Design Automation Conference*, Anaheim, CA, *1992*, pp. 524–529, June 1992.

23. H. Oprins, V. Cherman, M. Stucchi, B. Vandevelde, G. Van der Plas, P. Marchal, and E. Beyne. Steady state and transient thermal analysis of hot spots in 3D stacked ICS using dedicated test chips. In *27th Annual IEEE Semiconductor Thermal Measurement and Management Symposium (SEMI-THERM)*, San Jose, CA, *2011*, pp. 131–137, March 2011.

24. J.S. Pak, J. Kim, J. Cho, K. Kim, T. Song, S.-U. Ahn, J. Lee, H. Lee, K. Park, and J. Kim. PDN impedance modeling and analysis of 3D TSV IC by using proposed p/g TSV array model based on separated p/g TSV and chip-PDN models. *IEEE Transactions on Components, Packaging and Manufacturing Technology*, 1(2):208–219, February 2011.

25. M. Pedram and S. Nazarian. Thermal modeling, analysis, and management in VLSI circuits: Principles and methods. *Proceedings of the IEEE*, 94(8):1487–1501, August 2006.

26. M. Susa, R. Kojima Endo, and Y. Fujihara. Calculation of the density and heat capacity of silicon by molecular dynamics simulation, November 2006.

27. J. Singh and S.S. Sapatnekar. Congestion-aware topology optimization of structured power/ground networks. *IEEE Transactions on Computer-Aided Design of Integrated Circuits and Systems*, 24(5):683–695, May 2005.

28. K. Skadron, T. Abdelzaher, and M.R. Stan. Control-theoretic techniques and thermal-RC modeling for accurate and localized dynamic thermal management. In *Proceedings of the Eighth International Symposium on High-Performance Computer Architecture*, Cambridge, MA, *2002*, pp. 17–28, February 2002.

29. A. Todri, S. Kundu, P. Girard, A. Bosio, L. Dilillo, and A. Virazel. A study of tapered 3-D TSVs for power and thermal integrity. *IEEE Transactions on Very Large Scale Integration (VLSI) Systems*, San Jose, CA, 21(2):306–319, February 2013.

30. X.P. Wang, W.S. Zhao, and W.-Y. Yin. Electrothermal effects in high density through silicon via (TSV) arrays. *Progress in Electromagnetics Research*, 115:223–242, 2011.

31. K. Wang and M. Marek-Sadowska. On-chip power-supply network optimization using multigrid-based technique. *IEEE Transactions on Computer-Aided Design of Integrated Circuits and Systems*, San Jose, CA, 24(3):407–417, March 2005.

32. T.-Y. Wang and C.C.-P. Chen. Spice-compatible thermal simulation with lumped circuit modeling for thermal reliability analysis based on modeling order reduction. In *Proceedings of the Fifth International Symposium on Quality Electronic Design*, San Jose, CA, *2004*, pp. 357–362, 2004.

33. S.L. Wright, R. Polastre, H. Gan, L.P. Buchwalter, R. Horton, P.S. Andry, E. Sprogis et al. Characterization of micro-bump c4 interconnects for Si-carrier sop applications. In *Proceedings of the 56th Electronic Components and Technology Conference*, San Diego, CA, *2006*, 8pp., 2006.

34. J. Xie, D. Chung, M. Swaminathan, M. Mcallister, A. Deutsch, L. Jiang, and B.J. Rubin. Electrical-thermal co-analysis for power delivery networks in 3D system integration. In *IEEE International Conference on 3D System Integration, 2009. 3DIC 2009*, San Francisco, CA, pp. 1–4, September 2009.

35. H. Yu, J. Ho, and L. He. Simultaneous power and thermal integrity driven via stapling in 3D ICS. In *IEEE/ACM International Conference on Computer-Aided Design, 2006. ICCAD '06*, Dresden, Germany, pp. 802–808, November 2006.

36. C. Zhang, V.F. Pavlidis, and G. De Micheli. Voltage propagation method for 3-D power grid analysis. In *Design, Automation Test in Europe Conference Exhibition (DATE), 2012*, pp. 844–847, March 2012.

37. Y. Zhong and M.D.F. Wong. Thermal-aware IR drop analysis in large power grid. In *Ninth International Symposium on Quality Electronic Design, 2008. ISQED 2008*, San Jose, CA, pp. 194–199, March 2008.

38. P. Zhou, K. Sridharan, and S.S. Sapatnekar. Congestion-aware power grid optimization for 3D circuits using MIM and CMOS decoupling capacitors. In *Design Automation Conference, 2009. ASP-DAC 2009. Asia and South Pacific*, Yokohama, Japan, pp. 179–184, January 2009.

Section III

Reliability Concerns for 3D Integration

9 Live Free or Die Hard

Design for Reliability in 3D Integrated Circuits

Yu-Guang Chen, Yiyu Shi, and Shih-Chieh Chang

CONTENTS

Listen to the technology; find out what it's telling you.

Carver Mead (1981)

ABSTRACT

In this chapter, we discuss the reliability issues about 3D-ICs fabrication process under technologies nowadays, and propose possible solutions to address the problems. In Section 9.1, we lay down the reasons in detail as to why there are seldom 3D-IC products today versus the problem of low yield of TSVs. Section 9.2 shows two major fault-tolerant structures, double TSV and shared-spare TSV, to improve the chip yield. The fault-tolerant design constraints then are addressed in Section 9.3. In Section 9.4, we address the problem of spare TSV assignment, and proposed two heuristics to optimize area overhead under given constraints. A fault-tolerance clock network then is discussed in Section 9.5 with two different structures. Concluding remarks are given in Section 9.6.

9.1   RELIABILITY ISSUES IN 3D-IC DESIGNS

While the semiconductor industry is making every effort to make chips smaller and faster, further scaling of the current 45 nm technology has become prohibitively expensive. Accordingly, there has been a groundswell of interest in technologies that offer a path beyond the limits of device scaling. Among all the possible alternatives, the three-dimensional integrated circuit (3D-IC) is generally considered to be the most promising one, at least in the next decade, for its compatibility with the current technology. Instead of making transistors smaller, 3D-IC makes use of the vertical dimension, and stacks several tiers of circuits together to reduce chip area. It has been anticipated that by 2020 the worldwide market volume of 3D-ICs will exceed 50 billion [1].

A critical enabling technique for 3D-ICs is the through-silicon via (TSV), which can be classified into three types. Signal TSVs deliver signals from one tier to another; power/ground (P/G) TSVs connect the power delivery network (PDN) in each tier; finally, TSVs are the major vertical thermal channels. When signal and P/G TSVs are not enough, dummy TSVs (thermal TSVs) are inserted around hotspots to further reduce temperature [2–7]. In addition, a large number of TSVs are needed to deliver signal and power, to dissipate heat, and to provide redundancy. Moreover, foundries impose a minimum TSV density rule to maintain the planarity of the wafer during chemical and mechanical polishing (CMP). For example, Tezzaron requires at least one TSV in every 250 μm × 250 μm area [8]. Similar rules are imposed by Intel, TSMC, and other 3D foundries. To utilize these idle TSVs, Zhuo et al. [9–12] have explored the opportunistic utilization of them as 3D inductors in important circuit constructs such as resonant clocking and DC–DC converters.

However, despite their importance, TSVs have remained a bottleneck for yield boost of 3D-ICs. During fabrication, they may suffer from various reliability issues such as undercut, misalignment, or random open defects [13]. Undercut happens because of inaccuracy control of timing and speed during deep-reactive ion etching, and could happen below the mask at upper and wide edge of the TSVs, which can lead to mask overhang and TSVs open. Figure 9.1a shows the side view of undercut. Misalignment comes from unsuccessful wafer alignment prior to and during wafer bonding process as shown in Figure 9.1b, and is caused by shifts of bonding pads with respect to their nominal positions. Random defects comprise a variety of unpredictable physical phenomena related to the thermal compression process used in wafer stacking.

In addition, TSVs are subject to severe thermomechanical reliability hazards due to the coefficients of thermal expansion (CTE) mismatch between TSV and silicon substrate. Under normal operations, the 3D-IC experiences heavy thermal cycles and thus induces large thermomechanical stress. Consequently, the stress may initiate microcracks from the interface between a TSV and its dielectric liner, and further propagates them on the silicon substrate surface. Several studies have shown the existence of radial cracks adjacent to TSVs [4,14,15] due to thermomechanical stress. For example, Figure 9.2a shows the scanning electron microscopy (SEM) image of a TSV [14], where a radial micro crack can be clearly observed by zooming in its up-left corner as shown in Figure 9.2b. The cracks may cause bridging faults or delamination [16].

Because of the large number of TSVs in a chip, these issues in turn translate into low-chip yields [17]: For example, Bogaerts et al. [5] and Lee et al. [13] reported a 70% chip yield for a chip with 10,000 TSVs and only 20% yield for 55,000 TSVs in IMEC process technology. Moreover, Figure 9.3 shows limited yield of TSVs from three different process technologies: HRI [18], IMEC [5], and IBM [19]. From the figure, we can find that the chip yield significantly decreases while the number of TSVs increases for all three technologies, which would be a primary concern in 3D-ICs design.

(a)

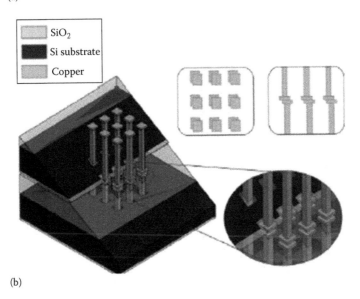

(b)

FIGURE 9.1 TSV failure due to (a) side view of the undercut (From Lung, C.L., Su, Y.S., Huang, S.H., Shi, Y., and Chang, S.C., Through-silicon via fault-tolerant clock networks for 3-D ICs, *IEEE Transactions on Computer-Aided Design of Integrated Circuits and Systems*, 32(7), 1100–1109, © 2013 IEEE.) and (b) worst-case misalignment scenario. (From Loi, I., Mitra, S., Lee, T.H., Fujita, S., and Benini, L., A low-overhead fault tolerance scheme for TSV-based 3-D network on chip links, in *Proceedings of IEEE/ACM International Conference on Computer-Aided Design (ICCAD)*, San Jose, CA, pp. 598–602, © 2008 IEEE.)

9.2 SPARE TSVs AND FAULT-TOLERANT STRUCTURES

Since TVS suffer from many possible defects during fabrication, many fault-tolerant mechanisms have been proposed in literature to enhance reliability. A simple but effective way is to add spare TSVs. Spare TSVs are redundant TSVs especially added to 3D-IC designs to fix failed TSVs should they occur, and therefore to increase chip yield and overall reliability with additional area and power overhead. In this section, we will briefly review some commonly used fault-tolerant structures of TSVs.

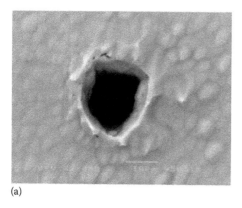
(a)

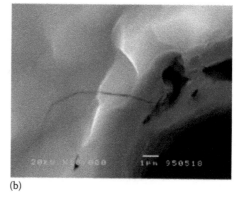
(b)

FIGURE 9.2 SEM image of (a) TSV and (b) zoom-in of the up-left corner of the TSV with radial crack. (From Chen, Y.-H., Lo, W.-C., and Kuo, T.-Y., Thermal effect characterization of laser-ablated silicon-through interconnect, in *Proceedings of Electronics System Integration Technology Conference*, Dresden, Germany, vol. 1, pp. 594–599, 2006.)

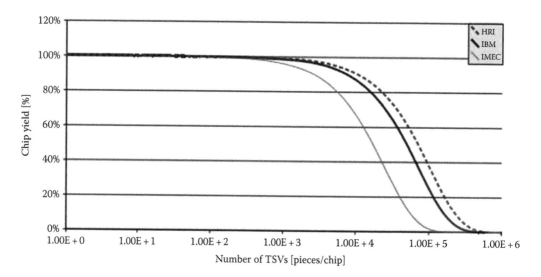

FIGURE 9.3 Yield trend for TSVs in three different processes: IBM, HRI, and IMEC. (From Lee, T.H. et al., *Proceedings of International Conference on Computer-Aided Design (ICCAD)*, 2008.)

9.2.1 DOUBLE TSV TECHNIQUE

The double TSV technique was first presented in a US patent [20]. The patent illustrated the systems and methods utilizing redundancy in semiconductor chip interconnects. To address the reliability issues, in the double TSV technique, each original TSV will be paired with an additional spare TSV, which is used to reroute the signal in the presence of TSV failure. Figure 9.4 illustrates the concept of the double TSV technique. In normal condition, the signal is transferred through the two TSVs simultaneously. Once a TSV is faulty, there is still a redundant TSV to pass the signal. By applying this technique, the only failure condition is that both TSVs in a pair are faulty, which has a much lower probability. Since the two TSVs are used to conduct the same signal, no additional circuits are required. While effective, this method uses up a lot of area and additional power consumption; given the large size of TSVs (10–125 μm^2) from industrial report [21–25], it is less practical.

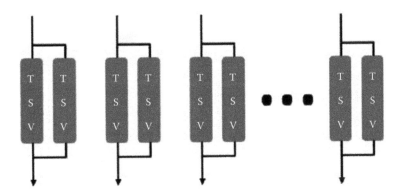

FIGURE 9.4 The concept of double TSV technique. (From Lung, C.L., Su, Y.S., Huang, S.H., Shi, Y., and Chang, S.C., Through-silicon via fault-tolerant clock networks for 3-D ICs, *IEEE Transactions on Computer-Aided Design of Integrated Circuits and Systems*, 32(7), 1100–1109, © 2013 IEEE.)

9.2.2 Shared-Spare TSV Technique

To lower the area cost induced by the redundant technique, another technique is proposed by grouping a set of original TSVs, and assigning spare TSVs to each group. By inserting multiplexers and arranging the reconfigurable routing properly, the spare TSVs can be used to deliver the signal in the presence of faulty TSVs if the number of faulty TSVs in a group is smaller than or equal to the number of spare TSVs. The recovery rate is decided by the ratio between the number of signal TSVs and the number of spare TSVs in each TSV group. The idea was first proposed in [26], where a 4:2 topology is used: four original TSVs and two spare TSVs are grouped together with appropriate MUXs and routing connection, as shown in Figure 9.5. If one TSV fails, the signal connecting to the faulty TSV can be rerouted through the spare TSV. Compared with double-TSV technique, the idea can significantly reduce area overhead caused by spare TSVs. However, a problem with this technique is that we have to cluster a few TSVs together physically, which introduces the potential long-distance rerouting and in turn translates into large delay overhead and even timing violations if not properly designed. It may also lead to routing difficulty, even unroutable.

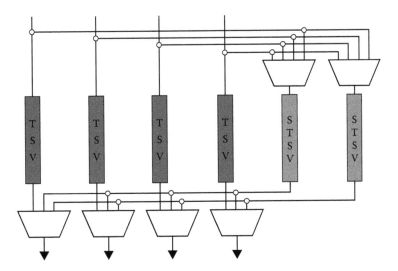

FIGURE 9.5 A fault-tolerant structure and the signal routing. (From Lung, C.L., Su, Y.S., Huang, S.H., Shi, Y., and Chang, S.C., Through-silicon via fault-tolerant clock networks for 3-D ICs, *IEEE Transactions on Computer-Aided Design of Integrated Circuits and Systems*, 32(7), 1100–1109, © 2013 IEEE.)

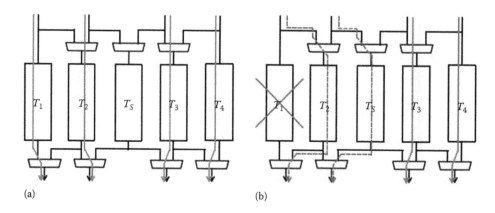

FIGURE 9.6 Another fault-tolerant structure and the signal routing when (a) all TSVs are good and (b) T_1 fails. (From Chen, Y.G., Wen, W.Y., Shi, Y., Hon, W.K., and Chang, S.C., Novel Spare TSV Deployment for 3D ICs Considering Yield and Timing Constraints, accepted in *IEEE Transactions on Computer-Aided Design of Integrated Circuits and Systems* (TCAD), © 2014 IEEE.)

To alleviate this issue, an alternative technique is proposed to simplify the complicated rerouting structure by inserting only one spare TSV in a group [2,13,27]. We use a simple example to explain how the concept works. Figure 9.6 shows one such structure with four original TSVs (denoted as T1, T2, T3, and T4) and one spare TSV (denoted as TS). Seven multiplexers serve as the control circuit. Figure 9.6a shows the original configuration where all TSVs are good and each signal passes through their corresponding TSV. When a TSV fails, the signal of the faulty TSV needs to be shifted. This in turn causes all signals between the faulty TSV and the spare TSV to be shifted. Figure 9.6b shows the operation when TSV T1 is faulty. In this approach, only one faulty TSV can be recovered in each group in this architecture, yet the simplified routing structure makes less possible for timing violations. However, when a signal is shifted, an unavoidable delay is still induced due to longer wires and extra components.

This concept was first presented in 2008. For both NOC-based and bus-based 3D interconnects, a simple and efficient defect-tolerant architecture is proposed in [13]. By taking advantage of the NOC switch architecture, the area overhead can be minimized. The physical level evaluation demonstrates the significant yield improvements and the lower cost.

Samsung realized this technique in an 8 GB 3D DDR3 DRAM [22]. Since there is no distinction between regular and redundant TSVs, in that design, they did not specify dedicated redundant TSVs. Different from conventional structures, which have dedicated and fixed-location of redundant TSVs, relaxing this constraint can decrease the detour path, reduce the routing complexity, and loading. In addition, they also evaluated two different configurations, 4:2 group (four signal TSVs with two redundant TSVs as shown in Figure 9.7) and 2:1 group (two signal TSVs with one redundant TSVs). The statistical analysis shows that, the yield rate of 4:2 group is higher than that of 2:1 group. The reason is that the chip has more repair flexibility and fewer TSV groups with the 4:2 configuration. In addition, each TSV will go through an open-short testing, and faulty TSVs will be repaired using e-fuses.

Fault-tolerant structures as spare TSVs, control logics, and rerouting wires incur significant area overhead [20,28,29]. However, inserting more spare TSVs would obtain higher yield rate. A trade-off between number of spare TSV inserted and additional overhead then is worth to be discussed. Recently a few efficient spare TSV allocation algorithms were proposed to reduce cost [28,29]. The early work in [28] tries to decide the optimal grouping ratio (number of spare TSVs/number of signal TSVs) in fault-tolerant structures to minimize the number of spare TSVs and related MUXs while achieving 100% yield. A more practical work is presented in [29], which tries to optimally cluster original TSVs and spare TSVs to minimize total wirelength overhead. To make these approaches become practical, it is important to solve the problem of timing violation.

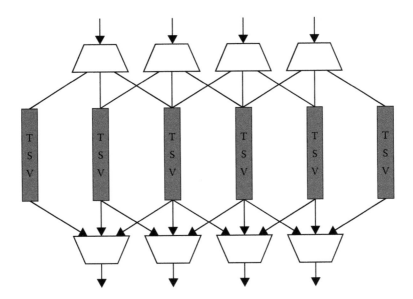

FIGURE 9.7 A 4:2 fault-tolerant structure without dedicated spare TSVs. (From Lung, C.L., Su, Y.S., Huang, S.H., Shi, Y., and Chang, S.C., Through-silicon via fault-tolerant clock networks for 3-D ICs, *IEEE Transactions on Computer-Aided Design of Integrated Circuits and Systems*, 32(7), 1100–1109, © 2013 IEEE.)

9.3 DESIGN CONSTRAINTS

To avoid timing violation to occur and to guarantee the efficiency of inserting spare TSVs, there are some design constraints when building such a fault-tolerant structure. It is suggested that there are two most important constraints: Yield and Timing [27,30]. First, the chip yield imposes a limit on the ratio of number of original TSVs and number of spare TSVs that can be in a group. Second, neighboring TSVs in a group should not be topologically far away to avoid timing violation when the signal is rerouted. For example, as shown in Figure 9.6b, the maximum distance between T1 and T2 depends on the slack of the signal that goes through T1. It also sets a lower bound on the distance between the TSVs in the same group due to the clustering effect of TSV faults (e.g., warpage) [28,31,32]. The clustering effect comes from the bonding quality of TSVs, which depends not only on the bonding technology but also the winding level of the thinned wafer and the surface roughness and cleanness of silicon dies. Consequently, if one TSV is defective during the bonding process, it is more likely that its neighboring TSVs are also faulty. Due to such clustering effect, earlier TSV repair techniques are less effective because a signal TSV and its neighboring redundant TSV may be defective at the same time [32]. These distance constraints prevent us from randomly grouping TSVs to form a TSV fault-tolerant structure.

Different literature may use different model to formulate these constraints. Here, we follow [27] to show a simple way to translate the constraints to a formal model.

9.3.1 Chip Yield Constraint

Assume the failure rate for a single TSV is F, and to simplify the discussion, we assume each TSV fails independently. It is important to note here due to clustering effect, the failure rate may not be independently. Since the clustering effect strongly depends on the location of TSVs, the accuracy model would take the location of TSVs as input and significantly increase the complexity of the model. Moreover, it is almost impossible to decided locations of TSVs before obtaining the number needed. Here we use a simple example that considers the fault-tolerant structure with N original TSVs and only one spare TSV. The structure will be good if all $N + 1$ TSVs are good, or if only one of them fails. The former has a probability of $(1 - F)^{N+1}$, while the latter has the probability of $(N + 1) \times (1 - F)^{N+1}$, so the yield of the structure can be calculated as

$$Y_{structure} = (1-F)^{N+1} + (N+1) \times (1-F)^N \times F = (1-F)^N \times (1+NF) \tag{9.1}$$

It is clear that a chip fails if any of the fault-tolerant structures fails. Therefore, the yield of the entire chip, Y_{chip}, can be calculated as

$$Y_{chip} = (Y_{structure})^{\# structure}, \tag{9.2}$$

where #*structure* denotes the number of fault-tolerant structures in a chip.

Given the yield constraint of a chip, $Y_{chip\_constraint}$, we can use Equations 9.1 and 9.2 to estimate N, the number of original TSVs in a fault-tolerant structure as

$$[(1-F)^N \times (1+NF)]^{M/N} \geq Y_{chip\_constraint}, \tag{9.3}$$

where M denotes the total number of original TSVs in the chip.

For example, if the failure rate of a TSV is 1%, given a chip with 500 original TSVs and the chip yield constraint is 90%, using Equation 9.3

$$[(1-0.01)^N \times (1+N \times 0.01)]^{500/N} \geq 0.9, \tag{9.4}$$

we can obtain $N \leq 3$. In other words, to achieve the yield of 90%, we may group no more than three original TSVs in a structure.

Note that the actual chip yield we obtain may be higher, because here we assume all fault-tolerant structures precisely contain the same number of original TSVs, N. However, the actual number of original TSVs in each fault-tolerant structure may be less than N due to other constraints (say, timing constraint which will be discussed later), which leads to higher real chip yield.

Again, due to traditional semiconductor manufacturing, TSV faults may have clustering effect rather than random distribution (e.g., warpage) [28,31,32]. Since we focus on the fault-tolerant structure with only one spare TSV, it is important to avoid grouping very close TSVs in the same structure [29,32]. Therefore, we impose a minimum distance constraint, D_{min}, to any two TSVs that can be in the same fault-tolerant structure, as suggested by [29,33].

9.3.2 Timing Constraint

As for the timing constraint, given the slack of the signal passing through a TSV, the maximum rerouting distance of the TSV i, $D_{max,i}$, is defined as the maximum distance between this TSV and its neighbor (the TSV used to reroute its signal when it fails). The maximum rerouting distance can be easily calculated, either through simple RC delay model, or through detailed SPICE simulation. The delay induced by control logics such as MUX should also be included when calculating maximum rerouting distance. Note that such maximum rerouting distance can be different for different TSVs, as shown in Figure 9.8.

The Built-In Self-Test (BIST) design can be helpful to detect the fault of TSVs and reroute the signal. Several BIST methods have been proposed in the literature such as [34–38] for detecting full-open, micro-voids, or pin-hole defects. TSVs are used as capacitive loads and the deviation of their expected RC parameters are detected by indirect measures, namely the delays required for charging or discharging the nets connected to their front ends—the only ends accessible before bonding [34].

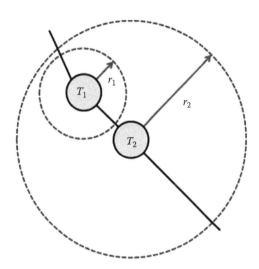

FIGURE 9.8 Asymmetry of maximum rerouting distance.

Note that one potential issue here is the asymmetry of the maximum rerouting distance. In Figure 9.8, T1 is within the maximum distance of T2, but not vice versa. Accordingly, T1 can be used to reroute the signal for T2, but T2 cannot be used for T1.

It is also worthwhile to point out here that this $D_{\max,i}$ also implicitly bounds the wirelength overhead induced by the fault-tolerant structures, as it specifies that the maximum wirelength overhead brought by each structure cannot exceed $\sum_{i=1}^{N} D_{\max,i}$.

9.3.3 TSV Placement

Since the distance could be a constraint for grouping TSVs to form a fault-tolerant structure, how original TSVs are placed may significantly influence the grouping results. Generally, there are two kinds of placement methods, regular or irregular structure in a tier.

As described in [2], due to manufacturing and physical design issues, TSVs are not recommended to be placed arbitrarily on a plane. From the aspect of manufacturing, a regular placement of TSVs improves the exposure quality of the lithographic process and therefore improves the yield. In real designs, TSVs are suggested to be placed regularly in TSV blocks which are determined in floor-plan stage. Inside each TSV block, TSVs are arranged in a grid-based structure to satisfy the pitch constraint of bond pads. Examples of TSV blocks are shown in Figure 9.9 [2,39].

On the other hand, some literature such as [27,29,40–43] suggest that TSVs should be placed irregularly instead of following an array structure in a TSV farm. The good points of this kind of placement can allow more flexibility when forming fault-tolerant group and show better wire-length [41,43].

We use Figure 9.10 as an example to demonstrate why for cases with irregularly placed TSVs, the appropriate assignment of spare TSVs can significantly reduce the number of fault-tolerant structures and therefore, the overall design cost. Figure 9.10a shows a tier in 3D-IC with eight original TSVs. A hollow circle stands for an original TSV and a solid circle for a spare TSV. For yield constraint, we assume that each fault-tolerant structure can have no more than four original TSVs. For timing constraint, a maximum distance is set between two neighboring TSVs. Figure 9.10b and c shows two different grouping results for inserting spare TSVs. Figure 9.10b uses a simple heuristic: Original TSVs are directly grouped based on distance, and then a spare TSV is inserted to each group. However, in Figure 9.10c, by properly inserting spare TSVs, the original TSVs that would

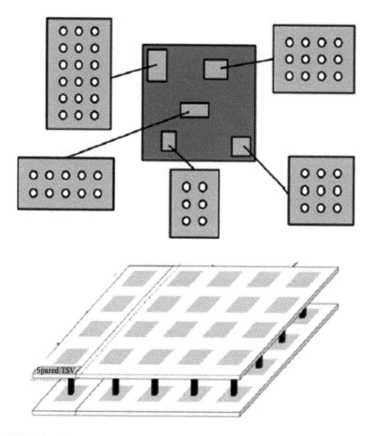

FIGURE 9.9 TSV blocks. (From Hsieh, A.C. et al. TSV redundancy: Architecture and design issues in 3-D IC, in *Proceedings of Design, Automation and Test in Europe (DATE)*, Dresden, Germany, pp. 1206–1211, 2010; Huang, H.-Y., Huang, Y.-S., and Hsu, C.-L., Built-in self-test/repair scheme for TSV-based three-dimensional integrated circuits, in *Proceedings of IEEE Asia Pacific Conference on Circuits and Systems (APCCAS)*, Kuala Lumpur, Malaysia, pp. 56–59, © 2010 IEEE.)

otherwise be in different groups due to the distance constraint can now be joined. Accordingly, a better solution is achieved: Only three spare TSVs are needed in Figure 9.10c instead of five used in Figure 9.10b, a 40% reduction.

As discussed earlier, it looks like a great benefit can be obtained with irregular placement. Therefore, in the following discussion we will focus on optimizing 3D designs with irregular TSV placement. Two important things should be noted here: First, actually it is impossible to place TSVs arbitrary due to white space constraint. In reality, the estimated location for TSVs may overlap with logic cells, IP blocks, or macro cells, and make the placement become illegality. This can be addressed by performing an extra step of local legalization, similar to the method used in [44].

The other problem is the routing complexity. Due to the routing congestion, we may face routing difficulty when actually forming a fault-tolerant structure (i.e., original TSVs and a spare TSV in a fault-tolerant structure cannot be connected as expected due to the routing congestion). One of the possible solutions is removing the formed structure and use double TSV technique [20] for each TSV in that structure. This step may increase the number of spare TSVs needed, however, it can guarantee the routability.

In the remainder of this chapter, we will explore two problems on 3D-IC reliability enhancement through fault-tolerant structures. The first one is fault-tolerance for signal tsvs; the second one is fault-tolerance for clock network.

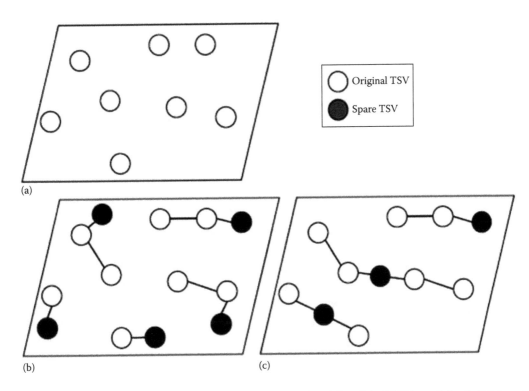

FIGURE 9.10 An example for spare TSVs assignment for (a) eight original TSVs, and the fault-tolerant structures formed with (b) five spare TSVs and (c) three spare TSVs.

9.4 FAULT-TOLERANCE FOR SIGNAL TSV

In this section, we address the problem of optimally inserting spare TSVs and forming fault-tolerant structures for signal TSVs. Section 9.4.1 describes the assumptions and problem formulation. The proposed algorithm is shown in Section 9.4.2. Experimental results are then proposed in Section 9.4.3.

9.4.1 Assumptions and Problem Formulation

Before we present our problem formulation, it is important to clarify the following five assumptions we use. First, the fault-tolerant structure could be formed during the standard cell placement stage when all the original TSVs are also placed. However, the stringent yield and timing constraints may result in significant distortion to the routing of signal nets. As such, in this paper, we assume that the original TSVs are placed without consideration of fault-tolerance. Second, we assume that in each fault-tolerant structure, only one spare TSV is inserted. This significantly simplifies the related control logic.

As such, our problem can be stated as follows: Given (1) a chip with all original TSVs are already placed, (2) the target yield of a chip, (3) the timing slacks of the signals that go through TSVs, determine an optimal spare TSV assignment such that total area overhead induced by the fault-tolerant mechanisms is minimized.

The area overhead is induced by the spare TSVs inserted, the rerouting wires, and the related control logics. To consider all of them during the optimization is quite complicated. Literature has pointed out that the size of TSVs is typically much larger than that of standard cells (about 5×–10× in

45 nm technology) [41]. Therefore, we simply reduce the objective from minimizing the total area overhead to minimizing the number of spare TSVs based on the second assumption described earlier. The reason is as follows. The control logic is formed by MUXs as shown in Figure 9.6, and the number of MUXs needed in a fault-tolerant structure with n original TSVs is $2n - 1$. Accordingly, assuming we cluster a total of N original TSVs into k fault-tolerant structures, then the total number of MUXs needed is $\sum_{i=1}^{k} (2n_i - 1)$, where n_i is the number of original TSVs in the ith structure. Considering $\sum_{i=1}^{k} (n_i = N)$, the total number of MUXs can then be expressed as $2N - k$. As such, the total area overhead is $(2N - k) \times Area_{MUX} + k \times Area_{TSV}$. Since $Area_{TSV} > Area_{MUX}$, we can express it as $k \times (Area_{TSV} - Area_{MUX}) + 2N \times Area_{MUX}$. It is now clear that minimizing the total area overhead is equivalent to minimizing k, which is the number of fault-tolerant structures and also the number of spare TSVs. Experimental results show that the latter objective can indeed significantly reduce the total area overhead.

Based on the earlier discussion, to form a TSV fault-tolerant structure while considering both yield and timing constraints, the maximum number of original TSVs in the group and distance constraints, $D_{max,i}$ and D_{min}, between neighboring TSVs must be held. As such, our problem can be stated as follows:

Formulation 9.1 (Yield and Timing Constrained Spare TSV Assignment): Given (1) the locations of all original TSVs, (x_i, y_i) on a tier from placement result, (2) the maximum number of original TSVs in a group, N, (3) the routing distance constraint, $D_{max,i}$, for each TSV, and (4) the minimum distance constraint, D_{min}, between neighboring TSVs, determine an optimal spare TSV assignment such that the total number of spare TSVs m is minimized.

9.4.2 Algorithm

It is interesting that Formulation 9.1 has a close relationship with graph theory, which shed light on the possible efficient solutions. In the following, we first describe the graph perspective, and then propose two efficient heuristic to address the graph problem.

9.4.2.1 Graph Perspective

To illustrate this point, we first define an assignment graph as follows:
Definition 9.1: An assignment graph is a directed graph constructed as follows:

1. Each vertex, s_i, represents an original TSV T_i.
2. There are two types of edges in the graph: Primary edge and second edge defined as follows.
3. There is a directed primary edge from vertices s_i to s_j if their distance is greater than D_{min} and smaller than $D_{max,j}$, that is, $D_{min} \le d_{ij} \le D_{max,j}$, otherwise
4. There is a directed secondary edge from vertices s_i to s_j if their distance is within the sum of maximum rerouting distance of each other, that is, $d_{ij} \le D_{max,i} + D_{max,j}$.

A primary edge from s_i to s_j implies that s_i can be used to reroute s_j while T_j is faulty. A secondary edge means that s_i can be used to reroute s_j only when a spare TSV is in between (e.g., the group in the lower left corner of Figure 9.10c). As an example, Figure 9.11 shows the assignment graph constructed from Figure 9.10a where primary edges are solid and secondary edges are dashed.

With the assignment graph, our problem can be readily cast as follows.
Formulation 9.2 (Constrained Decomposition of Assignment Graph): Given an assignment graph $G_{assignment}$ and an integer N, decompose the graph into a minimum number of disjointed sub-graphs, such that

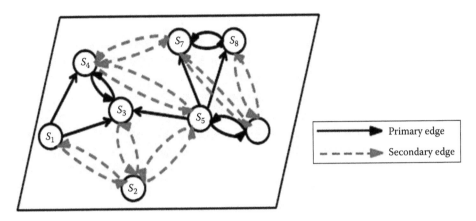

FIGURE 9.11   An example of assignment graph constructed from Figure 9.10a.

1. Each sub-graph contains no more than N vertices.
2. All the vertices in each sub-graph can be connected by a spanning tree.
3. The spanning tree contains no more than one secondary edge.

Each resulting sub-graph will be used to form a fault-tolerant structure by inserting a spare TSV. The first constraint limits the number of original TSVs in a structure to N. The second constraint guarantees that all the TSVs in the sub-graph can be connected through proper sequence. The last constraint ensures that only one spare TSV needs to be added.

While the formulation is quite self-explanatory, two things need to be noted here. First, we only require that all vertices in a sub-graph can be "connected by a spanning tree" instead of "connected in series." This can allow more flexibility and thus fewer spare TSVs. An example is shown in Figure 9.12, which shows different control circuit compared with the one in Figure 9.6, but with the same complexity.

Second, to actually form the fault-tolerant structure based on the decomposition, depending on whether a secondary edge is included, there are two different situations, as shown in Figure 9.13. If there is no secondary edge in the sub-graph, a spare TSV should be inserted with distance between $D_{max,i}$ and D_{min} from the TSV corresponding to the root of the spanning tree (e.g., S7, S8). If a secondary edge is used, then the spare TSV must be inserted between the two TSVs forming that secondary edge (e.g., between S2 and S3), satisfying $D_{max,i}$ and D_{min} constraints of both TSVs.

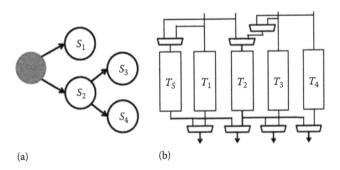

(a)   (b)

FIGURE 9.12   (a) A sub-graph with its spanning tree and (b) the corresponding fault-tolerant structure.

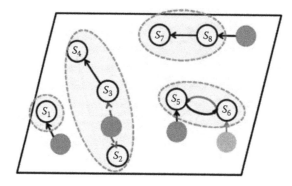

FIGURE 9.13   One of the decomposed results from Figure 9.11.

As the assignment graph decomposition formulation is clear now, the remaining problem is how to find its optimal solution. Simultaneously considering both graph decomposition and spanning tree formation may be very complicated for directed graphs. Trying to reduce the problem complexity, we will first solve it in an easier way by reducing the directed graph to an undirected one in Section 9.4.2.2. Actually, even with an undirected assignment graph, the decomposition problem can still be NP-hard. To prove NP-hardness, a commonly adopted technique is to show that it can be reduced from a well-known NP-complete problem in polynomial time. We choose to use Exact Cover by 3-Sets (X3C) problem to reduce to the undirected assignment graph decomposition problem. We omit the proof process since it may be out of range of topic of this chapter. Readers who are interested in the proof process can refer [27] for more details. However, doing so can make our algorithm much simpler. The reduction may prune some feasible and potentially better solutions, but will not injure the validity. Then, we will directly handle the directed assignment graph in Section 9.4.2.3.

9.4.2.2   Undirected Graph-Based Heuristic

We can reduce the directed graph to an undirected one by the following steps:

1. If there are a primary edge from vertices s_i to s_j and a primary edge from s_j to s_i, we merge them as one undirected primary edge between s_j and s_i.
2. If there are a secondary edge from vertices s_i to s_j and a secondary edge from s_j to s_i, we merge them as one undirected secondary edge between s_j and s_i.
3. Remove all the remaining directed edges in the graph.

As described in Section 9.3.2 and Definition 1, the directed edge on the graph comes from the asymmetry of the maximum rerouting distance. In Step 1, the condition implies that we will only put two TSVs as neighbors when they are within the maximum distance of each other. Such reduction skips the potential situation as shown in Figure 9.8, but will not injure the validity of the solution. In Step 2, we transform each pair of secondary edges between s_j and s_i into an undirected one since Definition 1 implies that secondary edges must appear in a pair. In other words, no secondary edges will be removed.

As an example, Figure 9.14 shows the undirected assignment graph constructed from Figure 9.11 where primary edges are solid and secondary edges are dashed.

Although the problem with undirected assignment graph is still NP-hard, an efficient algorithm can be developed. In our heuristic, we decompose the assignment graph by constructing disjointed sub-graphs iteratively, and eventually output all the sub-graphs and the spanning tree in each sub-graph. In the following, we will first provide an overview of the heuristic, and then discuss some implementation details.

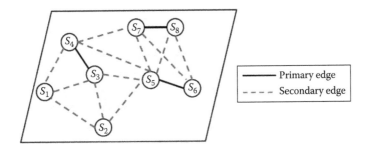

FIGURE 9.14   An example of undirected assignment graph constructed from Figure 9.11.

Our heuristic is based on two simple observations. First, a vertex with lower degree (fewer edges) has fewer choices to form a sub-graph. This implies that a higher priority should be given to vertices with lower degrees. Second, when forming sub-graphs, we should always try to first include the vertices that are connected to the sub-graph by primary edges, because once a secondary edge is included, we have less flexibility in selecting the remaining vertices.

Based on the two observations, we start with an empty sub-graph, find a vertex with the lowest degree in the graph, and add it to the sub-graph. We then search all the candidate vertices to find one with lowest residual degree, and add it to the sub-graph. The residual degree of a vertex is defined as the number of edges between the vertex and all vertices not in the sub-graph. The way to determine candidate vertices is explained later. This process continues until N vertices have been added, or no more candidate vertices can be found. The vertices in the sub-graph now form a fault-tolerant structure. These vertices, along with all the edges connected to them, are removed from the graph. And the same process repeats to form the next sub-graph.

To keep track of the spanning tree in the sub-graph, when expanding the sub-graph we will only keep the edges that form the spanning tree. Toward this, when a vertex is added, all the edges between that vertex and the vertices currently in the sub-graph are removed except one. This practice will make sure that a spanning tree is naturally formed as the sub-graph expands. The following rules are used to decide which edge to keep.

E1. A secondary edge will always be removed before a primary edge.
E2. When both edges are of the same type, the longer edge will always be removed before the shorter edge.
E3. Randomly select the edge to be removed if still tied after E1 and E2.

E2 attempts to minimize the delay overhead and rerouting wirelength whenever possible. Another advantage for this method is to keep track of the number of secondary edges that is essential to form the spanning tree, as we do not allow more than one secondary edges. Note that in Formulation 9.2, we do allow more than one secondary edges in the sub-graph, as long as they are not used in the spanning tree.

With the earlier edge deletion technique, a vertex is a candidate during sub-graph expansion if

C1. no secondary edge exists in the sub-graph, and the vertex is connected to at least one of the vertices in it; or
C2. one secondary edge exists in the sub-graph, and the vertex is connected to at least one of the vertices in it with a primary edge.

The last thing to note about our algorithm is the tie-breaking technique. Our algorithm iteratively selects the vertex of lowest residual degree. However, if more than one vertex has the same residual degree, the following tie-breaking rules will be used. They are all based on the observation that more flexibility should be left for the remaining vertices, and work pretty well in the experiments.

```
1: Function decomposition (G, N)
//decompose G to sub-graphs
2: sub_graph_collection{} ← φ;
3: While(G!= φ) do
4:    current_sub_graph_set[]← φ;
5:    candidate_vertices_set[]←φ;
//add lowest degree vertex to sub-graph
6:    add the vertex with the lowest degree in G to current_sub_graph_set[];
7:    update candidate_vertices_set[] according to C1;
//sub-graph expansion
8:    while (#current_sub_graph_set[]<N and
       candidate_vertices_set[]!= φ) do
9:       add the vertex with the lowest residual degree to current_sub_graph_set[]
          and apply T1-T4 for tie-breaking;
10:      add one edge to current_sub_graph_set[]
          according to E1-E3;
11:      update candidate_vertices_set[] according to C1-C2;
12:   endwhile
//sub-graph record
13:   add current_sub_graph_set[] to sub_graph_collection{};
//graph update
14:   update G by removing all the vertices in current_sub_graph_set[] and all the edges
       connected to them;
15: endwhile
16: return sub_graph_collection{};
```

FIGURE 9.15   The pseudo code of our heuristic for constrained graph decomposition.

T1. The vertex having one or more primary edges connected to the sub-graph will have higher priority over the vertex having only secondary edges connected to the sub-graph.

T2. When both vertices are connected to the sub-graph with primary edges, the one that has more secondary edges connected to the vertices not in the sub-graph has higher priority.

T3. When both vertices are connected to the sub-graph with secondary edges only, the one that has more primary edges connected to the vertices not in the sub-graph has higher priority.

T4. Randomly select the vertex to be added if still tied after T1, T2, and T3.

The algorithm is summarized in Figure 9.15.

Finally, we use the 8-vertex assignment graph in Figure 9.16 as an example to illustrate our algorithm with $N = 4$. The degrees of all vertices are labeled in Figure 9.16a. S1, S2, S6, and S8 are tied, and we choose S1 to start with. Now S2, S3, and S4 are the candidates. S2 has the lowest degree vertex and therefore be selected as shown in Figure 9.16b. Since no eligible candidates remain, the sub-graph is complete. The two vertices S1, S2, and all the edges connected to them are removed, as shown in Figure 9.16c.

With the updated graph in Figure 9.16c, we can construct the next sub-graph starting from S3, which has the lowest degree now. It can be easily verified that S4, S5, and S6 will then be added in sequence shown in Figure 9.16d. We then again remove these vertices (and the corresponding edges connected to them) from the graph. The updated assignment graph is shown in Figure 9.16e. Repeat the same process, S7 and S8 will form one sub-graph. Finally, the decomposition of the assignment graph is shown in Figure 9.16f.

9.4.2.3   Directed Graph-Based Heuristic

Previous section provides a simple yet efficient heuristic for undirected assignment graph, at the cost of pruning some feasible and potentially better solutions. In this section, we move a further step to adopt the heuristic to directly address the original problem in Formulation 9.2. In the following, we will first provide an overview of the heuristic, and then discuss some implementation details.

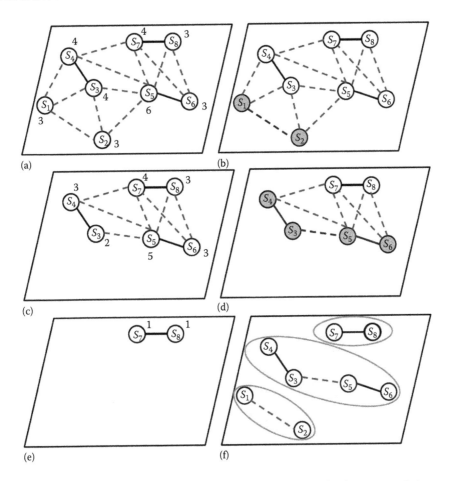

FIGURE 9.16 An example to illustrate the proposed heuristic for undirected assignment graph decomposition. (a) An example of 8-vertex undirected assignment graph with degrees of all vertices. (b) S1 and S2 are selected. (c) S1, S2, and all the edges connected to them are removed. (d) S3, S4, S5, and S6 are selected. (e) S3, S4, S5, S6, and all the edges connected to them are removed. (f) The final decomposition of the assignment graph.

Similar with the heuristic shown in the previous section, we again decompose the assignment graph by constructing disjointed sub-graphs iteratively and eventually output all the sub-graphs and the spanning tree in each sub-graph. The challenge here is to consider the direction of edges when forming a spanning tree. We solve the challenge by applying the following constraint when forming sub-graphs: All vertices except one should have at least one directed edge incident from at least one vertex in the subgraph. This constraint can guarantee that a spanning tree can be extracted from the sub-graph formed.

The heuristic is similar to the heuristic mentioned in the previous section, and the same complexity can be derived. The only differences are as follows:

1. When counting the degree or residual degree of a vertex, we count the number of vertices that have edges to/from it.
2. A vertex is a candidate during sub-graph expansion if no secondary edge exists in the subgraph, and the vertex has at least one incident edge; or at least one secondary edge exists in the sub-graph, and the vertex has at least one edge incident from other vertices with a primary edge.

We use the same example as previous section to demonstrate our algorithm in Figure 9.17. The degrees of all vertices are labeled in Figure 9.17a. We choose S1 to start with. Now S2, S3, and S4

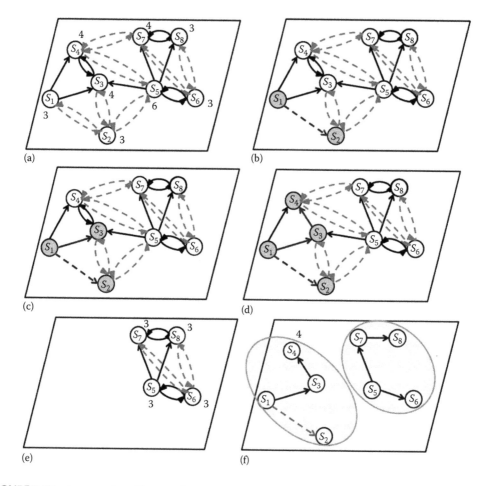

FIGURE 9.17 An example to illustrate the proposed heuristic for directed assignment graph decomposition. (a) An example of 8-vertex direct assignment graph with degrees of all vertices. (b) S1 and S2 are selected. (c) S3 is selected. (d) S4 is selected. (e) S1, S2, S3, S4, and all the edges connected to them are removed. (f) the final decomposition of the assignment graph.

are the candidates, but S2 has fewer degrees than S3 and S4. According to tie-breaking rule T3, S2 will be selected as shown in Figure 9.17b. S3 and S4 become candidates, but S3 have fewer degrees than S4. S3 be selected to the sub-graph as shown in Figure 9.17c. S4 is the only candidate because of C2. We added S4 into the sub-graph shown in Figure 9.17d. Since the sub-graph is full, it is complete. The four vertices S1, S2, S3, S4, and all the edges connected to them are removed, as shown in Figure 9.17e. It can be easily verified that S4, S5, S7, and S8 will then be added in sequence. Finally, the decomposition of the assignment graph is shown in Figure 9.17f.

9.4.3 EXPERIMENTAL RESULTS

In this section, experimental results are presented to demonstrate the efficacy of our proposed algorithms. For simplicity of presentation, in this section, we denote the heuristic for undirected assignment graph decomposition as UAGD while that for directed assignment graph decomposition as DAGD. We implement both in C, and perform experiments on 10 circuits, which are obtained by applying 3D-Craft [42] to five 2D industrial designs in 45 nm technology. Original TSVs and their locations in each case are extracted from layouts as our inputs. Target yields are set depending on the size of each circuit. Maximum distance constraint for each TSV is obtained from timing analysis results, and ranges between 20 μm and 50 μm. Minimum distance constraint is set to 10 μm, as [33,29] suggested. We also consider the Tezzaron design rule [8] for TSV density. After our algorithm completes, we check for routing congestions caused by pre-placed standard cells or other signal nets. If any fault-tolerance structure formed by our algorithm cannot be implemented, we remove it and use double TSV technique for each TSV in that structure.

Since no prior art exists, we implement another heuristic that sounds more intuitive and simpler. Instead of using an assignment graph, each time a sub-graph is constructed, we iteratively select the vertex that has the smallest distance to the current sub-graph, and add that vertex to it. The process is continued until no more vertices are within the maximum rerouting distance of any vertex in the sub-graph, or the limit of the vertices number N has reached. We call it nearest-neighbor-based heuristic (denoted as NN). Note that this heuristic excludes structures like group at the middle in Figure 9.16f. For the results from all the algorithms, we run static timing analysis with each signal going through the longest rerouting path, and no timing violation is reported. This convincingly shows that our maximum distance formulation used in Section 9.4.1 is effective.

We compare the area overhead induced by spare TSVs, control logics, and rerouting wires as well as the overall area overhead of the three algorithms as shown in Table 9.1. Columns 1 and 2 show benchmark information. Column 3 presents target chip yield. Columns 4–8 show the area of spare TSVs needed (as well as the number of spare TSVs), control logics for rerouting, rerouting wire, overall area overhead in NN. Columns 9–14 show the results for UAGD and Columns 15–20 show the results for GAGD. The area overhead reduction achieved by UAGD and GAGD compared with NN are shown in Columns 14 and 20, respectively. From the table we can find that the overall area overhead is reduced by up to 38% for UAGD and by up to 61% for DAGD. The average area overhead is also shown in the table. It is obvious that in most cases DAGD can obtain a significant improvement compared with UAGD, which is intuitive since UAGD prunes some potential solutions. However, in some cases such as circuit_8 and circuit_9, a better solution comes out from UAGD. It is because that although directed approach has a larger solution space, it might reduce the opportunity in forming other sub-graphs.

9.5 FAULT-TOLERANCE FOR CLOCK NETWORK

In previous section, we detailed describe two heuristics for spare TSV assignment and fault-tolerant structure formation. However, until now, we only focus on signal TSVs, that is, the TSVs we address are used to pass logic signals. When the circuit goes to vertical, the clock tree would also be constructed vertically with TSVs. The low-yield TSV would also influence the robustness of clock trees. We address this problem as follows: Section 9.5.1 detailed describes the challenge about reliability of 3D clock networks with necessary backgrounds of clock tree design. In Section 9.5.2, two kinds of fault-tolerant unit, their control units, and how they work are described. Experimental Results are shown in Section 9.5.3.

9.5.1 CHALLENGE OF RELIABILITY ABOUT 3D CLOCK NETWORKS

It is well known that an improperly synthesized 2D clock tree may introduce large skew and latency [7,45], degrading the system performance and even causing functional failures. The problem gets more involved in 3D, as the clock signals need to be delivered to different tiers through TSVs. There exist some researches to address the clock tree forming problems. A comparison between different clock distribution architectures is presented in [46]. A thermally adaptive 3D clocking scheme [47] is proposed to reduce clock skew by dynamically adjusting the driving strengths of the thermally adaptive clock buffers. BURITO [48] proposed to construct a 3D abstract tree under the wire-length versus via-congestion tradeoff considering thermal variations. By utilizing a dedicated clock tier, a global clock distribution network [3] was proposed to reduce the interconnect length and decrease the power consumption. Techniques for constructing prebond testable clock tree with

TABLE 9.1

Comparison between NN, UAGD, and GAGD in Terms of Area Overhead Induced by Spare TSVs, Control Logics, Rerouting Wires, the Overall Area Overhead (in μm²)

			NN				UAGD					DAGD				
Testbench	#TSV	Target Yield	Spare TSV	(Number)	Control	Wire	Overall	Spare TSV	(Number)	Control logic	Wire	Overall	Reduction	Spare TSV	(Number)	Control Logic	Wire	Overall	Reduction
circuit_1	186	95%	13973	(178)	1369	31	15373	7065	(90)	1990	545	9600	38%	4161	(53)	2272	766	7198	53%
circuit_2	188	95%	9106	(116)	1835	328	11269	4789	(61)	2223	699	7710	32%	1492	(19)	2554	956	5002	56%
circuit_3	256	90%	10127	(129)	2702	583	13412	5417	(69)	3126	999	9541	29%	1649	(21)	3464	1276	6389	52%
circuit_4	458	90%	25591	(326)	4163	587	30341	13188	(168)	5278	1537	20003	34%	3376	(43)	6181	2211	11768	61%
circuit_5	600	85%	23236	(296)	6379	1488	31102	13188	(168)	7282	3131	23601	24%	7458	(95)	7825	4256	19538	37%
circuit_6	721	85%	29124	(371)	7557	1622	38303	15151	(193)	8813	2833	26797	30%	4789	(61)	9744	3565	18098	53%
circuit_7	800	85%	35247	(449)	8121	1372	44740	20724	(264)	9427	3157	33308	26%	14601	(186)	10012	4216	28830	36%
circuit_8	1157	80%	18605	(237)	14655	5107	38367	10990	(140)	15340	6839	33169	14%	8321	(106)	15544	9667	33532	13%
circuit_9	1327	80%	20332	(259)	16899	6500	43730	12403	(158)	17612	8989	39004	11%	10127	(129)	17739	12534	40400	8%
circuit_10	1849	80%	73712	(939)	19468	4213	97392	41056	(523)	22403	9128	72586	25%	31479	(401)	23708	12060	67247	31%
Avg.	—	—	—	—	—	—	—	—	—	—	—	—	26%	—	—	—	—	—	40%

2D redundant clock tree for each die were proposed in [49–52] to facilitate the testing before die-stacking. However, none of these works considers the reliability of TSVs.

To alleviate the problem, TSV redundancy techniques utilizing spare TSVs and reconfigurable routing are proposed as shown in previous sections. Nevertheless, all these works explicitly or implicitly target at general signal deliveries, and should not be directly applied to the clock network for three reasons: First, clock signal is global in nature and needs to be distributed throughout the chip to all tiers. This provides the possibility to develop some special structures as the same signal is delivered through multiple TSVs. Second, clock signal has additional constraints such as skew/slew, which in turn imposes stringent requirements on the fault tolerance. Simply using spare TSVs or reconfigurable routing to accommodate TSV failure may cause potential skew or slew violations. Third, given the large sizes of TSVs (10V125 μm^2) from industrial reports [21–25], and their large numbers in the clock network, spare TSVs may bring significant area overhead, while reconfigurable routing may introduce much wire-length overhead.

We follow [21] to present a novel fault-tolerant framework specially tailored for clock networks. Different from any existing TSV redundancy techniques, our framework takes advantage of the unique property of clock delivery network by utilizing existing 2D redundant clock trees for pre-bond testing. The key concept is a TSV fault-tolerant unit (TFU), which can automatically re-route the clock signal using the 2D redundant tree in the presence of TSV failure. The number of TSVs Formula that forms a TFU (Formula-TFU) is adjustable to allow flexibility during clock network synthesis. It can be easily integrated with any bottom-up clock network synthesis algorithms.

Therefore, to realize TSV fault-tolerant clock network designs, the capability of propagation delay adjustment under various scenarios is important. In addition, in clock tree synthesis, utilizing a larger number of TSVs can effectively reduce the power consumption [52], but at the cost of higher yield loss. To mitigate the yield loss issue while keeping the power consumption low, we propose to use our TFU to reduce the number of extra TSVsXit applies to any bottom-up clock tree synthesis method. Finally, there exist certain structures in a pre-bond testable 3D clock network that can readily serve as redundant paths. To achieve TSV-fault-tolerant clock network designs, we proposed the TFU that balances the delay difference in different scenarios, reduces the number of total spare TSVs, and utilizes existing resources in the prebond testable clock tree.

Figure 9.18a illustrates the concept of a 3D clock tree, which utilizes TSVs to distribute the clock signal to different tiers. To lower the yield loss induced by stacking bad dies, 3D-ICs need to deal with the integration issue of known good die (KGD). Each bare-die is required to be tested before stacking, called pre-bond testing. However, the clock tree is incomplete before die-stacking. To deal with this issue, Figure 9.18b shows that a 2D redundant clock tree is needed in each die to distribute testing clock signal (TCLK) to the dies before stacking. By using transmission gates in Figure 9.18c, the 2D redundant tree is connected with the clock tree during the pre-bond testing stage, and is disconnected after die-stacking. Figure 9.18d shows the final 3D clock network, which is composed of the 3D clock tree and the 2D redundant tree.

The necessity of the prebond testable clock tree is first presented in [50]. Then, a pre-bond testable clock tree routing algorithm is proposed in [51,54]. The algorithm consists of two main steps: (1) 3D clock tree construction and (2) 2D redundant tree routing. The 3D clock tree is constructed with TSV-buffer insertion, and the objective is to minimize the total wire-length. The 2D redundant tree is synthesized with transmission gates, which are used to connect/disconnect the 2D redundant tree with the clock tree during different stages. To reduce the number of buffers and remove the global control signal of transmission gates, a comprehensive solution is provided in [49]. It utilizes a self-controlled clock transmission gate, thus eliminating the global control signals for the prebond testable clock tree.

For clarity, the 3D clock network in the context of this paper comprises the following components:

1. A 3D clock tree, which delivers the clock signal formula at the postbond stage.
2. 2D redundant trees, which deliver the testing clock signal formula at the prebond stage.

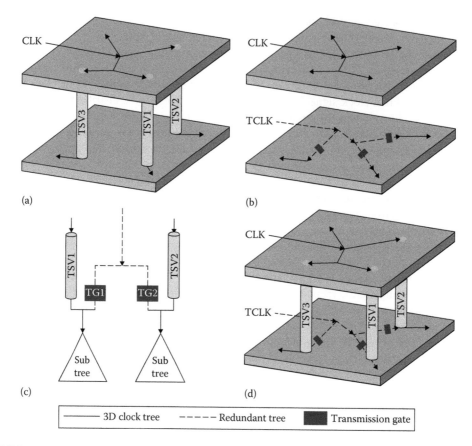

FIGURE 9.18 (a) 3-D clock tree after die-stacking. (b) 2-D clock tree of each die before die-stacking. (c) Connection structure of redundant tree. (d) 3-D clock network.

9.5.2 TSV Fault-Tolerant Unit

In Figure 9.18, the clock signal is delivered to different tiers through the 3D clock tree connected by TSVs. To keep the figures clear, in all related figures, we simply used a solid black triangle to represent the buffer chain, gray transmission gates to represent the off state, and gray TSVs with cross (X) to indicate faulty status. Once a TSV fails, the clock signal cannot be delivered to the subtree connected with the TSV, thus resulting in chip malfunction. This problem can be solved by providing a redundant path for each subtree. When a TSV fails, the corresponding redundant path can be used to deliver the clock instead. This idea is simple in concept, yet challenging in practice. There are many different ways of designing such a redundant path (e.g., double TSVs), each with different area penalties. We would like to choose the one with minimum area overhead.

The main idea of the TFU to be proposed in this section is to reuse a part of the 2D redundant tree as the redundant path for TSVs. In the following sections, we will first present the basic version of a TFU, formed by two TSVs (2-TFU), and explain how it works. Since a 2-TFU requires each TSV be paired with another, to enhance flexibility, we then propose an advanced version of TFU with three TSVs (3-TFU).

Note that 2-TFU and 3-TFU can be generalized to work with Formula TSVs (Formula-TFU). However, due to the complexity considerations, in this chapter, we limit our discussion to 2-TFU and 3-TFU. The details on how to actually form those TFUs during clock network synthesis are provided in Section 9.5.2.4.

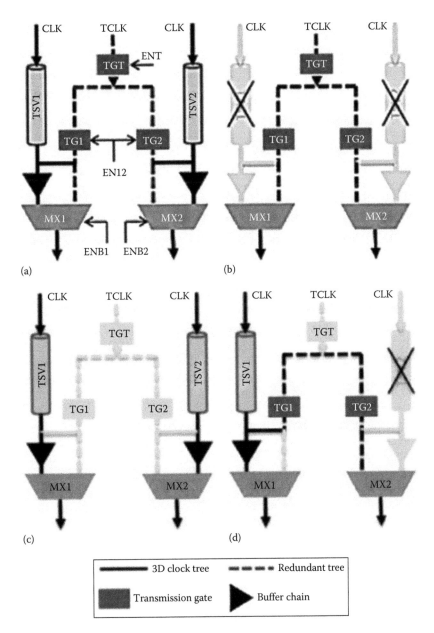

FIGURE 9.19 (a) Proposed 2-TFU. (b) Clock delivering path at the prebond stage. (c) Clock delivering path at the postbond stage when all TSV are valid. (d) Clock delivering path at the postbond stage when TSV2 fails.

9.5.2.1 Two-TSV-Based Fault-Tolerant Unit

Figure 9.19a shows the design of the proposed 2-TFU, which is composed of two TSVs. The three transmission gates (TGT, TG1, and TG2) and two multiplexers (MX1 and MX2) are configured to determine the clock signal paths in different situations. The two buffer chains are inserted for delay balance. Details are explained in the following.

In the pre-bond testing stage, the TSVs do not form yet, and the test clock signal is distributed through the redundant tree to each 2D die. After die-stacking, if both TSVs are good, the clock signal will go through the corresponding TSV to each subtree. Once a TSV is faulty, the clock signal can be delivered from the other TSVs through the redundant path.

To realize the earlier concept, proper configurations of the transmission gates and multiplexers are needed. In the prebond stage, the following configuration should be used: all the three transmission gates (TGT, TG1, and TG2) are turned on, and the two multiplexers (MX1 and MX2) should select the inputs corresponding to the redundant paths. The clock signal is distributed through the 2D redundant tree as shown in Figure 9.19b.

On the other hand, there are two configurations for the postbond stage. First, when both TSVs operate normally, TGT, TG1, and TG2 should be turned off and MX1 and MX2 are set to select the clock signal from the corresponding TSV. As such, the clock signal is delivered through TSV1 and TSV2 as shown in Figure 9.19c. Second, when one of the TSVs fails (assuming TSV2), TGT should be turned off, and the other two transmission gates (TG1 and TG2) should be turned on. This is illustrated in Figure 9.19d. In this case, MX1 selects the input corresponding to TSV1 while MX2 selects the input corresponding to the redundant path.

Although TFUs can provide tolerance against TSV failure, the latency of the clock changes when it alternates through the redundant path, causing potential skew violations. To tackle this problem, we can insert additional buffer chains to balance the skew in the presence of TSV failure.

9.5.2.2 Three-TSV-Based Fault-Tolerant Unit

Without utilizing any spare TSV, 2-TFU can guarantee the functionality of the entire clock network unless both TSVs in a 2-TFU fail. However, it would require each TSV be paired with another during clock network synthesis, which is not always possible. With details to be discussed in Section IV, in certain cases, it is necessary to group three TSVs and form a TFU in order to avoid stand-alone TSVs. Toward this end, we propose an extension of 2-TFU, namely, three-TSV-based fault-tolerant unit (3-TFU), which can deal with 3 TSVs.

As shown in Figure 9.20a, the major components in a 3-TFU are three TSV sets. Each set includes a TSV, a buffer chain, and a multiplexer. Two transmission gates, TGT1 and TGT2, are inserted into the redundant tree. Similar to 2-TFU, the automatic reroute scheme can be realized by properly arranging the transmission gates and multiplexers.

The operations of a 3-TFU are similar to that of a 2-TFU. In the prebond testing stage, the test clock signal passes through the redundant tree. If all TSVs function properly after die-stacking, the clock signal is delivered to the subtrees through the corresponding TSVs. If TSV1 fails, the subtree rooted in TSV1 receives the clock signal through TSV2 and redundant path, and the subtrees rooted in TSV2 and TSV3 get the clock signal through their corresponding TSVs. Figure 9.20b shows the clock delivery path when TSV2 is faulty. In this situation, the clock signal is delivered through TSV1 to the subtrees rooted in TSV1 and TSV2, and through TSV3 to the corresponding subtree rooted in TSV3.

The major difference between 2-TFU and 3-TFU is the condition when TSV3 is faulty. In Figure 9.20c, the subtree rooted in TSV1 receives the clock signal from TSV1, and the subtrees rooted in TSV2 and TSV3 receive the clock signal from TSV2.

9.5.2.3 Control Unit

While the general structures of 2-TFU and 3-TFU have been explained, one critical issue still needs to be addressed: the automatic detection of different scenarios (prebond, single TSV failure, etc.), and the generation of the corresponding control signals.

Figure 9.21a shows a schematic illustration of 2-TFU with control logics. Each TSV is followed by an on-chip detection circuit (OCDC), which can monitor the desired type of TSV failures. When the TSV is valid, the output signal EN (ENB) keeps high (low), otherwise, the EN (ENB) will be low; (high). The two OCDCs (OCDC1 and OCDC2) control the operation of the transmission gates (TG1 and TG2) and the multiplexers (MX1 and MX2). When both TSVs are valid, the control signal EN12 will be low; thus, transmission gates TG1 and TG2 are turned off and the multiplexers will

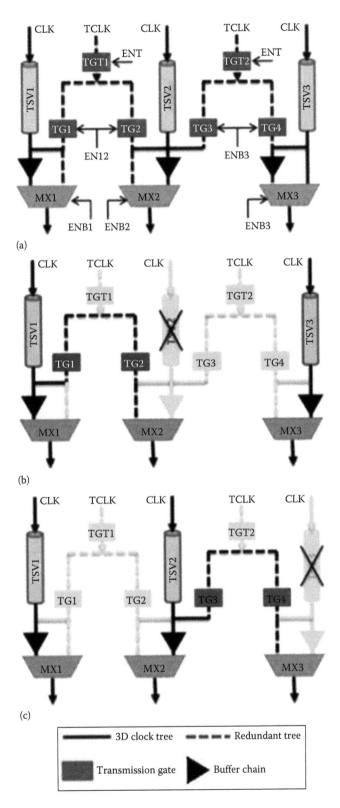

FIGURE 9.20 (a) Proposed 3-TFU. (b) Clock delivering path at the postbond stage when TSV2 fails. (c) Clock delivering path at the postbond stage when TSV3 fails.

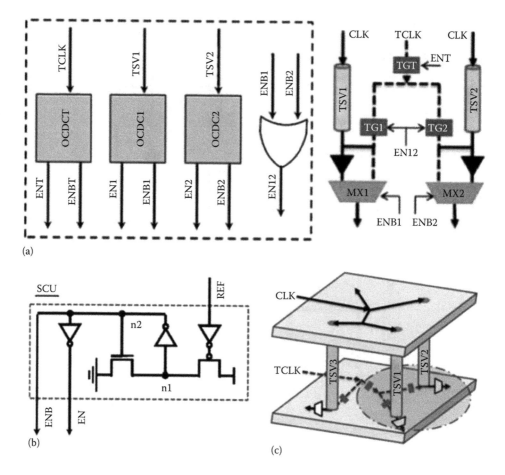

FIGURE 9.21 (a) 2-TFU with control signals. (b) Self-controlled unit (From Kim, T.-Y. and Kim, T., Clock tree synthesis with pre-bond testability for 3-D stacked IC designs, in *Proc. of Design Automation Conference (DAC)*, Anaheim, CA, pp. 184–190, 2010). (c) Fault-tolerant 3D clock network using TFUs.

deliver the clock signal from TSVs. The prebond stage detection and TGT controlling is same as the technique presented in [49].

One example of OCDC to monitor the random open defect of TSVs, named self-controlled units (SCUs), is discussed in [49]. Figure 9.21b shows the schematic of such an SCU, which is used to detect TCLK at input signal REF. When REF goes too high, the internal node $n1$ is charged (high) and $n2$ is discharged (low), which turns off the NMOS. When REF goes too low, the internal nodes $n1$ and $n2$ are gradually discharged by subthreshold leakage current. Once REF goes too high again, the internal nodes are recharged. If the switching of REF is fast enough, the internal nodes ($n1$ and $n2$) can be fully recharged before discharged; thus, the output signals EN remains high (and ENB remains low). Accordingly, the SCU can be used to automatically to detect a period signal at input REF. In the following, we use SCUs as OCDCs to demonstrate how to generate proper control signals in different scenarios. However, it is understood that other types of OCDCs can be applied as well. We will first explain our idea with 2-TFUs, and then extend it to 3-TFUs.

As shown in Figure 9.21a, each 2-TFU in the clock network is paired with OCDC1, OCDC2, and OCDCT (all three detection circuits are implemented by SCU), which monitor the clock signal from the two TSVs (TSV1 and TSV2) and the test clock (TCLK), respectively. Note that with this structure, all the wiring is local and the impact on the timing is minimum.

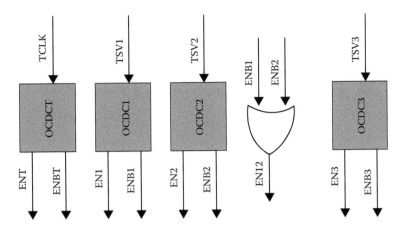

FIGURE 9.22   Control signals for a 3-TFU.

During pre-bond testing, the TSVs have not been formed yet and only TCLK passes through the redundant tree. The OCDCs detect this scenario and set ENT, ENB1, and ENB2 high. Thus, the three transmission gates in the 2-TFU are turned on, and TCLK can pass through the redundant tree. During the post-bond operations, ENT is low, which turns off the TGT, since no TCLK is detected. When both TSVs operate normally, the clock signal (CLK) is detected by OCDC1 and OCDC2, and ENB1, ENB2, and EN12 are low. Thus, the redundant path is disconnected, and the two multiplexers (MX1 and MX2) will choose the signal delivered from the corresponding TSV. Once a TSV fault is detected (assuming TSV1), the corresponding enable signal (ENB1) from OCDC1 is high, which in turn switches EN12 to high. Subsequently, the redundant path is connected, and the multiplexer MX1 will select the signal from the good TSV (TSV2) through the redundant path.

The resultant fault-tolerant 3D clock network with TFUs and aforementioned control signals is illustrated in Figure 9.21c.

To extend the idea to 3-TFU, only one extra detection circuit, namely OCDC3, is needed. OCDC3 is designed to monitor the functionality of TSV3. The output signal ENB3 keeps low (high) when TSV3 is good (faulty). During the prebond stage, ENT, ENB1, ENB2, ENB3, and EN12 keep high, and the redundant path is connected. During the postbond stage, if all three TSVs operate correctly, the five control signals keep low to disconnect the redundant path. The clock signal passes through all TSVs to the corresponding subtrees. The operations when TSV1 or TSV2 is faulty are similar to the operations in a 2-TFU. On the other hand, when TSV3 fails, ENB3 switches to high and connects the redundant path between TSV2 and TSV3. Therefore, the subtree rooted in the faulty TSV3 can still receive the clock signal through TSV2 and redundant path. The control signals of a 3-TFU are shown in Figure 9.22.

9.5.2.4   TFU-Embedded Clock Network Generation

In this section, we will present how to integrate the proposed TFUs with clock network synthesis. As a vehicle to demonstrate the efficacy of the proposed TFU structure, we extend the clustering algorithm in [53], mainly due to its simplicity, to synthesize the 3D clock network. However, it is understood that other bottom-up clock tree synthesis algorithms can be used as well. Since the number of TSV greatly affects the yield, choosing a better clock tree synthesis method can help to further reduce the total number of TSVs. In addition, the idea of integrated with top-down methods [54] can be an interesting topic to pursuit as our future research direction.

For the simplicity of discussion, we illustrate our algorithm using a simple example shown in Figure 9.23, which has two tiers and six sinks. Our clock network topology is generated by iteratively

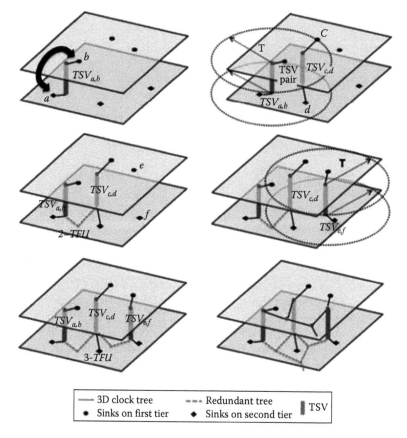

FIGURE 9.23 Example for the proposed fault-tolerant clock tree construction. (a) A TSV $TSV_{a,b}$ is generated to connect nodes a and b. (b) Nodes c and d are within the feasible range T of $TSV_{a,b}$. (c) $TSV_{c,d}$ is inserted to form a 2-TFU. (d) A 2-TFU can be found. (e) A 3-TFU is constructed. (f) The resultant clock tree.

forming a new parent node with the minimum cost in a bottom-up fashion. The cost of connecting two nodes N_a and N_b is based on the distance in 2D geometry and defined as follows:

$$\begin{cases} DistanceInGeometry(N_a, N_b), & \text{if } N_a, N_b \text{ in the same tier} \\ DistanceInGeometry(N_a, N_b) + \alpha \times TSV_{length}, & \text{otherwise} \end{cases} \quad (9.5)$$

where the parameter α denotes the weight factor which considers the overhead of using a TSV. For example in Figure 9.23a, a TSV $TSV_{a,b}$ is generated to connect nodes a and b since the two-node pair has the minimum cost. Different from [53,55], for level balancing purposes, a node will only be paired with another node in the same hierarchical level in the clock tree during the synthesis.

When a new TSV is created during the clock network synthesis, we start to find if another stand-alone TSV (pairing TSV) can be inserted within the feasible range T of the new TSV to form a 2-TFU. The decision of the feasible range will be detailed in next section. If there are multiple TSVs that can be inserted in the range, the one with the minimum distance to the new TSV will be chosen. For example in Figure 9.23b, nodes c and d are within the feasible range T of $TSV_{a,b}$, and accordingly, $TSV_{c,d}$ is inserted to form a 2-TFU shown in Figure 9.23c. The redundant path between the two TSVs is also determined and inserted at this time. Note that, since nodes e and f are the last two nodes without connection, $TSV_{e,f}$ is inserted to maintain the level balance of the resultant clock tree.

On the other hand, if no stand-alone TSVs can be inserted within the feasible range, we continue to search if any 2-TFU already exists within the range, which can be used to form a 3-TFUs structure. Similarly, we will select the 2-TFU with the minimum distance to the TSV. As shown in Figure 9.23d, no stand-alone TSVs can be inserted within the feasible range of Formula, but a 2-TFU can be found. As a result, a 3-TFU is constructed by connecting Formula and the 2-TFU, as shown in Figure 9.23e. If no stand-alone TSVs can be inserted and no 2-TFUs exist within the feasible range, the double TSV technique will be applied.

We then continue the bottom-up synthesis until we build the entire clock network. After the 3D clock network is synthesized, we synthesize the remainder of the 2D redundant tree in each tier according to the algorithm discussed in [53]. During the bottom-up synthesis process, each internal node located on the 3D clock tree and 2D redundant trees is generated considering skew balancing and latency controlling for all possible scenarios.

When the bottom-up merging process finishes, we can make the resultant clock tree depicted in Figure 9.23f satisfy the final constraint under any TSV failures. In addition, we also integrate the slew-aware buffer insertion technique [6,56] to address the issue of clock slew rate control. When the downstream capacitance of a node exceeds the maximum capacitance, denoted as Formula, a clock buffer chain is inserted, and proper adjustments at other buffers are necessary to ensure skew balance.

The TSVs in the same TFU act as the spare TSV for each other. If the distance between these two TSVs is too long, the redundant path will have excessive interconnect resistance and capacitance, which in turn increases the size of the buffer chains needed to balance the skew. Thus, we will only pair two TSVs within a feasible range Formula to form a TFU. Note that the range Formula cannot be too small as well. Otherwise, we might not be able to find any TSV pairs in the range.

In our experiments, we find out that setting the feasible range Formula to 100 Formula provides a good balance between the number of TFUs that can be formed and the delay overhead. Note that it is still possible that a TSV cannot find another TSV in the feasible range to pair with, this can be addressed either by forming a 3-TFU if a 2-TFU exists in the range, or otherwise by inserting a double TSV.

9.5.2.5   Experimental Results

We implement our algorithm in C++ and perform experiments on several 3D designs, including one industrial case with two tiers and 55.4 K clock sinks. Other designs are obtained by stacking benchmark circuits from IBM suite or ISCAS89. All designs are synthesized to an industrial 65 nm technology library, and the supply voltage is 1.0 V. At the implementation stage, all cell placements take the important keep-out-zone constraint [43] into consideration as well. Figure 9.24 shows the TSV model [57] used in this paper.

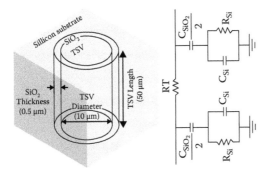

Symbol	Value
RT	4.39E – 03 (Ω)
C_{SiO_2}	1.08E – 13 (F)
C_{Si}	7.22E – 15 (F)
R_{Si}	1.46E + 03 (Ω)

FIGURE 9.24 TSV model. (From Shi, Y. et al., The effects of substrate doping density on the electrical performance of through-silicon-via (TSV), *Proceedings of Asia-Pacific EMC Symposium*, Jeju Island, Korea, May 2011.)

The wire resistance and capacitance per unit length are Formula and Formula, respectively. To control the slew rate, the maximum load capacitance allowed for each buffer is 200 fF. The feasible range Formula for our fault-tolerant unit is 100 Formula. The clock skew constraint is 100 ps. In other words, we enforce that the clock skew may not exceed 100 ps when only one TSV among all fails. Components such as buffers and multiplexers in TFUs are instantiated from the same library, and the area overheads of a 2-TFU and a 3-TFU are 11.7 Formula and 19.2 Formula, respectively. In addition, the diameter of a TSV is 10 Formula. The normal clock tree without these fault-tolerant techniques is denoted as the original design (Orig), and the tree utilizing the double TSV technique is denoted as Double 3D. We construct two fault-tolerant 3D clock networks, F3-D and A-F3-D (advanced F3-D), by utilizing different strategies: Only 2-TFUs are used in F3-D, whereas both 2-TFUs and 3-TFUs are used in A-F3-D. For both F3-D and A-F3-D, ungrouped TSVs are protected by the double TSV technique.

Table 9.2 summarizes the comparison between the original design, Double3-D, F3-D, and A-F3-D. Columns 1 and 2 show the name and number of sinks for each circuit. Columns 3–6 show the number of TSVs used in the original design, Double3-D, F3-D, and A-F3-D, respectively. Columns 7–9 show the number of TFUs in F3-D and A-F3-D. Columns 10–13 show the yield rate of the original design and all the three techniques based on the following facts: For original design without TSV redundancy, any faulty TSV results in a failure chip. On the other hand, for Double3-D, the design fails if both TSVs in a pair are faulty. For F3-D and A-F3-D, the design fails if more than one TSV in the same TFU fail.

From the table, we can see that the number of TSVs used by F3-D and A-F3-D is 44% and 64% less than that in case of Double3-D, which means that a large portion of the original TSVs can be used to form TFUs. In addition, compared with the double TSV technique, our F3-D and A-F3-D techniques result in similar yield rate.

We also analyze the power overhead and the total buffer count in different methods in Table 9.3. Columns 3 and 4 show the power and buffer count in the original design. Columns 5–7 report the power overhead using Double3-D, F-3-D, and A-F3-D, respectively. Columns 8–10 report the total buffer used in the clock network, and columns 11–12 report the percentage of the buffers used in

TABLE 9.2
Comparison of TSV Number and Yield Rate

			#TSVs			#TFUs F3-D	#TFUs A-F3-D		Yield Rate (%)			
Circuits		#Sinks	Orig	Double 3-D	F3-D	A-F3-D	2-TFUs	2-TFUs	3-TFUs	Orig	Double 3-D	F3-D	A-F3-D
2 tiers	Industrial	55410	586	1172	904	797	184	77	107	0.28	94.31	95.58	94.54
	2T_case1	3199	13	26	18	18	6	6	0	87.75	99.87	99.91	99.91
	2T_case2	3025	12	24	22	20	7	4	2	88.64	99.88	99.89	99.87
	2T_case3	3025	11	22	18	16	4	0	4	89.53	99.89	99.91	99.86
	2T_case4	1460	5	10	8	7	3	2	1	95.10	99.95	99.96	99.95
	2T_case5	2765	10	20	12	12	5	5	0	90.44	99.90	99.94	99.94
	2T_case6	5004	13	26	16	14	5	3	2	87.75	99.87	99.92	99.90
3 tiers	3T_case1	4496	40	80	72	55	18	7	11	66.90	99.60	99.64	99.56
	3T_case2	4753	39	78	70	56	19	5	14	67.57	99.61	99.65	99.51
	3T_case3	3363	17	34	24	24	10	10	0	84.29	99.83	99.88	99.88
	3T_case4	5866	23	46	38	29	14	8	5	79.36	99.77	99.81	99.78
Average ratio		–	1	2	1.56	1.36	–	–		1	1.42	1.42	1.42

TABLE 9.3

Normalized Power and Buffer Overhead Induced by Fault-Tolerant Techniques

Circuits		Orig		Power			Total Buffer Count			Buffer Used in TFU	
		Power (W)	Buffer Count	Double 3-D	F3-D	A-F3-D	Double 3-D	F3-D	A-F3-D	F3-D (%)	A-F3-D (%)
2 tiers	Industrial	1.04E–01	21795	1.09	1.16	1.23	1.09	1.15	1.22	13.02	20.18
	2T_case1	5.37E–03	465	1.22	1.27	1.27	1.21	1.27	1.27	20.27	20.27
	2T_case2	5.23E–03	451	1.22	1.26	1.31	1.22	1.31	1.34	19.43	29.79
	2T_case3	5.13E–03	443	1.21	1.26	1.33	1.20	1.29	1.33	39-36	41.30
	2T_case4	1.73E–03	99	1.24	1.30	1.34	1.24	1.31	1.37	36.81	34.53
	2T_case5	3.33E–03	189	1.22	1.36	1.36	1.21	1.38	1.38	49.14	49.14
	2T_case6	5.75E–03	326	1.19	1.29	1.30	1.17	1.30	1.33	47.51	29.79
3 tiers	3T_case1	8.23E–03	727	1.29	1.24	1.36	1.29	1.22	1.34	13.53	18.10
	3T_case2	8.20E–03	719	1.28	1.28	1.32	1.28	1.30	1.34	19.25	21.28
	3T_case3	4.06E–03	230	1.27	1.32	1.32	1.26	1.35	1.35	12.92	12.92
	3T_case4	7.20E–03	409	1.23	1.31	1.24	1.22	1.32	1.25	29.02	27.45
Average ratio		—	—	1.24	1.29	1.32	1.23	1.31	1.33	27.30	27.71

TFU structures. The power overhead and the buffer count are normalized against the original design. From the table, we can observe about 24% versus 32% power overhead on average caused by the TSV fault-tolerant techniques. Applying Double3-D, F-3-D, and A-F3-D needs 23%, 31%, and 33% more buffers, respectively. The buffers used in TFU structures account for 27% and 28% of the total buffers in F-3-D and A-F3-D, respectively.

We further compare the overhead in terms of clock network area and wire-length in Table 9.4. For fault-tolerant techniques, columns 2–4 report the area, which includes the cost of TFUs and all the clock TSVs. Columns 5–7 report the wire-length overhead. From the table, we can see that

TABLE 9.4

Area and Wire-Length Overhead Comparison

Circuits		Area (μm²)			Wire-Length (mm)		
		Double3-D	F3-D	A-F3-D	Double3-D	F3-D	A-F3-D
2 tiers	Industrial	117 200	92553	82655	1606.5	1630.0	1647.5
	2T_case1	2600	1870	1870	33.5	34.7	34.7
	2T_case2	2400	2282	2085	32.3	33.4	33.6
	2T_case3	2200	1847	1677	31.9	32.8	33.0
	2T_case4	1000	835	743	6.9	7.2	7.2
	2T_case5	2000	1259	1259	13.2	13.4	13.4
	2T_case6	2600	1659	1474	23.2	24.2	24.3
3 tiers	3T_case1	8000	741 i	5793	50.5	52.5	52.5
	3T_case2	7800	7222	5927	49.8	51.7	51.8
	3T_case3	3400	2517	2517	15.8	16.4	17.8
	3T_case4	2300	1664	790	27.6	29.7	29.5
Average ratio		1	0.80	0.70	1	1.02	1.03

TABLE 9.5

Skew Analysis

Circuits		Doublc3-D			F3-D			A-F3-D		
		Normal	Max	Avg	Normal	Max	Avg	Normal	Max	Avg
2 tiers	Industrial	88	93 (+5)	91 (+3)	85	94 (+9)	90 (+5)	87	96 (+9)	92 (+5)
	2T_case1	30	52 (+22)	37 (+7)	33	59 (+26)	38 (+5)	33	59 (+26)	38 (+5)
	2T_case2	25	54 (+29)	34 (+9)	25	65 (+40)	33 (+8)	24	68 (+44)	35 (+11)
	2T_case3	53	80 (+27)	59 (+6)	49	82 (+33)	58 (+9)	44	85 (+41)	54 (+10)
	2T_case4	24	43 (+19)	29 (+5)	22	43 (+21)	26 (+4)	17	45 (+28)	18 (+1)
	2T_case5	30	58 (+28)	35 (+5)	25	54 (+29)	26 (+1)	25	54 (+29)	26 (+1)
	2T_case6	33	78(445)	41 (+8)	55	70 (+15)	59 (+4)	59	71 (+12)	62 (+3)
3 tiers	3T_case1	38	64 (+26)	45 (+7)	52	62 (+10)	57 (+5)	56	65 (+9)	59 (+3)
	3T_case2	40	74 (+34)	43 (+3)	36	71 (+35)	46 (+10)	40	73 (+33)	48 (+8)
	3T_case3	30	46 (+16)	34 (+4)	35	46 (+11)	39 (+4)	35	46 (+11)	39 (+4)
	3T_case4	46	58 (+12)	51 (+5)	32	57 (+25)	37 (+5)	35	59 (+24)	41 (+6)
Average ratio		—			—	1.01	1.03	—	1.03	1.03

compared with Double3-D, our F3-D technique can reduce the area overhead by 20% and 30% on average, respectively. In terms of wire-length overhead, F3-D and A-F3-D result in a slight increase (2% and 3%, respectively). This convincingly demonstrates the area effectiveness of the proposed approach.

To demonstrate how the clock skew is affected by our techniques, we perform SPICE simulation to get the clock skews. The results are reported in Table 9.5. We report the normal skew where all TSVs are good, and perform Monte-Carlo simulation (assuming each TSV fails independently) to get the max skew and average skew under multiple TSV failures. The runs resulting in chip failure (e.g., both TSVs in a 2-TFU fail) are discarded. Columns 2–4, 5–7, and 8–10 correspond to the Double3-D, F3-D, and A-F3-D techniques, respectively. It is clear that Double3-D always has the smallest skew. Compared with it, F3-D has 1% and 3% increase in max skew and average skew, and A-F3-D has 3% and 3% increase. This is the penalty we have to pay to get the benefit in area. In all cases, we are able to suppress the skew below the 100 ps target. The skew can be further reduced at the cost of additional area for buffers.

9.6 CONCLUSIONS

In this chapter, we describe the problem and possible solutions about reliability issues in 3D-IC designs. The reliability issues majorly come from the fabrication defects of TSVs, and may cause serious timing violations and even a failed chip. To alleviate the problem, the concept of inserting spare TSVs and corresponding fault-tolerant structure are introduced. Then the problem of spare TSV assignment is detailed addressed. Finally, we address the problem about robustness of 3D clock tree, and described a fault-tolerant clock networks.

REFERENCES

1. http://www.prweb.com/releases/3d-4d-technology-market/forecasts-to-2020/prweb12140238.htm.
2. A. C. Hsieh, T. T. Hwang, M. H. Chang, M. H. Tsai, C. M. Tseng, and H. C. Li, TSV redundancy: Architecture and design issues in 3-D IC, in *Proceedings of Design, Automation and Test in Europe (DATE)*, Dresden, Germany, pp. 1206–1211, 2010.

3. V. Arunachalam and W. Burleson, Low-power clock distribution in a multilayer core 3-D microprocessor, in *Proceedings of the 18th ACM Great Lakes Symposium on VLSI (GLSVLSI)*, New York, pp. 429–434, 2008.

4. J. Auersperg et al., Nonlinear copper behavior of TSV and the cracking risks during BEoL-built-up for 3D-IC-integration, in *Proceedings of the International Conference on Thermal, Mechanical and Multi-Physics Simulation and Experiments in Microelectronics and Microsystems*, Cascais, Portugal, pp. 1–6, 2012.

5. B. Swinnen et al., 3D integration by Cu-Cu thermocompression bonding of extremely thinned bulk-Si die containing 10 µm pitch through-Si vias, in *Proceedings of International Electron Devices Meeting (IEDM)*, San Francisco, CA, pp. 1–4, 2006.

6. C. Albrecht, A. B. Kahng, B. Liu, I. I. Mandoiu, and A. Z. Zelikovsky, On the skew-bounded minimum-buffer routing tree problem, *IEEE Transactions on Computer-Aided Design of Integrated Circuits and Designs*, 22(7), 937–945, 2003.

7. C. L. Lung, Z. Y. Zeng, C. H. Chou, and S. C. Chang, Clock skew optimization considering complicated power modes, in *Proceedings of Design, Automation and Test in Europe (DATE)*, Dresden, Germany, pp. 1474–1480, 2010.

8. D. H. Kim et al., 3D-MAPS: 3D massively parallel processor with stacked memory, in *Proceedings of Solid-State Circuits Conference Digest of Technical Papers (ISSCC)*, San Francisco, CA, pp. 188–190, February 2012.

9. U. R. Tida, C. Zhuo, and Y. Shi, Novel through-silicon-via inductor based on-chip DC-DC converter designs in 3DICs, *ACM Journal on Emerging Technologies in Computing Systems*, 11(2), article 16, 2014.

10. U. R. Tida, C. Zhuo, and Y. Shi, Through-silicon-via inductor: Is it real or just a fantasy? in *Proceedings of Asia and South Pacific Design Automation Conference (ASPDAC)*, Singapore, pp. 837–842, 2014.

11. U. R. Tida, R. Yang, C. Zhuo, and Y. Shi, On the efficacy of through-silicon-via inductors, *IEEE Transactions on Very Large Scale Integration (VLSI) Systems*, 23(7), 1322–1334, 2014.

12. U. R. Tida, V. Mittapalli, C. Zhuo, and Y. Shi, "Green" on-chip inductors in three-dimensional integrated circuits, in *Proceedings of IEEE Computer Society Annual Symposium on VLSI*, Tampa, FL, 2014.

13. I. Loi, S. Mitra, T. H. Lee, S. Fujita, and L. Benini, A low-overhead fault tolerance scheme for TSV-based 3-D network on chip links, in *Proceedings of IEEE/ACM International Conference on Computer-Aided Design (ICCAD)*, San Jose, CA, pp. 598–602, 2008.

14. Y.-H. Chen, W.-C. Lo, and T.-Y. Kuo, Thermal effect characterization of laser-ablated silicon-through interconnect, in *Proceedings of Electronics System Integration Technology Conference*, vol. 1, pp. 594–599, Dresden, Germany, 2006.

15. Z. Zhang, Guideline to avoid cracking in 3D TSV design, in *Proceedings of IEEE Intersociety Conference on Thermal and Thermomechanical Phenomena in Electronic Systems (ITherm)*, Las Vegas, NV, pp. 1–5, 2010.

16. C. Zhang, M. Jung, S. K. Lim, and Y. Shi, Novel crack sensor for TSV-based 3D integrated circuits: Design and deployment perspectives, in *Proceedings of the International Conference on Computer-Aided Design (ICCAD)*, San Jose, CA, pp. 371–378, 2013.

17. J. H. Lau, TSV manufacturing yield and hidden costs for 3D IC integration, in *Proceedings of Electronic Components and Technology Conference (ECTC)*, Las Vegas, NV, pp. 1031–1042, June 2010.

18. N. Miyakawa, T. Maebashi, N. Nakamura, S.Nakayama, E. Hashimoto, and S. Toyoda, New multi-layer stacking technology and trial manufacture, in *Proceedings of the Architectures for Semiconductor Integration and Packaging*, 2007.

19. A. W. Topol et al., Enabling SOI-based assembly technology for three-dimensional integrated circuits, in *Proceedings of the IEEE International Electron Devices Meeting (IEDM) Technical Digest*, Washington, DC, pp. 352–355, 2005.

20. K. Arabi, T. Petrov, and M. Laisne, United State Patent Application Publication, US20100060310, 2010.

21. C. L. Lung, Y. S. Su, S. H. Huang, Y. Shi, and S. C. Chang, Through-silicon via fault-tolerant clock networks for 3-D ICs, *IEEE Transactions on Computer-Aided Design of Integrated Circuits and Systems*, 32(7), 1100–1109, July 2013.

22. Kang, U. et al., 8 Gb 3D DDR3 DRAM using through-silicon-via technology, in *Proceedings of the IEEE International Solid-State Circuits Conference (ISSCC)—Digest of Technical Papers*, pp. 130–132, 2009.

23. Kawano, M. et al., A 3D packaging technology for 4 Gbit stacked DRAM with 3 Gbps data transfer, in *Proceedings of the International Electron Devices Meeting (IEDM)*, San Francisco, CA, pp. 581–584, 2006.

24. Van der Plas, G. et al., Design issues and considerations for low-cost 3D TSV IC technology, in *Proceedings of the IEEE International Solid-State Circuits Conference Digest of Technical Papers (ISSCC)*, pp. 148–149, San Francisco, CA, 2010.

25. Yoshikawa, H. et al., Chip-scale camera module (CSCM) using through-silicon-via (TSV), in *Proceedings of the IEEE International Solid-State Circuits Conference (ISSCC)*, San Francisco, CA, pp. 476–477, 2009.

26. F. Wang, J. Kim, and M. Nowak, United State Patent Application Publication, US20100295600, 2010.

27. Y. G. Chen, W. Y. Wen, Y. Shi, W. K. Hon, and S. C. Chang, Novel spare TSV deployment for 3D ICs considering yield and timing constraints, *IEEE Transactions on Computer-Aided Design of Integrated Circuits and Systems (TCAD)*, 34(4), 577–588, 2014.

28. Y. Zhao, S. Khursheed, and B. M. Al-Hashimi, Cost-effective TSV grouping for yield improvement of 3D-ICs, in *Proceedings of Asian Test Symposium (ATS)*, New Delhi, India, pp. 201–206, November 2011.

29. F. Ye and K. Chakrabarty, TSV open defects in 3D integrated circuits: Characterization, test, and optimal spare allocation, in *Proceedings of ACM/IEEE Design Automation Conference (DAC)*, San Francisco, CA, pp. 1024–1030, June 2012.

30. Y.-G. Chen, K.-Y. Lai, M.-C. Lee, Y. Shi, W.-K. Hon, and S.-C. Chang, Yield and timing constrained spare TSV assignment for three-dimensional integrated circuits, in *Proceedings of Design, Automation and Test in Europe (DATE)*, Dresden, Germany, pp. 1–4, 2014.

31. F. X. Che, H. Y. Li, X. Zhang, S. Gao, and K. T. Teo, Wafer level warpage modeling methodology and characterization of TSV wafers, in *Proceedings of IEEE Electronic Components and Technology Conference (ECTC)*, Lake Buena Vista, FL, pp. 1196–1203, June 2011.

32. L. Jiang, Q. Xu, and B. Eklow, On effective TSV repair for 3D-stacked ICs, in *Proceedings of Design, Automation and Test in Europe Conference and Exhibition (DATE)*, Dresden, Germany, pp. 793–798, March 2012.

33. R. Glang, Defect size distribution in VLSI chips, in *Proceedings of International Conference on Microelectronic Test Structures (ICMTS)*, San Diego, CA, pp. 57–60, March 1990.

34. G. Di Natale, M.-L. Flottes, B. Rouzeyre, and H. Zimouche, Built-in self-test for manufacturing TSV defects before bonding, in *Proceedings of 2014 IEEE 32nd VLSI Test Symposium (VTS)*, Napa, CA, pp. 1–6, 2014.

35. L.-R. Huang, S.-Y. Huang, S. Sunter, K.-H. Tsai, and W.-T. Cheng, Oscillation-based pre-bond TSV test, *IEEE Transactions on Computer-Aided Design of Electronic Circuits (TCAD)*, 32(9), 1440–1444, September 2013.

36. S. Deutsch and K. Chakrabarty, Non-invasive pre-bond TSV test using ring oscillators and multiple voltage levels, in *Proceedings of the Design, Automation and Test in Europe Conference and Exhibition (DATE)*, Grenoble, France, pp. 1065–1070, March 2013.

37. Y. Fkih, P. Vivet, B. Rouzeyre, M.-L. Flottes, and G. Di Natale, 3D IC BIST for pre-bond test of TSVs using ring oscillators, in *Proceedings of the IEEE International New Circuits and Systems Conference (NEWCAS)*, Paris, France, pp. 1–4, 2013.

38. Y. Lou, Z. Yan, F. Zhang, and P. D. Franzon, Comparing Through-Silicon-Via (TSV) void/pinhole defect self-test methods, *Journal of Electronic Testing*, 28(1), 27–38, 2012.

39. H.-Y. Huang, Y.-S. Huang, and C.-L. Hsu, Built-in self-test/repair scheme for TSV-based three-dimensional integrated circuits, in *Proceedings of IEEE Asia Pacific Conference on Circuits and Systems (APCCAS)*, Kuala Lumpur, Malaysia, pp. 56–59, December 2010.

40. D. H. Kim, R. O. Topaloglu, and S. K. Lim, TSV density-driven global placement for 3D stacked ICs, in *Proceedings of International SoC Design Conference (ISOCC)*, Jeju, South Korea, pp. 135–138, November 2011.

41. H. Kim, K. Athikulwongse, and S. K. Lim, A study of Through-Silicon-Via impact on the 3D stacked IC layout, in *Proceedings of IEEE/ACM International Conference on Computer-Aided Design (ICCAD)*, pp. 674–680, San Jose, CA, November 2009.

42. J. Cong, G. Luo, and Y. Shi, Thermal-aware cell and through-silicon-via co-placement for 3D ICs, in *Proceedings of ACM/EDAC/IEEE Design Automation Conference (DAC)*, New York, pp. 670–675, June 2011.

43. M. Jung, J. Mitra, David Z. Pan, and S. K. Lim, TSV stress-aware full-chip mechanical reliability analysis and optimization for 3D IC, in *Proceedings of Design Automation Conference (DAC)*, New York, pp. 188–193, June 2011.

44. M. Cho, H. Ren, H. Xiang, and R. Puri, History-based VLSI legalization using network flow, in *Proceedings of ACM/IEEE Design Automation Conference (DAC)*, Anaheim, CA, pp. 286–291, June 2010.

45. C. L. Lung , H. C. Hsiao , Z. Y. Zeng, and S. C. Chang, LP-based multi-mode multi-corner clock skew optimization, in *Proceedings of the International Symposium on VLSI Design Automation and Test (VLSI-DAT)*, pp. 335–338, Hsin Chu, Taiwan, 2011.

46. V. F. Pavlidis , I. Savidis, and E. G. Friedman, Clock distribution networks for 3-D integrated circuits, in *Proceedings of IEEE Custom Integrated Circuits Conference*, pp. 651–654, San Jose, CA, 2008.

47. M. Mondal, A. Ricketts, S. Kirolos, T. Ragheb, G. Link, V. Narayanan, and Y. Massoud, Thermally robust clocking schemes for 3-D integrated circuits, in *Proceedings of the Design, Automation and Test in Europe Conference and Exhibition (DATE)*, Nice, France, pp. 1206–1211, 2007.

48. J. Minz, X. Zhao, and S. K. Lim, Buffered clock tree synthesis for 3-D ICs under thermal variations, in *Proceedings of Asia and South Pacific Design Automation Conference (ASPDAC)*, pp. 504–509, Seoul, South Korea, 2008.

49. T.-Y. Kim and T. Kim, Clock tree synthesis with pre-bond testability for 3-D stacked IC designs, in *Proceedings of the Design Automation Conference (DAC)*, Anaheim, CA, pp. 184–190, 2010.

50. D. L. Lewis and H.-H. S. Lee, A scan-island based design enabling pre-bond testability in die-stcked microprocessors, in *Proceedings of International Test Conference (ITC)*, Santa Clara, CA, pp. 1–8, 2007.

51. X. Zhao, D. L. Lewis, H.-H. S. Lee, and S. K. Lim, Pre-bond testable low-power clock tree design for 3-D stacked ICs, in *Proceedings of IEEE/ACM International Conference on Computer-Aided Design (ICCAD)*, San Jose, CA, pp. 184–190, 2009.

52. X. Zhao, D. L. Lewis, H.-H. S. Lee, and S. K. Lim, Low-power clock tree design for pre-bond testing of 3-D stacked ICs, *IEEE Transactions on Computer-Aided Design of Electronic Circuits (TCAD)*, 30(5), 732–745, 2011.

53. M. Edahiro, An efficient zero-skew routing algorithm, in *Proceedings of the Design Automation Conference (DAC)*, San Diego, CA, pp. 375–380, 1994.

54. X. Zhao, J. Minz, and S. K. Lim, Low-power and reliable clock network design for through silicon via based 3-D ICs, *IEEE Transactions on Components, Packaging and Manufacturing Technology (CPMT)*, 1(2), 247–259, 2011.

55. C. L. Lung , Y. S. Su , S. H. Huang , Y. Shi, and S. C. Chang, Fault-tolerant 3-D clock network, in *Proceedings of the Design Automation Conference (DAC)*, New York, pp. 645-651, 2011.

56. S. Hu, C. J. Alpert, J. Hu, S. K. Karandikar, Z. Li, W. Shi, and C. N. Sze, Fast algorithms for slew-constrained minimum cost buffering, *IEEE Transactions on Computer-Aided Design of Electronic Circuits (TCAD)*.

57. H. Wang, J. Kim, Y. Shi, and J. Fan, The effects of substrate doping density on the electrical performance of through-silicon vias, in *Proceedings of Asia-Pacific EMC Symposium*, Jeju, Korea, pp. 1–4, 2011.

10 Thermal Modeling and Management for 3D Stacked Systems

Tiansheng Zhang, Fulya Kaplan, and Ayse K. Coskun

CONTENTS

ABSTRACT

3D stacking technology allows integration of different technologies onto a single chip and provides performance benefits by decreasing communication latency. However, stacking resources vertically makes it harder to remove the generated heat, leading to elevated on-chip temperatures. High on-chip temperature is a limiting factor on the performance and reliability of the processor and incurs higher cooling costs. Thus, in order to unleash the true potential of 3D-stacked chips, accurate thermal modeling, design-stage thermal analysis, and development of efficient runtime management strategies are essential.

In this chapter, we elaborate on the state-of-the-art thermal modeling and management techniques targeting 3D-stacked systems. We focus on compact thermal modeling methods together with their integration into full system simulation. We then provide a selection of recently proposed dynamic thermal management techniques, which address the thermal challenges specific to 3D-stacked systems. These techniques include job scheduling, utilizing hardware control knobs, and active cooling using liquid microchannels. We conclude with a brief discussion of the open problems in the field.

10.1 INTRODUCTION

Three-dimensional (3D)-stacked architectures are getting increasingly attractive as they increase manufacturing yield, transistor density per chip footprint, and performance by reducing interconnect latency, while enabling the integration of layers manufactured with different technologies on the same chip. One of the major problems in 3D-stacked chips are the elevated temperatures due to increased thermal resistance brought by stacking multiple layers on top of each other. High temperatures impose a number of undesirable effects, including performance degradation, lifetime reduction due to reliability issues [26], and higher operational and cooling costs. This chapter discusses detailed thermal modeling approaches and the state-of-the-art techniques that address the high temperatures in 3D-stacked systems. We particularly focus on microchannel-based liquid cooling, which is a promising cooling solution for high-performance 3D systems.

Continued technology scaling with a stall in voltage scaling leads to increased on-chip power densities with each technology generation. In addition to that, 3D-stacked systems have higher power densities compared to their 2D counterparts, as they enable containing more transistors in the same footprint. Thus, the on-chip temperatures for 3D chips are higher than temperatures on 2D chips. The additional thermal resistance imposed by vertical layer stacking escalates the temperature problem, especially for the layers that are further away from the cooling infrastructure (i.e., heat spreader, heat sink or fans).

To address the thermal challenges and to efficiently utilize the true potential of 3D-stacked systems, novel design-time and run-time optimization techniques are essential. Job scheduling and dynamic task migration techniques that consider the strong thermal correlation between vertically neighboring on-chip components (i.e., core-to-core [6] and core-to-DRAM coupling [29]) provide significant reduction in thermal hot spots. Moreover, management through hardware control knobs such as dynamic voltage frequency scaling (DVFS) [11,17] provides finer temperature control at the cost of performance.

A detailed full system simulation framework is necessary for the design and accurate evaluation of such run-time management techniques. General simulation flow contains detailed modeling of the performance through microarchitectural simulation; modeling the power consumption of components such as cores, caches, and the network; and modeling the temperature behavior of each unit for a given processor geometry and power consumption levels. This chapter provides a step-by-step explanation of each of these modeling techniques.

Microchannel liquid cooling is also becoming an attractive solution to overcome temperature problems in 3D- stacked architectures due to the higher heat removal capability of liquids in comparison to air. Liquid cooling is performed by attaching a cold plate with built-in microchannels or by fabricating microchannels between the layers of the 3D chip. A coolant fluid is then pumped through the channels to remove the heat. The heat removal ability of this interlayer cooling approach scales with the number of layers. Current technology allows fabricating the infrastructure to enable interlayer liquid cooling. The IBM Zurich Research Laboratory has built a 3D chip that uses microchannel liquid cooling. Their cooling system can remove heat at a rate of 180 W/cm^2 per layer through 50 μm wide channels from a stack with 4 cm^2 footprint [9].

Despite its superior cooling performance, liquid cooling brings new management challenges. Large on-chip thermal gradients, which significantly deteriorate reliability [26], is one of them. The main reason for large thermal gradients is that the temperature of the coolant fluid rises as it flows along the channel and absorbs heat. Another challenge is the additional power consumption required to pump fluid into liquid-cooled systems. As the cooling requirement gets higher, coolant liquid needs to flow at a faster rate, resulting in an increase in the pump power. For energy-efficient computing purposes, it is desirable to have a cooling energy that constitutes a very small percentage of the total energy consumption. This section also discusses the proposed techniques to overcome the challenges faced in 3D systems with liquid cooling [5,20,21].

- We provide a detailed overview of approaches regarding thermal modeling of 3D-stacked systems (Section 10.2.1). We then discuss the steps of developing a full system simulation framework combining performance, power, and temperature simulation (Section 10.2.2). We finally show how to model the temperature behavior in a 3D-stacked system with microchannel liquid cooling (Section 10.2.3). The liquid cooling model includes fine-grained computation of the heat spread and considers the impact of through-silicon-vias (TSVs) and liquid flow within the microchannels.

- We carry out a detailed survey of the state-of-the-art techniques targeting improved energy efficiency under temperature and performance constraints in 3D-stacked systems. These techniques can be grouped into three main categories: (1) Job allocation and migration techniques (Section 10.3.1), (2) techniques implementing architectural modifications (Section 10.3.2), and (3) techniques improving the liquid-cooling efficiency (Section 10.3.3).

10.2 THERMAL MODELING

10.2.1 THERMAL MODELING OF 3D-STACKED SYSTEMS

Thermal challenges in 3D-stacked systems due to higher power density and increased thermal resistance should be considered while maximizing the performance gains of 3D integration. To enable early-stage design space exploration and avoid thermal issues in the manufactured chips, accurate thermal modeling of 3D-stacked systems is of great importance.

Chip-level thermal modeling has been studied extensively, which can be categorized into two types in general. The first type is finite element method (FEM), which is used in softwares such as COMSOL [1]. FEM divides a chip into many small subdomains and uses variational methods to model the thermal condition of the whole chip [19], as shown in Figure 10.1a. This method provides high accuracy; however, it is time-consuming and computation-intensive and thus, not suitable for system-level simulations [22]. The second commonly used modeling type is compact thermal modeling, such as HotSpot [23], which simulates the chip as a thermal resistor–capacitor (RC) network, as shown in Figure 10.1b. In this modeling method, the chip is represented by a floor plan and separated into small blocks, each of which acts as a heat dissipation source. Solving the thermal RC network for a given power distribution gives the temperature of each block [23]. Compact thermal modeling compromises between the simulation time and accuracy and is suitable for design space exploration. For various proposed compact thermal modeling techniques [10,23,28], FEM is usually used to verify the accuracy.

In 3D systems, the introduction of stacked layers, bonding techniques (e.g., TSVs, microbumps), and liquid cooling technique increases the complexity of thermal modeling. Other work [14] extends HotSpot with the ability to model multiple stacked layers inside a chip. Meng et al. [17] provide further improvement that enables modeling layers with blocks that have a heterogeneous set of thermal resistivity and specific heat capacity values (see Figure 10.2a). Their heterogeneous 3D thermal model was integrated as an extension to HotSpot-5.02, and it is currently included as a feature of the new release (HotSpot-6.0). Their approach allows finer modeling of blocks such as the TSVs, thus, provides better accuracy compared to assuming homogeneous layers. For example, TSVs are made of copper, which have much lower thermal resistivity than thermal interface material that lies between

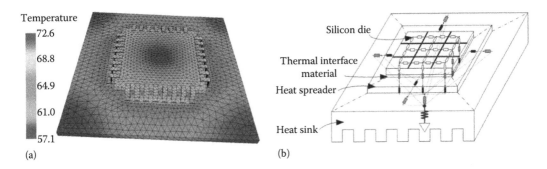

(a)

(b)

FIGURE 10.1 Thermal modeling examples of (a) finite element method (From Ellison, G., Thermal spreading and more using open-source FEA software, http://www.electronics-cooling.com/2013/06/thermal-spreading-and-more-using-open-source-fea-software/, accessed on August 31, 2015) and (b) compact thermal modeling. (From Skadron, K. et al., Temperature-aware microarchitecture, in: *ACM SIGARCH Computer Architecture News*, Vol. 31, ACM, New York, pp. 2–13, 2003.)

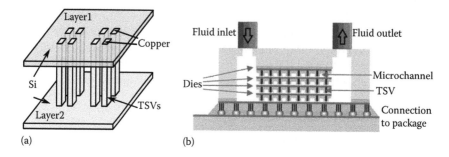

(a)   (b)

FIGURE 10.2 Thermal modeling of (a) heterogeneous material within one layer (Meng, J., Optimizing energy efficiency of 3D multicore systems with stacked DRAM under power and thermal constraints, in: *Proceedings of Design Automation Conference*, ACM, New York, pp. 648–655, 2012) and (b) liquid cooling in 3D-stacked system. (From Sridhar, A. Vincenzi, A., Ruggiero, M., Brunschwiler, T., and Atienza, D. 3D-ICE: Fast compact transient thermal modeling for 3D ICS with inter-tier liquid cooling, In *ICCAD*, pp. 463–470, © 2010 IEEE.)

the stacked layers. Thus, the area around the TSV groups generally has a lower temperature owing to better heat conduction. The ability to model heterogeneous materials on the same layer also allows us to explore novel technologies, such as silicon-photonic devices, implemented in 3D systems.

Since TSVs are used not only for data transfer among layers but also for power delivery, a 3D system usually needs a large quantity of TSVs. In order to investigate the impact of TSVs on on-chip thermal conditions, other compact thermal models [15,28] are proposed with FEM validation.

To address the thermal challenges in 3D-stacked systems, liquid cooling, which refers to pumping liquid through intertier microchannels to remove heat, has emerged as a promising solution (see Figure 10.2b) [5,25]. The main principles of thermal modeling of 3D chips with liquid cooling are presented in Section 10.2.3.

These existing thermal modeling tools allow for extensive design space exploration and thermal analysis in 3D-stacked systems with various techniques and integration choices.

10.2.2 Integration of Full-System Simulation and Thermal Modeling

On-chip thermal conditions are changing as applications and application phases change. A good chip design should prevent thermal violation during runtime as much as possible. For the purpose of investigating the runtime thermal conditions, thermal modeling needs to be integrated with performance and power simulations.

A typical workflow and integration of performance, power, and thermal simulation is presented in Figure 10.3. With a defined system architecture and application space, we can generate performance statistics traces (e.g., committed instructions per cycle (IPC), number of reads/writes and cache accesses) for the system and microarchitectural components within. Then, with the system architecture parameters, power simulator can convert these performance traces into power consumption traces for each component. In the end, thermal simulator takes the power traces and the corresponding chip floor plan as inputs and generates thermal profiles of the target system when running the defined applications. From the thermal simulation results, we are able to tell whether

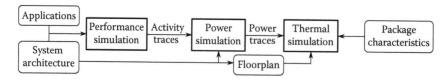

FIGURE 10.3 An example of performance, power, and thermal simulation framework.

the system operates within thermal constraints and design thermal management policies accordingly. The details for each stage in Figure 10.3 are elaborated as follows:

Performance simulation: Performance simulation is conducted using a full system simulator, which is a software-based tool that mimics the working flow of processor components to estimate outputs and performance metrics on a given architecture and an input application. A full system simulator (such as Gem5 [4]) includes detailed models for processor cores, memory system and peripheral devices, etc. The outputs of a full system simulator contain the activity traces for architectural components (e.g., load/store queues, branch predictor, register files, and caches), which can be used for performance analysis, architecture/application optimization as well as power/thermal simulations.

Power simulation: Power simulation takes the performance simulation outputs, architectural parameters, and technology node as inputs and generates estimated dynamic/static power values for the architectural components. Power simulators such as McPAT [13], CACTI [27], and ORION [12] are proposed to estimate power consumption for core, cache, and network, respectively. The output power traces from power simulators could be used for designing power management techniques and conducting thermal analysis.

Thermal simulation: Thermal simulations incorporate the thermal models described in Section 10.2.1 and the power traces from power simulations to generate temperature traces for the system. The framework shown in Figure 10.3 allows for choosing a static power trace for each round of thermal simulation. However, when designing thermal management techniques, the impact of thermal management decisions on the performance must also be considered and simulated. For example, when employing DVFS technique on a system, if the chip temperature threshold is violated, the system needs to switch to a lower voltage-frequency (V-F) setting so as to operate within the predefined temperature constraints. In such cases, a static power trace is not sufficient to meet the simulation requirements because each static trace corresponds to one performance simulation run under a fixed frequency. A framework, which allows for thermal simulations with either a fixed time interval or a fixed number of instructions, addresses this issue [17] (see Figure 10.4). In this framework, a database is constructed with performance/power traces based on a fixed instruction count interval. Thermal simulator dynamically polls this database for the power consumption values according to the operating V-F setting, as determined by a runtime policy. It takes the performance data (e.g., IPC) to convert the power traces from instruction-count-based samples to time-based samples as required by thermal simulations. Such an approach significantly speeds up dynamic thermal management development compared to using static power traces.

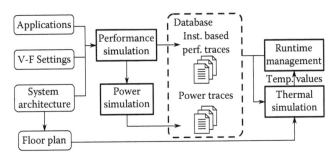

FIGURE 10.4   Improved performance, power, and thermal simulation frame-work with thermal impacts on performance. (From Meng, J., Optimizing energy efficiency of 3-D multicore systems with stacked DRAM under power and thermal constraints, in: *Proceedings of Design Automation Conference*, ACM, New York, pp. 648–655, 2012.)

10.2.3 THERMAL MODELING WITH ACTIVE COOLING

Active cooling refers to the techniques which require powered devices to remove heat from a computing system. These techniques include forced air cooling with fans, thermoelectric cooling, and liquid cooling through microchannels. Thermoelectric coolers use the Peltier effect for heat transfer such that when current passes through the device, heat is absorbed at one junction and rejected at the other, acting like a heat pump [18]. Liquid cooling with microchannels is a promising solution especially for 3D-stacked architectures, due to higher heat removal capability of liquids in comparison to air. Liquid cooling is performed by attaching a cold plate with built-in microchannels, and/or by fabricating microchannels between the layers of the 3D chip. A pump is then used to pump a coolant fluid (i.e., water or other fluid) through the microchannels to remove heat.

There are various thermal modeling approaches proposed for the design and evaluation of liquid-cooled 3D-ICs [5,8,20,25]. Thermal modeling of 3D-stacked systems with liquid cooling consists of three main steps: (1) constructing a grid level thermal R-C network of the 3D stack, (2) detailed modeling of the interlayer material including the microchannels and the TSVs, and (3) modeling the impact of pump and coolant flow rate. Coskun et al. propose a 3D thermal model with liquid cooling feature that they integrate into HotSpot-4.02 simulator as an extension [5]. HotSpot simulator constructs a 3D thermal RC network for a given physical geometry and grid size. Next step is the thermal characterization of the interlayer materials. Figure 10.5a shows an example layer with grid cells representing interface material, TSVs, and microchannels. Each of these different materials has different thermal resistivity and thermal capacitance values. By default, HotSpot assumes homogeneous thermal properties throughout the layer. In order to model the heterogeneous thermal characteristics of the interlayer materials, the proposed approach [5] (1) enables assigning various resistivity values to each grid cell and (2) allows the resistivity values to vary at runtime. Item (1) models the thermal impact of having different materials on specific locations of the layer and item (2) models the impact of changing the coolant flow rate at runtime.

Figure 10.5b depicts the resistive network that is used to compute the junction temperature of a 3D layer with liquid cooling. T_{S1} and T_{S2} represent the temperatures of the top and bottom layers, while $\dot{q}_1$ and $\dot{q}_2$ stand for the heat sources of the top and bottom layers, respectively. Three main resistances are combined to model the 3D stack: the thermal resistance of the back-end-of-line (BEOL) layers (R_{BEOL}), the silicon slab thermal resistance (R_{slab}), and the convective thermal resistance (R_{conv}). It is assumed that the top and bottom of the microchannel have the same temperature, and junction temperature $T_j = T_{S1}$ is computed accordingly.

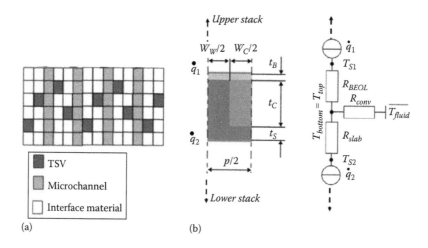

(a)   (b)

FIGURE 10.5   (a) Example layer with grid cells representing different interlayer materials; (b) cross-sectional view of the 3D layers and the resistive network.

Three components contribute to the rise of the junction temperature: (1) the thermal gradient due to conduction (ΔT_{cond}), (2) the rise in the coolant temperature due to heat absorption along the channel (ΔT_{heat}), and (3) the thermal gradient due to convection (ΔT_{conv}), which is independent of the flow rate when hydrodynamic and thermal boundary layers are reached. Thus, we can compute the total temperature rise at the junction as follows:

$$\Delta T_j = \Delta T_{cond} + \Delta T_{heat} + \Delta T_{conv} \tag{10.1}$$

Thermal gradient due to conduction is computed using Equations 10.2 and 10.3, where t_B and k_{BEOL} are the thickness and the conductivity of the BEOL layer, respectively.

$$\Delta T_{cond} = R_{th-BEOL} \cdot \dot{q}_1 \tag{10.2}$$

$$R_{th-BEOL} = \frac{t_B}{k_{BEOL}} \tag{10.3}$$

Equations 10.4 and 10.5 are used to compute the temperature rise due to heat absorption of the coolant through the channel. Here, A_{heater} is the total area that consumes power, c_p is the specific heat capacity of the coolant, ρ is the density of the coolant, and V (L/min) is the volumetric flow rate of the coolant.

$$\Delta T_{heat} = (\dot{q}_1 + \dot{q}_2) \cdot R_{th-heat} \tag{10.4}$$

$$R_{th-heat} = \frac{A_{heater}}{c_p \cdot \rho \cdot \dot{V}} \tag{10.5}$$

Similarly, Equations 10.6 and 10.7 are used to compute the convective portion of the temperature gradient. h is the heat transfer coefficient and it is dependent on the hydraulic diameter, Nusselt number, and conductivity of the coolant. w_c, t_c, and p are the width, thickness, and the pitch of the channel as shown in Figure 10.5b.

$$\Delta T_{conv} = (\dot{q}_1 + \dot{q}_2) \cdot h_{eff} \tag{10.6}$$

$$h_{eff} = h \cdot \frac{2 \cdot (w_c + t_c)}{p} \tag{10.7}$$

Final step in the liquid-cooling modeling is modeling the behavior of the pump and liquid flow rate. In a liquid-cooled 3D-stacked system, all the microchannels are connected to a pump that injects the coolant fluid at a required pressure difference. The fluid velocity is dependent on the applied pressure difference, and it is governed by Equations 10.8 through 10.10:

$$v_{bulk} = \frac{v_{darcy}}{\epsilon} \tag{10.8}$$

$$v_{darcy} = \frac{\kappa}{\mu} \cdot \nabla P \tag{10.9}$$

$$\nabla P = \frac{\Delta P}{L} \tag{10.10}$$

where
v_{bulk} is the actual coolant velocity in the channel
ϵ, κ, μ, and ΔP are the cavity porosity, cavity permeability, dynamic viscosity, and the pressure difference between the inlet and the outlet ports of the pump, respectively

Higher ΔP leads to higher liquid flow rate, and thus, provides smaller junction temperature according to Equations 10.4 and 10.5. However, there is a maximum limit for ΔP required by the system (i.e., prior work uses 1 bar [5]). The pump power also linearly increases with ΔP.

3D-ICE is another compact transient thermal model proposed for the simulation of 3D ICs with multiple intertier microchannel liquid cooling [25]. It uses a similar grid structure for modeling the heterogeneous thermal properties of TSVs and liquid microchannels as in prior work [5]. 3D-ICE provides further improvement by adding the ability to model the rise of temperature along the channel as the liquid absorbs heat. Other work proposes simulation of two-phase energy and mass balance (STEAM) [24], a compact simulator that models two-phase liquid cooling, focusing on the liquid-vapor phase change.

10.3 THERMAL MANAGEMENT

Thermal management is an essential part of any computing system today to prevent thermal-related reliability issues (e.g., thermal cycling, negative-bias temperature instability [26]) as well as improve system energy efficiency. There are three major ways of dynamic thermal management in 3D systems. In software-level, job allocation and migration can be used for changing the on-chip thermal conditions by adjusting the distribution of the workload. In hardware-level, techniques such as DVFS and clock gating can change the operating status of on-chip components and thus, their temperatures. In system-design level, cooling technique (e.g., liquid cooling) is another important control knob for thermal management.

10.3.1 JOB ALLOCATION AND MIGRATION

Job allocation and migration, or task scheduling, is extensively used for dynamic thermal management in 3D-stacked systems. Due to the strong temperature correlation between vertically neighboring on-chip components, when allocating/migrating jobs in the system, it is also important to take the cross-layer thermal interference into consideration. Researchers have proposed job allocation/migration policies for both logic-on-logic and memory-on-logic 3D systems.

Adapt3D job allocation policy [6] is proposed to address the thermal challenges for logic-on-logic 3D systems. This policy assigns a thermal index to each core based on its location in a 3D system and dynamically updates a probability value (P_t) for each core to determine the workload allocation at time t. The P_t values are computed at a given time interval as follows:

$$P_t = P_{t-1} + W \tag{10.11}$$

$$W_{diff} = T_{pref} - T_{avg} \tag{10.12}$$

$$W = \begin{cases} \beta_{inc} \cdot W_{diff} \cdot \dfrac{1}{\alpha_i} & : T_{pref} \geq T_{avg} \\ \beta_{dec} \cdot W_{diff} \cdot \dfrac{1}{\alpha_i} & : T_{pref} \leq T_{avg} \end{cases} \tag{10.13}$$

where
   W is the weight factor
   T_{pref} is the preferred operating temperature
   T_{avg} is the average temperature observed in the history window
   α_i ($0 < \alpha_i < 1$) is the thermal index of core i
   β is an empirically determined constant to decide the rate of change in the probability values

Using the equations above, *Adapt3D* computes the probabilities for each core at every scheduling time interval. This policy prefers to allocate jobs to the cores that are less likely to heat up in the near future so as to balance temperatures across the chip. If the temperature of a core exceeds the temperature threshold value, its probability value is set to 0 to avoid heating up further and to prevent reliability failures. The test system floor plans are shown in Figure 10.6, and the experimental results are presented in Figure 10.7. The experimental comparison group includes widely used and state-of-the-art thermal management policies for 2D systems: dynamic load balancing (*Default*), clockgating (*CGate*), temperature-triggered DVFS (*DVFS TT*), utilization-based DVFS (*DVFS Util*), floorplan-induced DVFS (*DVFS FLP*), task migration (*Migr*), and random policy with temperature-aware adaption (*AdaptRand*). The results demonstrate that, compared to 2D system-based thermal balancing techniques, *Adapt3D* reduces thermal hot spots by up to 32% with much lower spatial and temporal temperature gradients.

For logic-on-memory 3D stacking system, Meng and Coskun [16] provide a simulation framework with joint performance, power, and thermal models. This work also proposes a memory management policy for the memory-intensive applications that have varying access patterns to DRAM banks. The policy performs temperature-aware virtual address to physical address mapping in order to decrease the peak temperature and thermal variations. However, this work does not consider the

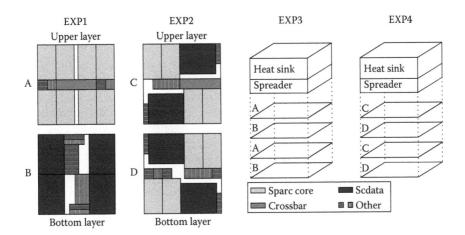

FIGURE 10.6 Floorplans for experimental evaluations. (From Coskun, A.K., Ayala, J.L., Rosing, T.S., Leblebici, Y., Dynamic thermal management in 3D multicore architectures, in: *Proceedings of the Conference on Design, Automation and Test in Europe*, Leuven, Belgium, pp. 1410–1415, © 2009 IEEE.)

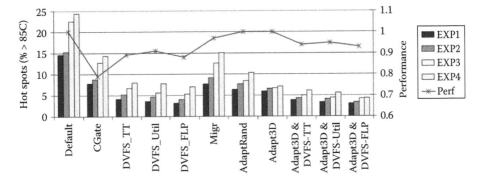

FIGURE 10.7 Thermal hot spots and performance comparison across temperature management techniques. (From Coskun, A.K., Ayala, J.L., Rosing, T.S., Leblebici, Y., Dynamic thermal management in 3D multicore architectures, in: *Proceedings of the Conference on Design, Automation and Test in Europe*, Leuven, Belgium, pp. 1410–1415, © 2009 IEEE.)

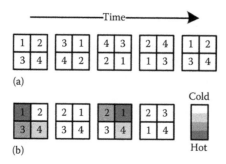

(a)

(b)

Cold

Hot

FIGURE 10.8 Thermal hot spots and performance comparison across temperature management techniques. (a) Rotation and (b) pair-wise 1–4 migration. (From Zhao, D., Homayoun, H., Veidenbaum, A.V., Temperature aware thread migration in 3D architecture with stacked DRAM, in: *Proceedings of IEEE International Symposium on Quality Electronic Design*, Santa Clara, CA, pp. 80–87, © 2013 IEEE.)

thermal interference between logic and memory layers and the thermal conditions under multi-program workloads. Recent work [29] demonstrates through simulations that for 3D systems with stacked DRAM, logic layer components have strong impacts on DRAM layer's temperature. They propose using dynamic thread migration to reduce the peak temperature and thermal variations in a 4-core 3D system for multi-program workloads. There are three baseline migration algorithms used as the comparison group (see Figure 10.8). *Rotation* algorithm migrates the threads clockwise every time interval. This algorithm does not require on-chip temperature sensors; however, rotating all four threads every time interval has a relatively high cost compared to the remaining ones. *Pair-wise* algorithm, on the other hand, requires temperature sensors for each core on the chip. During each time interval, this algorithm reads temperature sensors; sorts the cores according to their temperature values and swaps a pair of cores to adjust the temperature map. For example, *Pair-wise 1–4* swaps the hottest core with the coolest core. As for the proposed dynamic thread migration algorithm, the authors set up a thermal variation threshold to prevent unnecessary migrations so as to reduce performance loss due to migrations. The experimental results demonstrate that *Dynamic Pair-wise 1–4* algorithm reduces the number of migrations by 43% while maintaining similar thermal variation as *Rotation* provides.

Job migration could significantly improve the on-chip thermal conditions since it adapts to system runtime status. A proper migration frequency based on the migration overhead is as important as the migration policy itself.

10.3.2 Thermal Management through Hardware Control Knobs

Thermal management using hardware control knobs include the employment of dynamic voltage and frequency scaling (DVFS), clock gating, fetch gating, as well as dynamic reconfiguration of architectural resources. Compared to job allocation and migration, these techniques provide more flexible and finer control knobs for thermal management but at a higher sacrifice of performance at the same time.

Systems with DVFS features provide several available V-F settings for cores to operate at, and switching from a high V-F setting to a low V-F setting helps lower the temperature of a core. When utilizing DVFS for dynamic thermal management, determining which V-F setting a core should operate at is quite important, because the thermal constraint may be violated with a high V-F setting and a low V-F setting may decrease performance dramatically. A recent paper [17] proposes a novel runtime policy to select V-F setting to meet power and thermal constraints while maximizing performance for a 3D multicore system with stacked DRAM. The proposed policy utilizes instruction per cycle (IPC) and memory access per instruction (MA) to construct an offline regression

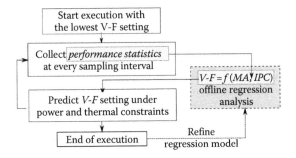

FIGURE 10.9 The flowchart of the runtime policy. (From Meng, J., Optimizing energy efficiency of 3D multicore systems with stacked DRAM under power and thermal constraints, in: *Proceedings of Design Automation Conference*, ACM, New York, pp. 648–655, 2012.)

model in order to predict the appropriate V-F setting for the 3D system (see Figure 10.9). The applied regression model is as follows:

$$VF = c_0 + c_1 \cdot MA + c_2 \cdot IPC + c_3 \cdot MA \cdot IPC \qquad (10.14)$$

IPC and MA are used since they are good indicators for the power of the logic layer and the DRAM layer, respectively. This regression model is trained with power and performance statistics from simulations across PARSEC [3] and NAS Parallel benchmarks [2]. Since MA and IPC vary with the V-F setting, a set of coefficients is computed for each V-F setting [17].

Figure 10.10 shows an example temperature/V-F setting trace when running *ua*, a NAS benchmark. For this application, 1.4GHz/1.02V is the reliable static operating point with a temperature threshold of 85°C. The proposed policy takes advantage of the temperature slack and switches to 1.7GHz during periods of low temperature in order to achieve the highest performance. In comparison to temperature-triggered DVFS policy, the proposed policy improves EDP by up to 61.9% and IPS by 32.2% on average across all benchmarks.

In spite of its advantages in thermal management, DVFS induces performance degradation due to lower operating frequency and the overhead of scaling the frequency/voltage. Skadron et al. [23] firstly introduce the concept of adapting architectural resources for temperature management.

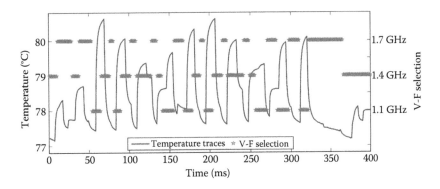

FIGURE 10.10 Temperature trace of application *ua* from NAS Parallel Benchmark Suite [2] on the 3D system running at 1.4GHz/1.02V and the V-F setting selected by proposed runtime management policy. (From Meng, J., Optimizing energy efficiency of 3D multicore systems with stacked DRAM under power and thermal constraints, in: *Proceedings of Design Automation Conference*, ACM, New York, pp. 648–655, 2012.)

Jayaseelan and Mitra [11] propose a method to tailor the architectural parameters to track the available instruction-level parallelism of the program in order to reduce temperature with minimal performance loss. Their framework is composed of (1) a neural-network-based classifier that filters out the thermally unsafe configurations, (2) a fast performance prediction model for any hardware configuration, and (3) an efficient configuration space search algorithm (see Figure 10.11). At runtime, for each time interval, processor parses performance counter information to the framework and gets the optimal configuration in return. Based on the statistics from the processor side, the neural network classifier decides whether the current architecture configuration is thermally safe or unsafe. If it is thermally unsafe, the configuration search algorithm will search for the best performing configuration that is within the thermal limit. The performance prediction model then takes this configuration, predicts issued IPC and feeds back to the classifier for thermal safety check.

The results, as shown in Figure 10.12, demonstrate the performance benefits of the proposed method (*AdaptiveDTM*). In this figure, *DVS* refers to a PI-controller based voltage scaling scheme that has a threshold of 81.8°C for voltage transitions; *HybridDTM* combines fetch gating and *DVS* for thermal management. On average, *AdaptiveDTM* has 11.68% performance slowdown while *DVS* and *HybridDTM* have 24.4% and 19.37%, respectively.

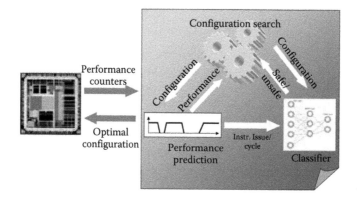

FIGURE 10.11 Framework for dynamic thermal management using the hardware control knobs. (From Jayaseelan, R. and Mitra, T., Dynamic thermal management via architectural adaptation, in: *Proceedings of Design Automation Conference*, New York, pp. 484–489, © 2009 IEEE.)

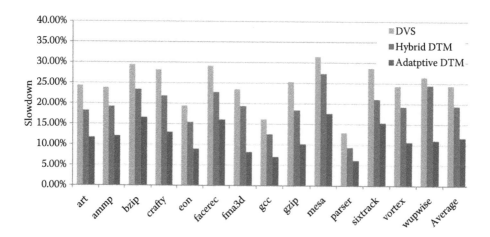

FIGURE 10.12 Performance comparison of various dynamic thermal management policies. (From Jayaseelan, R. and Mitra, T., Dynamic thermal management via architectural adaptation, in: *Proceedings of Design Automation Conference*, New York, pp. 484–489, © 2009 IEEE.)

10.3.3 MANAGEMENT OF 3D-STACKED SYSTEMS WITH ACTIVE COOLING

A number of dynamic management strategies improve the energy efficiency of 3D-stacked systems by optimizing for cooling efficiency [5,20,21]. These strategies include liquid flow rate adjustment [5], job allocation combined with DVFS [20], and microchannel width modulation [21].

Coskun et al. propose adjusting the liquid flow rate at runtime to save pump power [5]. In multicore systems, the workload varies over time and the system is generally not 100% utilized. Thus, setting the coolant flow rate based on worst-case temperature creates inefficiency by consuming higher pump power than needed. In order to avoid that, a novel controller was developed, which adjusts the liquid flow rate to minimize the pump power while operating under a temperature threshold. They monitor the temperature and predict the maximum temperature of the next interval by using autoregressive moving average (ARMA) technique. The input to the flow rate controller is the predicted maximum temperature and the output is the required flow rate for the next interval. While designing the controller, they first carry out offline analysis to find the minimum flow rate that should be applied to cool down the system from a maximum temperature of $T_{\max}$ to the target operating temperature of 80°C. They then create a look-up table where each entry corresponds to a flow rate associated with a $T_{\max}$ value. At runtime, they pick the appropriate flow rate from the table based on the predicted maximum temperature. Due to the time delay in adjusting the flow rate, forecasting the temperature provides the ability to adapt to the changes in the system on time. They combine their liquid flow rate control mechanism with temperature-aware load balancing (*TALB*) for scheduling jobs. Their liquid flow control algorithm combined with *TALB* achieves 10% energy savings on average in comparison to using worst-case flow rate. For lower utilization workloads, it achieves up to 12% total energy savings (chip and pump energy) and 30% cooling energy reduction.

Fuzzy controller approach was proposed to decide on the V-F settings of the cores and the liquid flow rate at runtime [20]. Before designing the controller, they carry out extensive analysis of the individual impacts of (1) task scheduling, (2) variable fluid flow rate, and (3) DVFS on the resulting temperature profile of the chip. Their analysis shows that the cores that are located closer to the fluid outlet port are hotter than the ones located near the inlet port due to the heat absorption of the liquid. Thus, the cores closer to the outlet ports benefit more from higher liquid flow rates. Based on the same observation, they also show that lower peak temperatures and lower thermal gradients can be achieved by allocating jobs to the cores according to their distance from the inlet port. Their flow-aware 3D load balancing technique (*FALB*) adjusts the workload queue length of the individual cores based on their distance to the fluid inlet port (i.e., the cores that are away from the inlet port have shorter queues, and thus, are utilized less). Next, they design a rule-based fuzzy controller which takes the temperature (*T*), workload utilization (*U*), and the relative distance of the core from the inlet port (*D*) as inputs and gives the liquid flow rate and V-F setting as the output. Figure 10.13 illustrates the proposed fuzzy thermal management technique (*LC FUZZY*), where the fuzzy controller operates as an interrupt-driven software routine. At every temperature sampling interval (100 ms), the routine is triggered and the values for *T* and *U* are acquired from temperature sensors and from the workload allocated to each core's queue, respectively. Fuzzy controller then selects the appropriate V-F setting for each core as well as the flow rate.

They implement various dynamic thermal management policies and compare them against the proposed *LC FUZZY* policy. *Liquid cooling with LB (LC LB)* applies maximum flow rate and schedules jobs with load balancing. *LUT-based flow rate control with LB (LC VAR)* uses the look-up table-based flow rate adjustment introduced by Coskun et al. [5], and uses *LB* scheduler. Experimental evaluation shows that for a 4-tier 3D-stacked system, *LC FUZZY* provides 21% and 18% peak system energy savings in comparison to *LC LB* and *LC VAR*. Moreover, applying *FALB* scheduler together with *LC FUZZY* reduces the frequency of intralayer thermal gradients (i.e., larger than 15°C) to 50% in comparison to 80% when using *LC FUZZY* only.

Other work tackles the problem of large on-chip thermal gradients brought by microchannel-based liquid cooling. GreenCool was proposed as an optimal design method to minimize the cooling

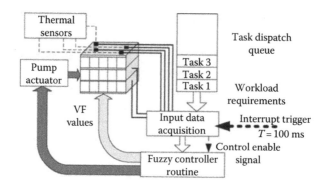

FIGURE 10.13 The flowchart of the proposed management scheme with fuzzy controller. (From Sabry, M.M., Coskun, A.K., Atienza, D., Rosing, T.S., and Brunschwiler, T., Energy-efficient multiobjective thermal control for liquid-cooled 3-D stacked architectures, *IEEE Transactions on CAD of Integrated Circuits and Systems*, 30(12), 1883–1896, © 2011 IEEE.)

energy while maintaining the thermal gradients and peak temperatures under safe limits [21]. They achieve this through adjusting the heat transfer characteristics of the microchannels by channel width modulation. Two main factors contribute to the thermal gradients in liquid-cooled ICs: (1) nonuniform on-chip heat distribution, and (2) sensible heat absorption by the coolant, creating uneven heat sinking from inlet to outlet. The latter is far more dominant than the former one. Figure 10.14 illustrates this effect for a single microchannel cooling a strip of silicon chip with uniform heat flux distribution. The coolant flows along the z-axis and Figure 10.14a plots the junction temperature along the z-axis showing the individual components as discussed in Section 10.2.3 for a uniform nonmodulated channel. ΔT_{heat} (i.e., represents sensible heat absorption) increases along the z-axis and it is the primary contributor of thermal gradients. The thermal gradient in Figure 10.14a can be reduced by modifying one of the contributors. GreenCool considers modifying ΔT_{conv}, as it does not incur major changes in the fabrication process and does not require additional pump power. ΔT_{conv} depends on the channel aspect ratio, and thus, it is possible to create a decreasing ΔT_{conv} profile by modulating the width of the channel from inlet to outlet (see Figure 10.14b).

Drawing upon these observations, *GreenCool* finds the optimal channel width profile that minimizes the pumping energy under thermal gradient (ΔT_{max}) and peak temperature (T_{max}) constraints. The authors evaluate the benefit of *GreenCool* with channel width modulation on a 3D MPSoC with stacked on-chip DRAM through detailed simulations. *GreenCool* with channel width modulation improves power usage effectiveness (i.e., PUE = Total power consumption/Computational power consumption) by up to 54% and on average by 11.4% when $T_{max}= 60°C$ and $\Delta T_{max} = 10°C$. Similarly, when $\Delta T_{max} = 12°C$, 6% average and 35% peak PUE enhancement is observed.

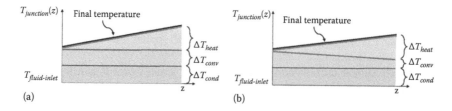

FIGURE 10.14 Junction temperature distribution for (a) uniform nonmodulated channel width and (b) modulated channel width. (From Sabry, M.M., Sridhar, A., Meng, J., Coskun, A.K., and Atienza, D., Greencool: An energy-efficient liquid cooling design technique for 3-D MPSoCs via channel width modulation, *IEEE Transactions on CAD of Integrated Circuits and Systems*, 32(4), 524–537, © 2013 IEEE.)

10.4 SUMMARY AND CONCLUSIONS

3D stacking technology promises significant performance benefits, however, it also raises thermal challenges due to higher thermal resistance along the heat transfer path. In order to avoid hot spots and get the most out of 3D-stacked processors, detailed modeling and optimization of these systems are necessary. This chapter has reviewed detailed thermal modeling approaches for 3D-stacked systems including consideration of the TSVs and liquid microchannels. The chapter has also provided a description of the full system simulation framework that integrates the performance, power and thermal simulation. Finally, it has demonstrated an analysis of the state-of-the-art thermal management techniques targeting improved energy-efficiency in 3D systems with and without microchannel liquid cooling.

There are many open problems regarding the thermal management and energy efficiency optimization of 3D-stacked architectures. 3D-stacking allows integration of multiple layers manufactured using different technologies. This ability leads the way to heterogeneous designs where CPUs, GPUs, accelerators, different memories such as DRAM and Phase Change Memory (PCM) as well as silicon photonic devices are integrated into a chip stack to achieve higher performance and better energy efficiency. The performance, power and temperature characteristics of these various components must be considered in the design and optimization of future processors.

REFERENCES

1. Comsol multiphysics. http://www.comsol.com.
2. D.H. Bailey, E. Barszcz, J.T. Barton, D.S. Browning, R.L. Carter, L. Dagum, R.A. Fatoohi et al. The NAS parallel benchmarks. *International Journal of High Performance Computing Applications*, 5(3):63–73, 1991.
3. C. Bienia, S. Kumar, J.P. Singh, and K. Li. The PARSEC benchmark suite: Characterization and architectural implications. In *Proceedings of 17th International Conference on Parallel Architectures and Compilation Techniques*, New York, pp. 72–81. ACM, 2008.
4. N.L. Binkert, R.G. Dreslinski, L.R. Hsu, K.T. Lim, A.G. Saidi, and S.K. Reinhardt. The m5 simulator: Modeling networked systems. *IEEE Micro*, 26(4):52–60, 2006.
5. A.K. Coskun, D. Atienza, T.S. Rosing, T. Brunschwiler, and B. Michel. Energy-efficient variable-flow liquid cooling in 3D stacked architectures. In *Proceedings of the Conference on Design, Automation and Test in Europe,* Leuven, Belgium, pp. 111–116. IEEE, 2010.
6. A.K. Coskun, J.L. Ayala, D. Atienza, T.S. Rosing, and Y. Leblebici. Dynamic thermal management in 3D multicore architectures. In *Proceedings of the Conference on Design, Automation and Test in Europe,* Leuven, Belgium, pp. 1410–1415. IEEE, 2009.
7. G. Ellison. Thermal spreading and more using open-source FEA software. http://www.electronics-cooling.com/2013/06/thermal-spreading-and-more-using-open-source-fea-software/, accessed on August 31, 2015.
8. A. Fourmigue, G. Beltrame, and G. Nicolescu. Efficient transient thermal simulation of 3D ICS with liquid-cooling and through silicon vias. In *Proceedings of the Conference on Design, Automation and Test in Europe*, DATE'14, pp. 74:1–74:6, 3001 Leuven, Belgium. European Design and Automation Association, 2014.
9. W. Gruener. IBM cools 3D chips with integrated water channels. http://www.tomshardware.com/news/IBm-research,5604.html, June 5, 2008.
10. A. Jain, R.E. Jones, R. Chatterjee, S. Pozder, and Z. Huang. Thermal modeling and design of 3D integrated circuits. In *Proceedings of 11th Intersociety Conference on Thermal and Thermomechanical Phenomena in Electronic Systems*, Orlando, FL, pp. 1139–1145. IEEE, 2008.
11. R. Jayaseelan and T. Mitra. Dynamic thermal management via architectural adaptation. In *Proceedings of 46th Annual Design Automation Conference*, New York, pp. 484–489. IEEE, 2009.
12. A.B. Kahng, B. Li, L.-S. Peh, and K. Samadi. Orion 2.0: A power-area simulator for interconnection networks. Institute of Electrical and Electronics Engineers, 2011.
13. S. Li, J.H. Ahn, R.D. Strong, J.B. Brockman, D.M. Tullsen, and N.P. Jouppi. McPAT: An integrated power, area, and timing modeling framework for multicore and manycore architectures. In *Proceedings of 42nd Annual IEEE/ACM International Symposium on Microarchitecture*, New York, pp. 469–480. IEEE, 2009.

14. G.M. Link and N. Vijaykrishnan. Thermal trends in emerging technologies. In *Proceedings of the 7th International Symposium on Quality Electronic Design*, pp. 625–632. IEEE Computer Society, Washington, DC, 2006.

15. Z. Liu, S. Swarup, and S.X.-D. Tan. Compact lateral thermal resistance modeling and characterization for TSV and TSV array. In *Proceedings of the International Conference on Computer-Aided Design*, pp. 275–280. IEEE Press, Piscataway, NJ, 2013.

16. J. Meng and A.K. Coskun. Analysis and runtime management of 3D systems with stacked DRAM for boosting energy efficiency. In *Proceedings of the Conference on Design, Automation and Test in Europe*, San Jose, CA, pp. 611–616. EDA Consortium, 2012.

17. J. Meng, K. Kawakami, and A.K. Coskun. Optimizing energy efficiency of 3-D multicore systems with stacked DRAM under power and thermal constraints. In *Proceedings of the 49th Annual Design Automation Conference*, pp. 648–655. ACM, New York, 2012.

18. F. Paterna and S. Reda. Mitigating dark-silicon problems using superlattice-based thermoelectric coolers. In *Proceedings of the Conference on Design, Automation and Test in Europe*, pp. 1391–1394, San Jose, CA. EDA Consortium, 2013.

19. J.N. Reddy. *An Introduction to the Finite Element Method*, Vol. 2. McGraw-Hill, New York, 1993.

20. M.M. Sabry, A.K. Coskun, D. Atienza, T.S. Rosing, and T. Brunschwiler. Energy-efficient multiobjective thermal control for liquid-cooled 3-D stacked architectures. *IEEE Transactions on CAD of Integrated Circuits and Systems*, 30(12):1883–1896, 2011.

21. M.M. Sabry, A. Sridhar, J. Meng, A.K. Coskun, and D. Atienza. Greencool: An energy-efficient liquid cooling design technique for 3-D MPSoCs via channel width modulation. *IEEE Transactions on CAD of Integrated Circuits and Systems*, 32(4):524–537, 2013.

22. C. Santos, P. Vivet, D. Dutoit, P. Garrault, N. Peltier, and R. Reis. System-level thermal modeling for 3D circuits: Characterization with a 65 nm memory-on-logic circuit. In *3D Systems Integration Conference (3DIC), 2013 IEEE International*, pp. 1–6. IEEE, San Francisco, CA, 2013.

23. K. Skadron, M.R. Stan, W. Huang, S. Velusamy, K. Sankaranarayanan, and D. Tarjan. Temperature-aware microarchitecture. In *ACM SIGARCH Computer Architecture News*, Vol. 31, pp. 2–13. ACM, New York, 2003.

24. A. Sridhar, Y. Madhour, D. Atienza, T. Brunschwiler, and J. Thome. Steam: A fast compact thermal model for two-phase cooling of integrated circuits. In *ICCAD*, pp. 256–263, Nov 2013.

25. A. Sridhar, A. Vincenzi, M. Ruggiero, T. Brunschwiler, and D. Atienza. 3D-ICE: Fast compact transient thermal modeling for 3D ICS with inter-tier liquid cooling. In *ICCAD*, pp. 463–470. IEEE, 2010.

26. J. Srinivasan et al. Ramp: A model for reliability aware microprocessor design. Technical Report IBM-RC23048(W0312-122), December 2003.

27. S. Thoziyoor, N. Muralimanohar, J.H. Ahn, and N.P. Jouppi. Cacti 5.1. Technical Report HPL-2008-20, *HP Laboratories*, April 2, 2008.

28. H. Xu, V.F. Pavlidis, and G. De Micheli. Analytical heat transfer model for thermal through-silicon vias. In *Proceedings of Design, Automation and Test in Europe Conference and Exhibition*, pp. 1–6. IEEE, Grenoble, France, 2011.

29. D. Zhao, H. Homayoun, and A.V. Veidenbaum. Temperature aware thread migration in 3D architecture with stacked DRAM. In *Proceedings of IEEE International Symposium on Quality Electronic Design*, pp. 80–87. IEEE, Santa Clara, CA, 2013.

11 Exploration of the Thermal Design Space in 3D Integrated Circuits

Sumeet S. Kumar, Amir Zjajo, and Rene van Leuken

CONTENTS

ABSTRACT

The complex nature of heat flow within 3D integrated circuits (IC) results in nonuniform operating temperatures throughout the die stack, and, consequently, in the formation of performance-degrading hotspots. The mitigation of such issues is predicated upon the accurate characterization of the physical design parameters of 3D ICs, and their specific thermal impact. This chapter presents a high-level flow that uncovers the influence of stack composition, physical construction, power density, and the design of the vertical interconnect structure on thermal behavior. These findings provide new insights into the thermal design space of 3D ICs, and the specific role of individual design parameters in reducing operating temperature gradients.

11.1 INTRODUCTION

Three-dimensional integrated circuits (3D-ICs) are composed of multiple stacked dies interconnected using *Through-Silicon-Vias (TSVs)*. These enable the high-density integration of devices without increasing the area footprint of the system. However, the number of devices that can be integrated within the system is limited by its thermal behavior. This is largely a consequence of the complex nature of heatflow in 3D-ICs, where each stacked die adds up to 12 material layers to the conduction path between power dissipating elements and the heatsink [14]. A number of studies in literature examine specific thermal characteristics of die stacks. Oprins et al. [15] investigated the thermal coupling between memory and logic dies in a two-tier stack in order to determine operating temperatures and thermal profiles in the memory. Their study used a thermal model calibrated with a 130 nm stacked-die silicon test chip and examined the influence of microbumps and underfill thermal conductivity on peak temperatures. Most recently, Clarke et al. [1] characterized the influence of stacked memory on hotspot formation in underlying logic tiers. Their work revealed the nontrivial nature of heat flow within 3D-ICs and uncovered the benign influence of symmetric metallization in memories on hotspot temperatures. Matsumoto and Taira [10] performed an experimental measurement of thermal resistance of multi-layer die stacks interconnected with C4 microbumps. Their work highlighted the dependence of thermal resistance on microbump pitch and size. Similarly, Vaisband et al. [17,20] investigated thermal conduction paths within a two-tier die stack in order to evaluate the magnitude of heat transfer between dies and to characterize the temperature dependence of thermal conductivity.

Nagata [13] determined the relationship between power dissipation by circuit elements and thermal constraints as

$$\frac{\alpha N_G E}{t_{pd}} \leq g \cdot \Delta T \tag{11.1}$$

where N_G is the number of gates within the system operating with a clock period t_{pd}. The relation dictates that the maximum number of gates that can be integrated within a single IC is limited by the average thermal conductance g and temperature gradient ΔT between the dissipating elements and the ambient air. In order to increase integration density, either energy dissipation E of the gates, or their activity rate α must be decreased. Alternatively, the thermal conductance of the system must be improved so as to efficiently dissipate the larger amount of heat generated (Q) in the left hand side of the relation. Equation 11.1 essentially specifies the thermal design space for 3D-ICs in terms of its most significant parameters. In this chapter, we explore the influence of these parameters on the steady-state temperature profiles of dies in 3D-ICs using a high-level flow. This exploration is performed using a physical model of a real 3D-stacked test chip [14], and provides insights into the formation of hotspots, and the implications of physical design parameters on operating temperatures of dies.

11.2   SIGNIFICANCE OF DESIGN PARAMETERS

The heat flow between power dissipating elements and sink surfaces in 3D-ICs follows the Fourier heat transfer equation:

$$c_v \frac{\delta T}{\delta t} = \nabla \cdot g (\nabla T)^T + Q \tag{11.2}$$

$$\nabla T = \left[\frac{\delta T}{\delta x}, \frac{\delta T}{\delta y}, \frac{\delta T}{\delta z} \right] \quad \text{and} \quad Q \propto \frac{\alpha N_G E}{t_{pd}} \tag{11.3}$$

where Q is the heat flowing from a power dissipating element toward multiple sink surfaces, at a rate $\delta T/\delta t$ through a material of volumetric heat capacity c_v. The conductance matrix g defines the thermal conductance along the orthogonal x, y, and z axes of the material as a function of effective thermal conductivity (κ_{eff}) of the materials encountered along those respective axes. Equation 11.2 can be rewritten in the one-dimensional steady-state form as

$$Q = g \cdot \Delta T \quad \text{where } g = \kappa_{eff} \frac{A}{l_{x,y,z}} \tag{11.4}$$

$$\kappa_{eff} = A_{tsv}\kappa_{tsv} + A_{mat}\kappa_{mat} \tag{11.5}$$

This equation indicates the relationship between thermal conductance and effective thermal conductivity of the material. The conductance g across a die layer is thus a function of its κ_{eff}, its area (A), and its material thickness ($l_{x,y,z}$). Although κ_{eff} is material-dependent, it is significantly influenced by the presence of structures, such as TSVs, whose excellent thermal conductance improves the overall conductivity of the material surrounding them. For a material layer with TSVs, the κ_{eff} may be computed as the weighted average of the thermal conductivities of the TSV material (κ_{tsv}) and the layer material (κ_{mat}), respectively, as given in Equation 11.5. A_{mat} is the area of the material layer in a die with length h_{die} and width w_{die}, and A_{tsv} represents the total area of n TSVs of radius r_{tsv} on this layer.

$$A_{tsv} = n\left(\pi r_{tsv}^2\right) \quad \text{and} \quad A_{mat} = (h_{die}w_{die}) - A_{tsv} \tag{11.6}$$

From Equation 11.4, the magnitude of thermal gradients observed at a given point across any layer of the stack is determined by effective thermal conductivity (κ_{eff}) of that layer, its thickness, and the power density at that point in the die stack. Lateral thermal gradients in stacked dies are predominantly influenced by the effective thermal conductivity of the bulk silicon. As κ_{eff} increases, more heat is conducted away from the power dissipation site resulting in shallower temperature profiles. Low κ_{eff} on the other hand results in the stagnation of heat and thus, in the formation of zones with high temperature, that is, hotspots. The sustained operation of devices at elevated temperatures can accelerate their degradation and give rise to reliability issues [6,16]. Therefore, it is essential that potential hotspot zones are spotted early in the design process, and suitably addressed.

The spread of temperatures is also dependent on die thickness ($l_{x,y,z}$). The use of die thinning to improve integration density causes a decrease in the thickness of bulk material, which hampers the lateral spreading of heat, and leads to the formation of high-temperature hotspots. The spread of temperatures is however limited by lateral thermal resistance beyond a certain die thickness. Therefore, dies thinner than this threshold can be expected to experience hotspots of greater magnitude than thicker dies. In 3D-ICs, $l_{x,y,z}$ can also be considered as the effective length of the heat flow path between the power dissipating elements and the sink. Consequently, the thermal conductance of the heat flow path decreases as stack depth increases.

Power density refers to the ratio of generated heat (Q), and the surface area (A) across which it can be conducted to the heatsink surface. In 3D-ICs, power dissipated by vertically stacked components can result in the generation of large quantities of heat within a small area, and yield highly localized hotspots. Since vertical heatflow is determined by κ_{eff}, the temperature gradient of a die is predominantly affected by both local power dissipation as well as dissipation in the tiers directly above and below it. Consequently, increasing power density causes temperature profiles to shift to higher ranges.

11.3 EXPLORATION FLOW

A high-level exploration flow is used to characterize the influence of these parameters on the thermal behavior of die stacks. The flow comprises of two stages: thermal conductivity estimation and thermal simulation, as shown in Figure 11.1. In the first stage, input design and technology parameters are translated into a physical model of the die stack. This describes the dies and their constituent layers, materials and dimensions, floor plans, and peak power dissipation values for components. The model also describes the physical characteristics of the TSV-based vertical interconnect structure determined through an exploration of the design space. In the second stage of the flow, an

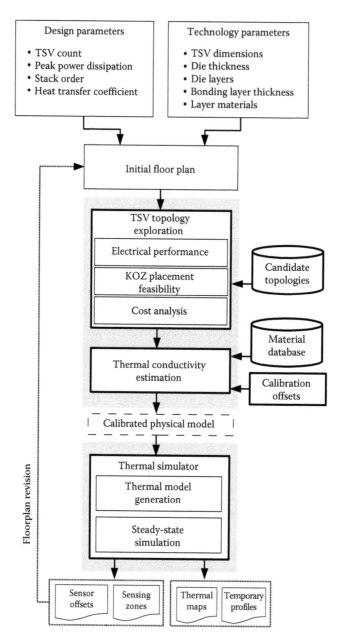

FIGURE 11.1 High-level flow for the thermal design space exploration of 3D-ICs.

electrical resistor–capacitor (RC) equivalent of the thermal relationships between power-dissipating elements is generated from the physical model. This electrical model is used to accurately simulate the thermal behavior of the 3D-IC. The results of the simulation provide insights into the thermal characteristics of stacked dies, and the influence of power density, vertical interconnect topology, and other physical parameters on operating temperatures.

11.3.1 TSV Topology Exploration

The vertical interconnect forms a critical part of the 3D-IC, enabling the interconnection of components in stacked dies. TSVs essentially act as high conductance thermal pathways that facilitate the flow of heat from power-dissipating elements toward heatsink surfaces. The difference in the coefficient of thermal expansion of TSVs and silicon means that the two expand by different amounts when heated. As a consequence, when power is dissipated within 3D-ICs, TSVs expand more than their surrounding silicon, and create regions of stress in their immediate vicinity. In order to insulate circuit elements from potential damage due to this stress, TSVs are enveloped by a Keep Out Zone (KOZ) within which no active devices are placed. The dimensions of the KOZ, however, are found to vary based on the placement topology of TSVs [12]. Similarly, electrical noise performance considerations can necessitate the inclusion of shield TSVs to minimize noise induced due to capacitive coupling. Together, the electrical performance requirements, KOZ dimensions, and area constraints determine the number and density of TSVs constituting the vertical interconnect [5,7]. Since these decisions affect κ_{eff} as indicated in Equation 11.5, it is essential to account for their influence on the thermal behavior of 3D-ICs.

Our flow integrates a methodology for the exploration of TSV topologies based on electrical performance, signal integrity characteristics, KOZ feasibility, and area cost [7]. This enables the selection of an optimal candidate from several TSV topologies on the basis of performance and feasibility, and subsequently evaluates its thermal performance in the 3D-IC.

11.3.2 THERMAL CONDUCTIVITY ESTIMATION

For each candidate topology, the effective thermal conductivity of each stack layer due to the presence of TSVs is computed using the weighted average equation (Equation 11.5). The conductivity data is embedded into a stack descriptor as part of the physical model. This provides an accurate model of the heatflow paths between power dissipating elements and the heat sink surfaces.

11.3.3 THERMAL SIMULATION

In the second stage of the flow, the thermal behavior of this model is simulated in order to obtain fine-grained steady-state temperature maps corresponding to the peak power dissipation of components. In order to do this, a thermal simulator is used [19]. Thermal simulation begins with the discretization of the die stack into a grid of thermal cells, each representing a unit of heat generation (Q), and temperature (T), as illustrated in Figure 11.2. The thermal conductance (g) between cells essentially determines the heatflow across the die, and is derived from Equation 11.4. The *Resistance-Capacitance* (*RC*) network within each cell represents the electrical equivalent of that cell's thermal relationship with its neighbors. In order to simulate thermal behavior, power values

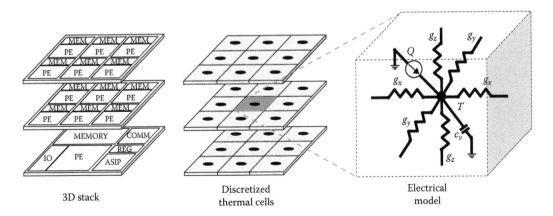

3D stack   Discretized
thermal cells   Electrical
model

FIGURE 11.2 Illustration of the equivalent electrical model used to determine the temperature of thermal cells.

are inserted into the appropriate thermal grid current sources corresponding to the locations of power dissipating elements. This results in a rise in local temperature, and depending on the magnitude of conductance ($g_{x,y,z}$), the temperature of neighboring thermal cells as well. The inclusion of a capacitance representing the specific heat capacity (c_v) of the silicon allows modeling of both steady-state as well as transient thermal behavior. The accuracy of the model, however, is largely dependent on the dimensions of the thermal grid, which in turn has a considerable influence on runtime. Since power and temperature are considered to be uniform within a thermal cell, large dimensions can result in loss of accuracy. For instance, if a small circuit element dissipates 100% of the power within a thermal cell, large-cell dimensions would result in that total power being considered as having been dissipated over the entire cell's area—thereby erroneously modeling a lower power density. Furthermore, this would result in overly optimistic temperature estimates. On the other hand, very small cell dimensions result in the generation of a massive thermal model, increasing system memory usage considerably, and resulting in long runtimes [8]. More recent thermal simulators eliminate this problem by using a nonuniform grid that uses variable-sized thermal cells to better model the spatial differences in floor plan complexity and power dissipation [4,21].

11.4   EXPERIMENTAL SETUP AND VALIDATION

In order to validate our exploration flow, we created a physical model of the 3D-stacked test chip described in [14] and carefully calibrated the heat transfer coefficient of the connection to ambient using available data. Thereafter, we carried out a series of thermal simulations using test floorplans containing heating elements of size and power dissipation identical to those in [14]. Our setup accurately reproduced the same temperature profiles, with a maximum temperature deviation of 6% at hotspot peaks.

The experimental setup utilizes a 3D stack consisting of a 250 µm base die with multiple thinner dies bonded above using 1 µm thick bonding layers, as illustrated in Figure 11.3. Each die in the stack contains five configurable power-dissipating elements of 50 µm × 50 µm to generate heat. The complete stack is connected to ambient air through an interface with the previously calibrated heat transfer coefficient. The material composition and layer dimensions for dies are identical to the 3D test chip described in [14]. The dies used in the characterization have a size of 2000 µm × 2000 µm, and the thermal simulations use a grid width of 10 µm, which provides good accuracy with acceptable simulation times. Simulations are carried out using the 3D-ICE thermal simulator [19] and are run until steady-state temperatures are reached. To quantify the variation in temperatures with changing parameter values, the normalized temperature profile along the horizontal line at $y = 500$ µm is plotted.

11.5   EXPLORATION RESULTS

The following sections explore the effect of thermal conductivity, TSV topologies, die thickness, stack depth, and power density on the steady-state thermal behavior of 3D-ICs.

11.5.1   Influence of Thermal Conductivity

The total number of TSVs in the vertical interconnect is effectively a function of the system architecture, electrical noise performance requirements, and the area constraints for the design, and influences the effective thermal conductivity (k_{eff}) of silicon dies. Since hotspot temperatures are directly related to the rate of heatflow toward the heatsink surfaces, k_{eff} has an impact on temperature profiles. Figure 11.4a through d depicts the temperature rise across an observation die in the middle of a five-tier die stack, mounted on a 250 µm base die, with different values of k_{eff}, for a power dissipation of 120 mW at heater 3. As effective conductivity increases, the heat produced due to this localized power dissipation is conducted more efficiently to adjacent dies in the stack, and in turn to

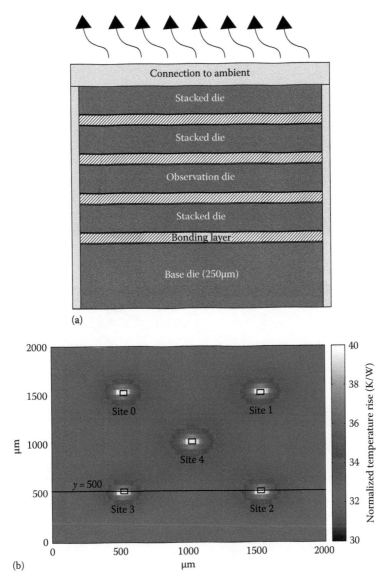

FIGURE 11.3 (a) Composition of the die stack and (b) normalized map of temperature across the die with all heaters dissipating 25mW.

the sink surfaces. The generated heat is thus prevented from stagnating in the die, thereby resulting in a decreased hotspot temperature, and smaller thermal gradients.

Figure 11.4 also illustrates the improved heat spread due to decreased lateral thermal resistance resulting from the higher κ_{eff}. The lateral spread of heat holds a number of implications for system planning. Firstly, the accuracy of temperature sensors is influenced by their distance from the hotspot, and owing to their size, sensors cannot always be placed at close proximity to the region of interest. This necessitates the use of calibration techniques to offset induced measurement errors [11,21]. We observe that the radius of the zone within which hotspot temperatures can be tracked with 100% accuracy increases with κ_{eff}. Consequently, temperature sensors can be placed farther away from the hotspot without compromising on measurement accuracy [9]. Secondly, the spread of heat across the die can result in the power dissipation of one component negatively influencing

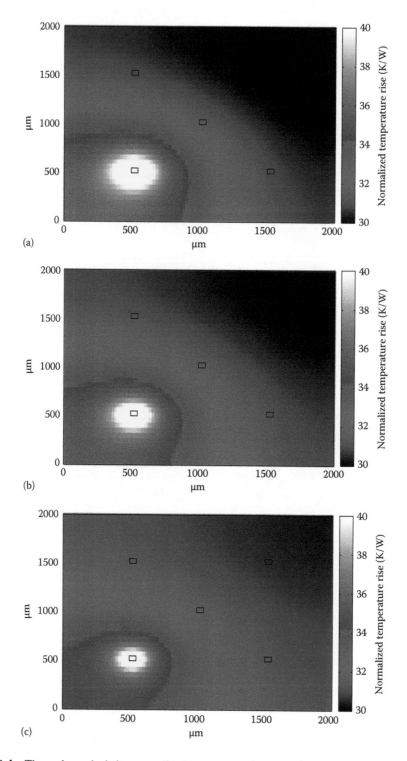

FIGURE 11.4 Thermal map depicting normalized temperature rise across the observation die with increasing effective thermal conductivity relative to conventional silicon. (a) 1×, (b) 2×, and (c) 4×.   (*Continued*)

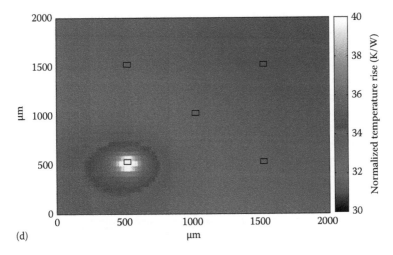

(d)

FIGURE 11.4 (*Continued*)   Thermal map depicting normalized temperature rise across the observation die with increasing effective thermal conductivity relative to conventional silicon. (d) 8×.

the performance or reliability of another. This issue can necessitate thermal-aware floor planning to mitigate temperature-related interactions between components [2,3].

In order to demonstrate the impact of k_{eff} due to the vertical interconnect, we compare two TSV topologies determined to be optimal in terms of their electrical performance in the context of a 300 μm × 300 μm on-chip communication block in the 45 nm technology node [7]. The *Shielded* topology intersperses signal and shield TSVs in order to improve electrical performance and signal integrity. In addition to their electrical benefits, these shield TSVs improve thermal conductance toward the heatsink surfaces, as well as into the silicon bulk. The *Isolated* topology on the other hand spaces TSVs far apart to prevent capacitive coupling. However, due to its fewer number of TSVs and comparatively lower density, it yields an effective thermal conductance 1.8× lower than that of the shielded topology. Figure 11.5 presents the heatmaps corresponding to power dissipation at all five heaters on the observation die of the stack. Due to its superior thermal conductance, the shielded topology produces lower peak temperatures, and smaller hotspot dimensions. The exact difference in their temperature profiles is depicted in Figure 11.5c. Although the difference between the two is small, it can yield appreciable performance improvements in real systems.

To illustrate this, the two topologies are deployed within a three-tier multiprocessor with 12 processing elements (PEs), and a temperature-aware power manager. During thermal emergencies, voltage and frequency levels are scaled down in order to limit power dissipation and prevent hotspots from forming [18]. The additional temperature margin realized with the shielded topology enables the operation of PEs at higher frequencies, and improves overall system performance. This is observed in Figure 11.5d where the execution of the application program completes 38 ms faster with the shielded topology, than with the isolated topology. Essentially, the additional thermal conductance afforded by the shield TSVs yields a 11% improvement in the multiprocessor's execution performance without any modifications to the architecture.

11.5.2   INFLUENCE OF DIE THICKNESS AND STACK DEPTH

Figure 11.6a shows the temperature profile on a single die of varying thickness, stacked above a 250 μm base die. The relatively shallow depth of bulk silicon in thin dies inhibits the spread of heat away from the hotspot, and thus causes the highest peak temperatures. While increasing the thickness improves lateral spreading, this benefit diminishes as die thickness exceeds 100 µm. The influence of die thickness can also be emulated by a stack of dies. Figure 11.6b reports the temperature profiles resulting from the stacking of multiple 25 µm thick dies on top of the dissipating die used in the previous case. The additional dies stacked above the dissipating die in this case serve as bulk material and facilitate the lateral spreading of generated heat, as evidenced by the increased temperature rise registered at the eastern edge of the die. The heat-spreading effect is observed for stacks with up to two dies above the dissipating die, and translates to an effective thickness ceiling of 75 µm (excluding the base die), which is lower than the 100 µm ceiling observed in the previous case. This observation is explained by the fact that die stacks contain a range of materials such as silicon dioxide, which have a thermal conductivity much lower than that of silicon. Consequently,

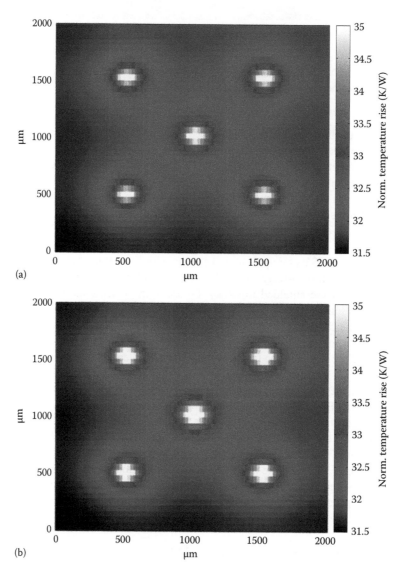

FIGURE 11.5 Heatmaps corresponding to power dissipation at five heaters on the middle die of a five-tier stack with (a) Shielded topology and (b) Isolated topology.   (*Continued*)

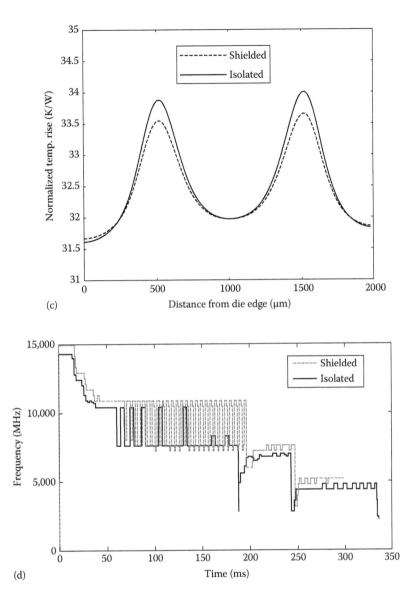

FIGURE 11.5 (*Continued*) Heatmaps corresponding to power dissipation at five heaters on the middle die of a five-tier stack with (c) Steady-state peak temperature profile corresponding to shielded and Isolated TSV topologies. (d) Aggregated frequency profile corresponding to execution of computational workload on a three-tier multiprocessor with shielded and isolated topologies. *Note:* The trace for shielded ends 38 ms earlier than that of isolated due to faster completion of execution.

the thermal characteristics of a die stack with effective thickness of 100 μm are certain to differ from those of a single die of the same thickness.

The composition of layers within the stack can also positively influence thermal behavior. For instance, the density of interconnect wiring in memory elements cause their metallization layers to act as heat spreaders. Hotspots arising due to high-power density are thus dissipated due to the lateral conduction in the metallization layers [1]. This observation hints at the potential for using memories within die stacks to mitigate thermal issues.

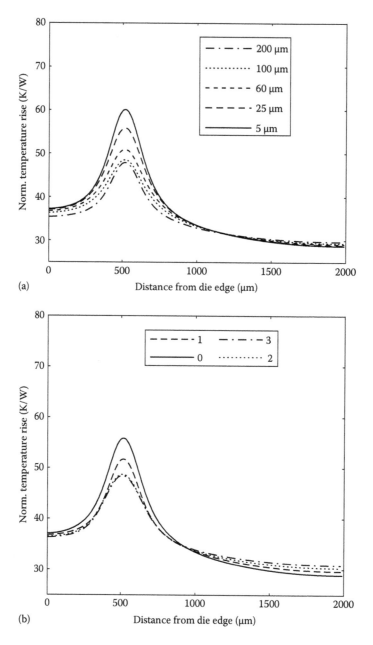

FIGURE 11.6 (a) Influence of varying die thickness on normalized temperature rise. (thickness of base die: 250 μm.) (b) Influence of stack depth on normalized temperature rise on an active observation die mounted on a 250 μm thick base die. (thickness of observation/stacked dies: 25 μm.) (Results adapted from Kumar, S.S. et al., Physical characterization of steady-state temperature profiles in three-dimensional integrated circuits, in: *Proceedings of the International Symposium on Circuits and Systems (ISCAS)*, May 2015, pp. 1969–1972.)

11.5.3 INFLUENCE OF POWER DENSITY

Figure 11.7 shows the temperature profile of an observation die located in the middle of a five-die stack mounted on a 250 μm base die. The graph reports the temperature rise experienced in the observation die due to power dissipated at Site 3 heaters on other tiers of the stack, that is, changing stack power density. Since the primary sink in the stack is located at the surface of the topmost die,

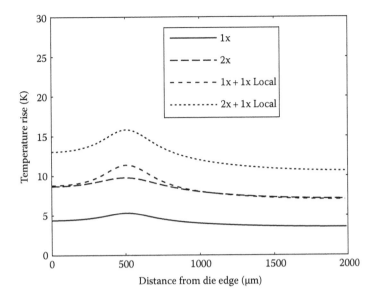

FIGURE 11.7 Influence of stack power density on the actual temperature rise in an observation die. Note that the case marked +*Local* uses an active Site 3 heater on the observation die. Each 1× represents a Site 3 heater which dissipates 120 mW of power. (Results adapted from Kumar, S.S. et al., Physical characterization of steady-state temperature profiles in 3D integrated circuits, in: *Proceedings of the International Symposium on Circuits and Systems (ISCAS)*, May 2015, pp. 1969–1972.)

generated heat must flow through all intermediate dies. This is evidenced by the higher temperature rise noted when the dissipating dies are located below the observation die. However, it is interesting to note that similar behavior is observed even when the dissipating dies are placed above the observation die. This indicates that in addition to the heat flowing toward the primary heatsink, conduction also occurs toward regions of the 3D-IC with available temperature margin, and secondary heat sinking surfaces. Consequently, the heat flowing from the site of power dissipation follows multiple complex paths, not necessarily in the direction of the primary heatsink, resulting in elevated temperature profiles on all dies within the stack. The overall steady-state temperature profile is observed to be notably influenced by the stack power density and the heat transfer coefficient of the sink. The magnitude of hotspots on the other hand depends on the k_{eff} between the power dissipating dies.

It is important to note that in this work, we focused only on the steady-state thermal behavior of 3D-ICs. Since the volumetric heat conductivity (c_v) of bulk silicon is affected by the inclusion of metal TSVs, we expect the transient behavior of die stacks, and the dynamics of hotspot formation to vary with the stack construction and power density. This is an important topic for future investigation.

11.6 CONCLUSIONS

The performance and integration density of modern ICs is constrained by their physical characteristics and the thermal efficiency of their cooling interfaces. The complex thermal characteristics of 3D-ICs further aggravate this issue due to the nonuniform nature of heat transfer from stacked dies to the ambience. In this chapter, we investigated the steady-state thermal behavior of 3D-ICs using a high-level exploration flow and determined the impact of various physical design parameters on temperature profiles. The exploration uncovered the critical influence of power density on hotspot magnitudes, the significance of thermal conductivity in determining temperature gradients, and the impact of multidie stacks on steady-state temperatures. In addition, we also examined how the choice of vertical interconnect topology influences thermal performance, and execution performance of 3D systems. The work presented in this chapter is important in the context of design space exploration of 3D-ICs and provides a basis for evolving thermal-aware design methodologies for stacked die architectures. An application of the presented flow in the context of a larger exploratory framework can be found in our other papers [7,8].

ACKNOWLEDGMENTS

The work presented in this chapter was supported in part by the CATRENE programme under the Computing Fabric for High Performance Applications (COBRA) project CA104. The authors would like to thank Arnica Aggarwal (ASML), Radhika Jagtap (ARM), Yash Joshi, and Michel Berkelaar (Delft University of Technology) for their help in the development of the flow.

REFERENCES

1. R. Clarke, P. Jacob, O. Erdogan, P. Belemijian, S. Raman, M.R. Leroy, T.G. Neogi, R.P. Kraft, D. Borca-Tasciuc, and J.F. McDonald. Thermal modeling of 3-D stacked DRAM over SiGe HBT BICMOS CPU. *IEEE Access*, 3:43–54, 2015.
2. D. Cuesta, J.L. Risco-Martin, J.L. Ayala, and D. Atienza. 3D thermal-aware floorplanner for many-core single-chip systems. In *Proceedings of the Latin American Test Workshop (LATW)*, Porto de Galinhas, Brazil, pp. 1–6, 2011.
3. D. Cuesta, J.L. Risco-Martin, J.L. Ayala, and J.I. Hidalgo. 3D thermal-aware floorplanner using a moea approximation. *Integration, the VLSI Journal*, 46(1):10–21, 2013.
4. W. Huang, S. Ghosh, S. Velusamy, K. Sankaranarayanan, K. Skadron, and M.R. Stan. Hotspot: A compact thermal modeling methodology for early-stage VLSI design. *IEEE Transactions on Very Large Scale Integration Systems*, 14(5):501–513, 2006.
5. R. Jagtap, S.S. Kumar, and R. van Leuken. A methodology for early exploration of TSV placement topologies in 3D stacked ICS. In *Proceedings of the 15th Euromicro Conference on Digital System Design*, Izmir, Turkey, pp. 382–388, September 2012.
6. H. Kufluoglu and M. Ashraful Alam. A computational model of NBTI and hot carrier injection time-exponents for mosfet reliability. *Journal of Computational Electronics*, 3(3–4):165–169, 2004.
7. S.S. Kumar, A. Aggarwal, R. Jagtap, A. Zjajo, and R. van Leuken. System level methodology for interconnect aware and temperature constrained power management of 3-D MP-SOCS. *IEEE Transactions on Very Large Scale Integration (VLSI) Systems*, 22(7):1606–1619, July 2014.
8. S.S. Kumar, A. Zjajo, and R. van Leuken. Ctherm: An Integrated Framework for Thermal-Functional Co-simulation of Systems-on-Chip. In *Proceedings of the Euromicro International Conference on Parallel, Distributed and Network-Based Processing*, Turku, Finland, pp. 674–681, March 2015.
9. S.S. Kumar, A. Zjajo, and R. van Leuken. Physical characterization of steady-state temperature profiles in three-dimensional integrated circuits. In *Proceedings of the International Symposium on Circuits and Systems*, Lisbon, Portugal, pp. 1969–1972, May 2015.
10. K. Matsumoto and Y. Taira. Thermal resistance measurements of interconnections, for the investigation of the thermal resistance of a three-dimensional (3D) chip stack. In *Proceedings of the Semiconductor Thermal Measurement and Management Symposium*, San Jose, CA, 2009, pp. 321–328, March 2009.
11. S.O. Memik, R. Mukherjee, M. Ni, and J. Long. Optimizing thermal sensor allocation for microprocessors. *IEEE Transactions on Computer-Aided Design of Integrated Circuits and Systems*, 27(3):516–527, March 2008.
12. A. Mercha et al. Comprehensive analysis of the impact of single and arrays of through silicon vias induced stress on high-k/metal gate CMOS performance. In *IEEE International Electron Devices Meeting*, San Francisco, CA, pp. 2.2.1–2.2.4, December 2010.
13. M. Nagata. Limitations, innovations, and challenges of circuits and devices into a half micrometer and beyond. *IEEE Journal of Solid-State Circuits*, 27(4):465–472, April 1992.
14. H. Oprins et al. Fine grain thermal modeling and experimental validation of 3D-ICS. *Microelectronics Journal*, 42(4):572–578, April 2011.

15. H. Oprins et al. Numerical and experimental characterization of the thermal behavior of a packaged dram-on-logic stack. In *Proceedings of the Electronic Components and Technology Conference*, San Diego, CA, pp. 1081–1088, May 2012.
16. K. Ramakrishnan et al. Variation impact on ser of combinational circuits. In *International Symposium on Quality Electronic Design*, San Jose, CA, pp. 911–916, March 2007.
17. I. Savidis and E. G. Friedman. Thermal coupling in TSV-based 3-D integrated circuits. In *Proceedings of the Workshop on 3D Integration—Design, Automation and Test in Europe*, Dresden, Germany, March 2014.
18. K. Skadron, M.R. Stan, K. Sankaranarayanan, W. Huang, S. Velusamy, and D. Tarjan. Temperature-aware microarchitecture: Modeling and implementation. *ACM Transactions on Architecture and Code Optimization*, 1(1):94–125, March 2004.
19. A. Sridhar et al. 3D-ICE: Fast compact transient thermal modeling for 3D ICS with inter-tier liquid cooling. In *International Conference on Computer-Aided Design*, San Jose, CA, pp. 463–470, November 2010.
20. B. Vaisband, I. Savidis, and E.G. Friedman. Thermal conduction path analysis in 3-D ICS. In *Proceedings of the International Symposium on Circuits and Systems*, Melbourne, Australia, pp. 594–597, June 2014.
21. A. Zjajo, N. van der Meijs, and R. van Leuken. Dynamic thermal estimation methodology for high-performance 3-D MPSOC. *IEEE Transactions on Very Large Scale Integration (VLSI) Systems*, 22(9):1920–1933, September 2014.

12 Dynamic Thermal Optimization for 3D Many-Core Systems

Nizar Dahir, Ra'ed Al-Dujaily, Terrence Mak, and Alex Yakolev

CONTENTS

ABSTRACT

Despite the numerous advantages of substantial silicon density in 3D VLSI, such as small form size, low power consumption, and high operating frequency, it introduces many serious challenges. The most alerting ones is the thermal threat, which would lead to various faults and system failures. This chapter introduces a new strategy to effectively diffuse heat from NoC-based 3D CMPs. A runtime dynamic programming network (DPN) is proposed to optimize routing directions and provide temperature moderation across the whole chip. Both on-chip reliability and computational performance have been improved by 63% and 27%, respectively, with the DPN approach. This work enables a new avenue to explore the adaptability for future large-scale 3D integration.

12.1 INTRODUCTION AND MOTIVATION

Semiconductor manufacturing processes are approaching the physical limits. This motivated the exploration of the use of 3D VLSI design, which could have many advantages including shorter global interconnect lengths, less delay, better scalability, and smaller form factors. On the other hand, the NoC has been proposed as a promising communication paradigm for SoC and CMP systems, which could overcome the limitations associated with on-chip bus connectivity. NoCs provide a scalable, flexible, and power-efficient solution to integrate many cores in a single chip [3,13].

In 3D NoCs, benefits can be gained from both 3D integration and NoCs [2,40,49], with shorter interconnects, smaller form factors, and reduced delay leading to major performance increase in 3D SoCs and 3D CMPs compared to 2D systems [16,50]. However, future 3D VLSI systems in general, and 3D NoCs in particular, are prone to thermal challenges due to decreased transistors junction temperature, exacerbated spatial temperature gradients, and increased device density. For these reasons, worst-case cooling system design will not be feasible. Instead, RTM techniques at various levels of optimization are indispensable, especially at the network level where the NoC communication power budget takes up a significant portion of overall chip power, and may dominate logic as a source of heat [43]. NoCs power consumption would increase in the future due to advances in both technology and architecture.

Technology scaling is causing interconnects to consume more power than logic [25]. This is mainly because smaller technology reduces delay of logic gates and their power consumption, but results in relatively slower and more power-hungry wires. This is due to the fact that wires do not scale in the same way as logic. It is expected that in future technology nodes, interconnect power would take up to 65%–80% of total chip power [38].

Meanwhile, in terms of architecture, an important trend in the microarchitecture of many-core systems advocates integrating many (hundreds or thousands of) simple cores rather than integrating few complex cores. This has many advantages, such as the higher performance and finer control of these simple cores [4]. As a result, as the complexity of the individual core is decreasing their number increases, causing the communication power budget to increase relative to computation power.

As a result of the aforementioned reasons, in future many-core CMP design will become communication-centric, causing NoCs to contribute significantly to power consumption and chip heating. Even in present designs, the NoC is shown to make a contribution to heat generation either comparable or greater than that of processors, particularly for communication-intensive applications. Examples are the MIT-RAW chip [43] and Intel single chip cloud computer [39]. Furthermore, for Intel's 80-core CMP, the power density of the NoC routers is nearly double that of other units such as the floating-point and memory units in the tile [49]. Thus, the NoC would make a greater contribution to chip heating compared to these units.

Moreover, controlling the NoCs workload offers a unique opportunity to control a workload spanning the whole chip. This implies that the thermal-aware control of routing paths could achieve better thermal distribution over the entire chip. In this chapter, an adaptive RTM scheme for 3D NoCs is proposed. This scheme is based on a routing strategy, which tries to reduce thermal hotspots by effectively migrating routing load towards the coolest regions in the 3D geometry in order to achieve thermal optimization. Figure 12.1 gives an example depicting how the proposed routing strategy works. In this figure, there are several shortest paths between the source node $S1$ and the destination node $D1$. However, paths that involve thermal hotspots are avoided and the proposed scheme chooses the coolest path among those available to reduce the power density at these nodes

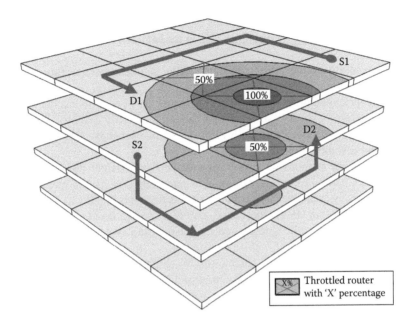

FIGURE 12.1 Illustration of thermal-aware routing paths.

and thus moderate the temperature. Similarly, in another example for inter-layer paths between source node *S*2 and destination node *D*2, a path which avoids thermal hotspots is selected. The contributions of this chapter can be summarized as follows:

1. A new runtime thermal-adaptive routing strategy is proposed to effectively diffuse heat from the NoC-based 3D CMPs, and a distributed Dynamic programming (DP)-based control architecture, DPN, is introduced to implement the proposed strategy.
2. The DPN is improved such that costs are computed and propagated in compliance with deadlock-free routing algorithms. Furthermore, 3D deadlock-free adaptive routing algorithms are improved in order to achieve higher path diversity, more balanced adaptiveness, and better performance.
3. The proposed methodology is evaluated through experimental studies and comparisons with state-of-the-art NoCs RTM techniques using various synthetic traffic scenarios and real benchmarks. Results for temperature, reliability, energy, and performance are compared and discussed.
4. The hardware implementation of the proposed method is discussed in detail and area and power overheads are evaluated.

12.2   RELATED WORK

Network-on-chip design space exploration may involve many design-time and runtime optimization techniques [34]. Dynamic and static application mapping and scheduling, which aim to maximize performance or minimize energy are examples of these techniques [21,51]. When the application mapping and traffic characteristics are known, the communication paradigm can be optimized to meet various objectives. The key to this optimization is designing a static or dynamic routing algorithm that determines the path taken by data packets.

The routing algorithm is important as it impacts on all network metrics such as latency, throughput, and power consumption. It can also affect thermal issues, since temperature distribution is highly correlated with power consumption. This is particularly so when the NoC takes up a substantial portion of system power, for example, 40% in the MIT-RAW chip [47] and 30% in the Intel 80-core teraflop chip [48].

On the other hand, due to the greater thermal challenge in current and future VLSI systems, thermal modeling and management has gained a lot of attention in recent years [6,14,30,44]. For example, a generic modeling of the thermal behavior of VLSI chip has been proposed [44]. This model starts from basic; a resistance and capacitance (RC) dynamic compact model in modeling the main heat transfer path with typical package settings. A thermal modeling tool called *HotSpot* has also been released [22].

Thermal management methods have also been studied by different researchers. For example, distributed task migration is proposed for thermal management [17]. This strategy relies on distributed agents to proactively exchange tasks among neighboring cores in order to balance the workload and avoid thermal emergency. In another work, DVFS is used to avoid exceeding the emergency temperature [14]. On-line task allocation to reduce hotspots and avoid exceeding the thermal limit has also been proposed [31].

Some runtime thermal management schemes for on-chip networks that mainly use traffic throttling to avoid thermal emergencies have been proposed [30,42,43]. However, exploiting NoC routing to better control chip heat distribution has received limited attention in existing literature. A characterization of the thermal profile in the MIT-Raw chip [42] revealed that the NoC can surpass processors in heat generation for highly parallelizable and communication-centric applications. For example, for the 802.11a encoding application, the chip temperature reached 53°C when only processors workload was considered, while with NoC workload alone the chip temperature reached 60°C. Similar proportions were found for the 8b_10b encoding and the FIR applications [42]. Motivated by these results, a routing-based NoC RTM strategy has also been proposed. In the proactive phase of this strategy, neighboring nodes exchange traffic counters as a mean of thermal balancing. When the thermal limit of the chip is violated, throttling is proposed as a reactive strategy [43]. In another work [37], thermal-aware application-specific routing path allocation was proposed for 2D-mesh MPSoCs. The authors propose using linear programming (LP) to allocate routing paths at design time such that thermal variations among the cores are compensated and thermal hotspots are minimized. However, this scheme is an offline technique and cannot adapt to application dynamics at runtime.

In 3D NoCs, significant thermal variations among the 3D layers can occur due to longer heat paths to the heat sink. In a recent study [7], the authors proposed a nonminimal routing called downward XYZ routing (dw_xyz) for 3D NoCs to migrate routing load, and thus power consumption, to layers closer to the heat sink. This would improve the efficiency of heat diffusion. The downward level is determined by traffic counters in order to minimize the impact of downward routing on performance and to avoid congestion. However, the implementation of this approach requires an H/W overhead for holding, updating, and communicating these counters. Other shortcomings of this scheme are that cool paths within the layer are not exploited, and furthermore it is tailored for a particular cooling system and cannot adapt to different cooling systems.

The present work attempts to provide a routing scheme, which is flexible in maneuvering packets away from hot paths. The proposed routing exploits cool paths wherever they are and whenever they become available in the chip, using a DP-based distributed control architecture.

12.3   PROBLEM DEFINITION AND BACKGROUND

This section presents definitions and the necessary background for this chapter. This includes a discussion of the motivation behind RTM techniques and an introduction to temperature-related failure mechanisms.

12.3.1 THERMAL OPTIMIZATION AND MANAGEMENT

Due to continuous shrinking of feature size, severe thermal challenges emerge. Thus, design-time thermal optimization is becoming increasingly difficult. Furthermore, 3D die stacking results in higher spatial temperature gradients over different strata due to longer main heat-flow paths. Thus, worst-case cooling system designs are becoming infeasible due to the prohibitive packaging cost associated with such designs [43]. Alternatively, RTM techniques can be used. These techniques would diffuse heat and regulate the system's operating temperature at runtime before the thermal limit is exceeded, in order to keep it within a safe range. In this scenario, chip and package design for typical cases would be possible.

Techniques that use RTM monitor the temperature at runtime and alter system behavior accordingly. These techniques can be divided into two categories: reactive and proactive. Reactive techniques work when the thermal limit is exceeded and sacrifice performance in order to achieve thermal regulation (e.g., DVFS). On the other hand, proactive techniques try to reduce thermal hotspots and minimize the temperature at runtime. This reduces, and may alleviate, the need for reactive action, thus improving both chip performance and reliability. Examples of these techniques are dynamic task scheduling and allocation in CMPs [10,31].

12.3.2 TEMPERATURE-RELATED FAULTS

Higher temperatures can lead to slower devices and increased leakage current as well as interconnect delay due to higher resistivity. Furthermore, a higher thermal gradient over different chip regions may lead to failure of timing closure and increase in soft errors. However, the most prominent impact of higher temperatures and thermal gradients is on lifetime reliability [5]. It has been reported that a 10°C increase in chip temperature would cause the lifespan of the device to be shortened by half [41]. As a result, increased temperature and spatial thermal variation (or temperature gradients) increases the MTTF, reducing chip reliability and shortening the device lifespan. Moreover, it has been reported that over 50% of electronic products failures are temperature-related [26,43].

The impact of a failure mechanism is usually expressed in terms of the MTTF. Failure mechanisms include the following [5,26,45]:

1. *Electromigration* failure is caused by the displacement of interconnect mass due to the flow of electrical current. This will lead to thinner wires and higher resistance in interconnects. Eventually, it can result in interconnect faults due to an open circuit. The MTTF for electromigration is given by [26]:

$$MTTF_{EM} = \frac{A_{EM}}{J^n} e^{E_{aEM}/kT} \tag{12.1}$$

   where
   T is temperature
   A_{EM} is a constant
   J is current density
   k is Boltzmann's constant
   E_{aEM} is the activation energy of electromigration
   n is an empirically determined constant
2. *Time-dependent dielectric breakdown* is caused by the breakdown of the gate oxide dielectric, which results in a conductive path in this dielectric. This failure becomes more prominent with technology scaling due to lower dielectric thickness, lower operating voltages,

and higher operating temperatures. The MTTF due to time-dependent dielectric break-down is given by [26]

$$MTTF_{TDDB} = A_{TDDB} \left(\frac{1}{V} \right)^{(a-bT)} e^{(A+B/T+CT)/kT} \qquad (12.2)$$

where

A_{TDDB} is a constant
V is the supply voltage
a, b, A, B and C are fitting parameters

3. *Stress migration* is similar to electromigration but is caused by the migration of intercon-nect mass atoms due to mechanical stress caused by mismatches in thermal expansion for different materials. The MTTF due to stress migration is given by [26]

$$MTTF_{SM} = A_{SM} \left| T_0 - T \right|^{-n} e^{E_{aSM}/kT} \qquad (12.3)$$

where

A_{SM} is a constant
T_0 is the metal deposition temperature during fabrication
n is an empirically determined constant
E_{aSM} is the activation energy of stress migration

4. *Thermal cycling* is IC fatigue failure which accumulates every time there is a cycle in tem-perature. It is also caused by thermal expansion mismatch and occurs mainly in adjacent die and package metal layers (e.g., solder joints). The MTTF due to thermal cycling is given by the Coffin–Manson equation as follows [26]:

$$N_f = C_o \frac{1}{(T_{average} - T_{ambient})^q} \qquad (12.4)$$

where

N_f is the number of cycles to failure
C_o is a constant
$T_{average}$ is the average chip temperature
$T_{ambient}$ is the ambient temperature
q is the Coffin–Manson exponent constant

Now MTTF due to thermal cycling can be expressed as the product of N_f and the average time of a thermal cycle, t_{TC}, as follows:

$$MTTF_{TC} = A_{TC} \frac{1}{(T_{average} - T_{ambient})^q} \qquad (12.5)$$

where, $A_{TC} = t_{TC} \cdot C_o$.

To calculate the total MTTF, the effects of various failure mechanisms need to be combined. A model such as the sum-of-failure-rates model (SOFR) can be used to obtain the resultant failure rate due to different fault mechanisms. This model is based on two assumptions: (1) system failure is a series of failures, where any failure due to any mechanism will cause the entire system to fail; and

(2) all failure mechanisms have constant failure rates. Under these assumptions, the total failure rate (λ_{tot}) can be expressed as follows:

$$\lambda_{tot} = \frac{1}{MTTF_{tot}} = \sum_{\forall f \in \{F\}} \sum_{\forall k \in \{K\}} \lambda_{fk} \tag{12.6}$$

where
{F} is the set of faults
{K} is the set of components in the system
λ_{fk} is the fault rate of component k due to fault f

The fault rate is the reciprocal of MTTF.

The conventional way of expressing failure rates for electronic devices is in terms of FIT. The FIT value is the number of failures expected in one billion (10^9) device-hours. Thus, the total FIT can be expressed as

$$FIT_{tot} = \lambda_{tot} \times 10^9 \tag{12.7}$$

FIT is used to report improvements in reliability as a result of temperature reductions achieved by the thermal optimization methods discussed in this chapter.

12.4 DPN-BASED THERMAL OPTIMIZATION IN 3D NoCs

In this section the proposed DPN-based RTM is presented. All aspects associated with the DPN are discussed such as dynamic programming, routing algorithm deadlock-freeness, routing algorithm adaptiveness, and convergence time.

12.4.1 SHORTEST PATH COMPUTATION USING DYNAMIC PROGRAMMING

Dynamic programming (DP) is an efficient optimization method that is suitable for problems that can be broken down into subproblems. Problems which can be solved with decisions that span several points in time recursively, and where *Bellman's principle of optimality* can be applied, are said to have optimal substructures and can be solved using DP [9].

One such problem is the *shortest path* problem in graph theory. This problem is essential in NoCs runtime management. Runtime dynamic routing with congestion avoidance, fault tolerance, or thermal management can be formulated as shortest path problems [33]. Figure 12.2 illustrates the shortest path as a bold line between a source node (S) and a destination node (D).

The shortest path problem can be described as follows: Given a directed graph $G = (\mathcal{V}, \mathcal{A})$ with $N = |\mathcal{V}|$ nodes, $m = |\mathcal{A}|$ edges, and a cost $C_{u,v}$ associated with each edge $u, v \in \mathcal{A}$. The total cost of a path of length l, $p = \langle n_0, n_1, \ldots, n_{l-1} \rangle$ is the sum of the costs of its constituent edges: $Cost(p) = \sum_{i=1}^{l-1} C_{i-1,i}$. The shortest path from a source node s to a destination node d is then defined as any path p with minimum cost, $\min\{Cost(p)\}, \forall p \in P_{s,d}^l$, where $P_{s,d}^l$ is the set of all paths between s and d.

The shortest path problem can be readily formulated and solved using a standard LP solver. Alternatively, it can be simplified by breaking it down into simpler subproblems and then solved recursively using DP [19]. Solving the shortest path problem using DP involves stating this problem in the form of Bellman equations, which define a recursive procedure in step k and can lead to a simple parallel architecture to speed up the computation.

In DP, finding the shortest path from a source node s to a destination node d requires computing the DP value or namely the *cost-to-go*, which is the expected cost from s to d. This cost is updated recursively

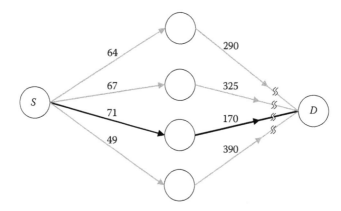

FIGURE 12.2 Illustration of finding the shortest path in a graph: a straight line indicates a single edge; a discontinuous line indicates a shortest path between the two nodes it connects (other nodes on these paths are not shown); the bold line is the overall shortest path from source, S, to destination, D.

until it reaches its optimal value. The DP value from s to d at the kth iteration is denoted as $V^{(k)}(s, d)$ here. After the algorithm converges the optimal DP value, $V^*(s, d)$, will hold the minimum cost from s to d. For any intermediate node u between s and d, the Bellman equation can be written as follows:

$$V^{(k)}(s,d) = \min_{\forall u \in \mathcal{V}} \left\{ V^{(k-1)}(u,d) + C_{s,u} \right\}$$
(12.8)

Starting from $u = d$ and $V(d, d) = 0$, the recursion can be expanded for a path of length l between nodes s and d, that is, $s = n_0$ to $d = n_{l-1}$. The optimal DP-value can then be expressed as the total cost of the optimal path from node s to node d:

$$V^*(s,d) = \min_{p \in P_{s,d}^l} \left\{ Cost(p) \right\}$$
(12.9)

where $V^*(s, d)$ is the optimal (minimum) cost for the path between s to d and $P_{s,d}^l$ is the set of all paths of length l from s to d.

From this minimum cost path (or shortest path), the optimal decision (direction) to the destination node can be readily obtained from the argument of the minimum operator, as follows:

$$\mu(s,d) = \arg\min_{\forall u \in \mathcal{V}} \left\{ V^*(u,d) + C_{s,u} \right\}$$
(12.10)

where $\mu(s, d)$ is the optimal decision, or direction, to be taken in order to reach destination node d with the minimum cost.

Cost can be associated with nodes rather than edges. This is the case in this chapter, since the cost is defined as the router temperature. In such cases, the costs of all directed edges entering a node are equal to the cost associated with this node, which is the router's local temperature (T_{local}).

12.4.2 DPN-GUIDED 3D NoC ROUTING

The proposed routing strategy relies on distributed DP units to guide the routing load to the coolest path in the chip. These units are connected via a dynamic programming network (DPN).

The DPN is tightly coupled with the NoC communication fabric, and consists of distributed computational units. At each NoC router, there is a DP unit that implements the DP algorithm and propagates the solution to the neighboring units. Each computational unit locally exchanges control and system parameters with the corresponding NoC router.

Assuming a multisource single destination, each DP unit receives the cost of the neighboring units as input, and computes and propagates the minimum cost to the neighbors after adding its local temperature. The temperature is used as the node cost and, thus, the shortest path from a source to a destination is the one with minimum total temperature (i.e., the coolest). In this scenario, thermal hotspots are avoided whenever possible and the coolest paths are always exploited to minimize the thermal effect of the routing workload.

Considering the costs of all neighbors in the DP unit's decision implies that packets can be relayed in any direction towards the destination. However, this is only possible for fully adaptive routing, which cannot guarantee deadlock-freeness. One alternative is to use partially adaptive routing, which guarantees deadlock-freeness by prohibiting some turns in order to break waiting cycles. Thus, cost propagation and computation must only consider *possible* directions, which are those allowed by the routing algorithm. Algorithm 12.1 presents the operations required for updating the routing directions using the DP unit. The router's local temperature, which is used as the cost in the DP unit computation, comes from the distributed embedded sensor in the chip. The main algorithm is outlined in lines 1–12. If the current node is the destination, the DP-unit outputs zero as a cost and the routing decision is the local port (lines 1–4). For other destinations, the local cost is computed as shown in line 5. Each DP unit takes the optimal cost of each neighbor node as input. However, as mentioned earlier, the DP unit should consider only candidate neighbors returned by the routing function *ROUTE* as shown in line 6.

Algorithm 12.1 Operations performed by the DP-unit for thermal optimization.

Define: -
  n_c: Current node,
  $V'(n_c,k,n_d)$: Cost of sending packet from n_c to n_d through channel k,
  $\mathcal{N}(n_c)$: All neighbor nodes of node n_c,
  $ROUTE(n_c,n_d)$: Routing function that takes n_c and n_d and return candidate routing directions
      $K_{n_c,n_d} \subset \mathcal{N}(n_c)$,
  par: denotes parallel operations.

Inputs: -
  n_d: Destination node,
  $V(i,n_d)$: $\forall i \in \mathcal{N}(n_c)$, Costs for all neighbors of n_c to n_d,
  T_{local}: Router local temperature.

Outputs: -
  $\mu(n_c,n_d)$: Optimal direction from node n_c to n_d,
  $V^*(n_c,n_d)$: Optimal cost from node n_c to n_d.

1: **if** $n_d = n_c$ **then**
2:   $V^*(n_c,n_d) = 0$
3:   $\mu(n_c,n_d) = LOCAL$
4: **else**
5:   $Cost_c = T_{local}$
6:   $K_{nc,nd} = ROUTE(n_c,n_d)$
7:   **par for all** directions $k \in K_{nc,nd}$ **do**
8:     $V'(n_c,k,n_d) = V(k,n_d) + Cost_c$
9:   **end par for all**
10:   $V^*(n_c,n_d) = \min_{\forall k} V'(n_c,k,n_d)$
11:   $\mu(n_c,n_d) = \arg\min_{\forall k} V'(n_c,k,n_d)$ {Update optimal directions}
12: **end if**

Given a destination n_d and a direction k, the expected cost is computed for all routable directions (lines 7–9). Then the minimum cost is selected in line 10. The optimal routing direction is selected and used to update the routing directions in line 11. The outputs of the unit at node n_c, for a given destination n_d, are the updated expected cost $V^*(n_c, n_d)$ and the best direction $\mu(n_c, n_d)$. $V^*(n_c, n_d)$ is propagated to all neighboring nodes (to perform a similar operation), while $\mu(n_c, n_d)$ is sent to the local router to update the routing table.

Although Algorithm 12.1 has a loop, it can be performed in hardware using parallel architecture and the computational delay reduces to linear. Computational delay in the DP unit and its convergence are discussed in Section 12.4.5, while the hardware realization of the DP unit is detailed in Section 12.6.6.

12.4.3   DEADLOCK-FREENESS AND ADAPTIVENESS

In Algorithm 12.1, the degree of adaptiveness offered by the routing function *ROUTE* plays a crucial role in DPN performance. For instance, if the routing is deterministic XYZ, the DPN has no impact on routing paths. Conversely, for fully adaptive routing, the routing paths are completely determined by the DPN. However, fully adaptive routing is prone to deadlocks and requires deadlock detection and recovery techniques [28,29]. These techniques have their power and area overhead and would impact performance. Thus, in this work, a turn model is improved and adopted to ensure deadlock-freeness of the proposed routing.

12.4.3.1   Balanced 3D Routing

In this work, an improved 3D turn model is used to ensure the deadlock-freeness of the proposed adaptive routing. The proposed 3D turn model is based on a 2D *odd–even* turn model. The odd–even routing is considered to be an improvement in terms of its degree of adaptiveness compared to other turn model routing algorithms [8]. The odd–even routing differs from other turn models in prohibiting different turns for odd and even columns in a 2D.

To describe the proposed 3D routing, the rules of odd–even in 2D meshes are first summarized as follows [8]:

- **Rule 1**: *In odd columns packets are allowed to take neither North-West (NW) nor South-West (SW) turns,*
- **Rule 2**: *In even column packets are allowed to take neither East-North (EN) nor East-South (ES) turns.*

The deadlock-freeness of the odd–even turn model is proven by contradiction [8]. These rules are illustrated in Figure 12.3. For a waiting cycle to exist in a 2D mesh NoC, both ES and SW have to occur in the same column (for clockwise cycles), or both EN and NW turns have to occur in the same column (for counterclockwise cycles). Both scenarios contradict rules 1 and 2, since these rules ensure that the column of an NW turn cannot have an EN turn and the column of an SW turn cannot have an ES turn. Thus, the odd–even turn model defined by rules 1 and 2 is deadlock-free.

Extending the 2D odd–even model (described by rules 1 and 2) to 3D meshes requires the application of a rule to ensure deadlock-freeness and prohibit waiting cycles that consist of vertical turns (turns involving Up or Down directions). The following rule ensures this:

- **Rule 3**: *xy—Down turns are not allowed in an odd xy-plane and Up—xy turns are not allowed in an even xy-plane.*

In other words, packets travelling upward cannot enter an even xy-plane (turn North, East, South, or West) and packets travelling within an odd xy-plane cannot leave this plane in the downward direction (as illustrated in Figure 12.4). The 3D odd–even routing that is described by rules 1, 2, and 3

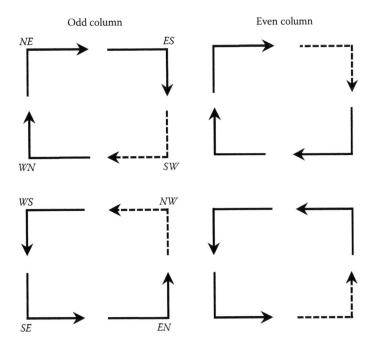

FIGURE 12.3 Illustration of prohibited turns for odd–even routing (rules 1 and 2). Dashed lines represent prohibited turns.

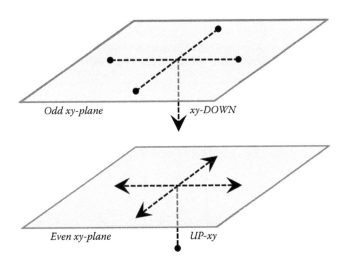

FIGURE 12.4 Illustration of prohibited vertical turns for the 3D odd–even routing (rule 3).

can then be called *conventional odd–even* or, for short, oe. Other versions of 2D odd–even routing can be defined. Here a modified odd–even routing is defined with turn prohibitions that are applied according to the row, and not the column, of a 2D mesh. The rules of the modified odd–even can be stated as follows:

- **Rule 4**: *In odd row, packets are allowed to take neither West-North (WN) nor East-North (EN) turns.*
- **Rule 5**: *In even row, packets are allowed to take neither South-West (SW) nor South-East (SE) turns.*

Similar to the conventional odd–even turn model defined by rules 1 and 2, the odd–even turn model defined by rules 4 and 5 is deadlock-free and the proof is similar to that of the conventional odd–even. Looking at the waiting cycles row-wise instead of column-wise, the row of the SW turn cannot have a WN turn (prohibiting any clockwise cycles) and the row of the SE turn cannot have an EN turn (prohibiting any counterclockwise cycles). Thus, a 2D NoC that applies rules 4 and 5 is deadlock-free, since no waiting cycles can exist (Figure 12.5).

The proposed extension of 2D partially adaptive routing to 3D NoCs is based on the following corollary.

Corollary 12.1 In 3D NoCs deadlock-freeness is still guaranteed when different layers have different turn prohibition rules if these rules guarantee intra-layer deadlock-freeness and a rule is applied to guarantee freeness from deadlocks that involve vertical turns.

Proof 12.1 *Corollary 12.1 is proven by contradiction. Assume that there are a set of packets that form a deadlock cycle in a 3D mesh. Then this cycle must be either on the same plane (planar deadlocks) or span two or more planes (3D deadlocks). The first case contradicts Corollary 12.1 since it states that intra-layer rules must exist to guarantee deadlock-freeness within the layer. The second case cannot occur since vertical links cannot be part of a cycle if a rule is applied to guarantee freeness from deadlocks that involve vertical turns as stated by the corollary. Thus, any 3D routing algorithm that satisfies Corollary 1 is deadlock-free.*

For conventional 3D odd–even routing, odd–even rules within the xy-plane are the same for all planes. They are applied along the *column*. Based on Corollary 12.1, different rules can be used for different layers to achieve more balanced adaptiveness. Thus, the conventional 3D odd–even can be modified such that the rules for the odd xy-plane (where the z coordinate is odd) are different from those for the even xy-plane (where the z coordinate is even). The 3D routing algorithm proposed in this work *uses rules 1 and 2 in an odd plane, and rules 4 and 5 in an even plane.* These rules are applied to prohibit

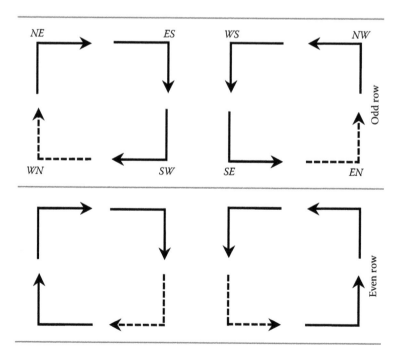

FIGURE 12.5   Illustration of prohibited turns in modified odd–even routing (rules 4 and 5).

planar deadlocks. Rule 3 is used to prohibit 3D deadlocks. The resulting 3D odd–even algorithm balances the adaptiveness among different planes in a 3D mesh, as will be seen in the following, and is called *balanced odd–even* (boe). Both oe and boe are deadlock-free since they satisfy Corollary 1.

12.4.3.2 Degree of Adaptiveness

The degree of adaptiveness is one of the metrics that is used to evaluate adaptive routing algorithms [18]. It can be defined as the number of different allowable paths from a source to a destination. For a 3D mesh, let the coordinates of the source node be (x_s, y_s, z_s) and the coordinates of destination node are (x_d, y_d, z_d). Also, in the following, let $d_x = |x_d - x_s|$, $d_y = |y_d - y_s|$ and $d_z = |z_d - z_s|$.

For fully adaptive routing, the degree of adaptiveness is the number of *all* shortest paths from source to destination and is given by [8]

$$P_{fully\_adaptive} = \frac{(d_x + d_y + d_z)!}{d_x! d_y! d_z!} \tag{12.11}$$

As a result of applying Rules 1 and 2 in all planes, the degree of adaptiveness of the conventional oe (P_{oe}) can be expressed as follows [8]:

$$P_{oe} = \frac{(h + d_y + k)!}{h! d_y! k!} \tag{12.12}$$

where h is equal to $\lceil d_x/2 \rceil$ or $\lceil (d_x-1)/2 \rceil$ depending on the column at which x_s and d_x lie. Similarly, k is equal to $\lceil d_z/2 \rceil$ or $\lceil (d_z-1)/2 \rceil$ depending on the layer at which z_s and d_z lie. It can be noted that, for the conventional oe, the constrained directions are x and z while the y direction is relaxed.

As opposed to the conventional oe (Equation 12.12), which has the x direction constrained in all planes, the balanced oe constrains direction x in an odd plane and direction y in an even plane. This results in different restrictions on the odd xy-plane from those on the even xy-plane. Consequently, the regularity of traffic patterns (and the resulting communication workload) which occur in adjacent layers (due to similar restrictions) is broken as shown in Figure 12.6. This results in a more balanced adaptiveness among the planes which enhances the performance of runtime adaptive selection strategies [52].

12.4.4 Coupling DP with 3D-NoC

The 3D NoC, with the DP network coupled, is illustrated in Figure 12.7. DP network is a network of distributed computational units. The topology of the DP network resembles that of the communication structure of the NoC. At each node there is a computation unit to implement the DP shortest path computation. The solution is propagated to the neighboring units. The DP network is tightly coupled with the NoC and each computational unit locally exchanges control and system parameters with the corresponding NoC router (as detailed in Section 12.6.6).

The DP network converges to the optimal solution in a time period which depends on the network structure (the diameter of the network) and the clock frequency of the DP network [32]. This is detailed in the following section. After convergence, the DPN passes the control decisions to the NoC routers. The cost can be communicated across the DP network by dedicated links. This enables fast convergence of the network in response to rapid changes in the cost function. Another option is to use the existing NoC structure to propagate the DP cost. This increases DPN convergence delay. For the present, case cost is defined as the local router temperature, which changes slowly compared to the system clock since the thermal time constant of the chip is usually in the order of milliseconds or seconds while the frequency period is in the order of nanoseconds. This enables the use of the existing NoC structure for DP cost propagation.

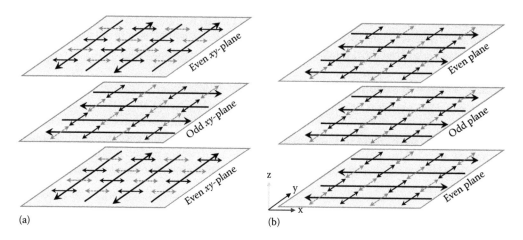

(a)   (b)

FIGURE 12.6 Illustration of path diversities for both, (a) conventional 3D odd–even, and (b) the proposed balanced 3D odd–even.

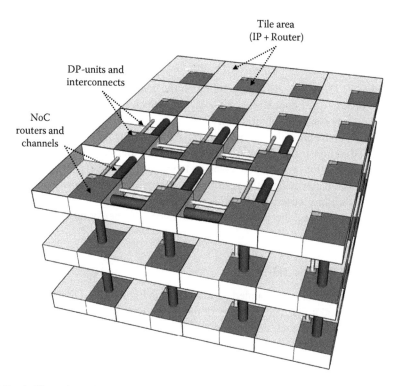

FIGURE 12.7 A 3D mesh NoC with DPN for coupled.

12.4.5 DP-Network Convergence Time and Complexity

The DP-network converges to an optimal routing solution after a delay which is determined by the network topology, the delay of data propagation, and the computational delay of the DP-unit. Each unit involves $\mathcal{O}(|\mathcal{A}|)$ additions and comparisons where $|\mathcal{A}|$ is the number of edges in the unit. Note that the number of additions corresponds to the number of adjacent nodes. Hence, the worst-case solution time is $\mathcal{O}(i\,|\mathcal{A}|)$, where i is the number of iterations evaluated by each unit. In software computation, the number of iterations, i, which guarantees that all nodes have been updated, equals the number of nodes in the network, that is, $i = |\mathcal{V}|$ [9]. However, in hardware implementation with parallel execution, i is determined by the network structure and $\mathcal{A}$ additions can be executed in parallel. Each computational unit can simultaneously compute the new expected cost for all neighboring nodes. The network convergence time is proportional to the network diameter, which is the longest path in the network. To determine the minimum clock frequency of the DPN that guarantees the convergence, the following condition must be met:

$$N_{dim} \times N \times (t_{link} + t_{unit}) < t_{temp\_sampling}, \tag{12.13}$$

where
N_{dim} is the NoC diameter
N is the number of NoC nodes (number of destinations)
t_{link} is the delay time of the DP-interconnect
t_{unit} is the delay time of the DP-unit computation
$t_{temp\_sampling}$ is the time period for temperature sampling

This condition can yield an upper bound of t_{unit} from which the minimum frequency of the unit can be computed.

12.5 DYNAMIC THERMAL MODELING FOR 3D NoCs

A traffic and thermal co-simulation tool is developed in this work for the dynamic thermal modeling of the 3D NoC in order to evaluate the proposed routing strategy. This tool comprises traffic, power, and thermal models. These models are integrated in an automated flow. Figure 12.8 illustrates the input configuration files and parameters required by the model. This figure also depicts the model components and the computational flows among them. The technology, architecture, and packaging parameters are used to configure the tool. The results, in terms of cycle-accurate temperature variations over a discrete sampling interval, are saved in computer storage. The models used in this traffic and thermal co-simulation tool are described in this section.

12.5.1 TRAFFIC MODEL

The network of concern in this work is a collection of router nodes connected by channels. Each router is connected to a single IP core that injects and consumes packets. A wormhole flow control technique [12] is used and the configurations of Intel's TeraFLOPS chip [49] is adopted with an input buffer size of 16 flits and a packet size of 3 flits.

Traffic simulation is performed using a modified version of *Noxim* [15]. The router architecture is modified by adding additional ports for communicating the DP costs, and a DP-based routing-table updating algorithm (Algorithm 12.1) is introduced. Moreover, the 2D NoC routing algorithms and traffic patterns are modified to support the proposed 3D NoC routings and traffic patterns. The power model of the original simulator is updated with power for both the router and the DP unit. Router power and area are evaluated using a NoC power simulator, while DP unit power and area are evaluated using a hardware synthesis tool.

12.5.2 AREA AND POWER MODEL

Router power and area are computed using the NoC power and area model *ORION 2.0* [27]. The power traces of the computational units and floorplan are taken from the Intel TeraFLOPS chip [49]. In general, the energy dissipated by the NoC is divided into the following categories: (1) routing

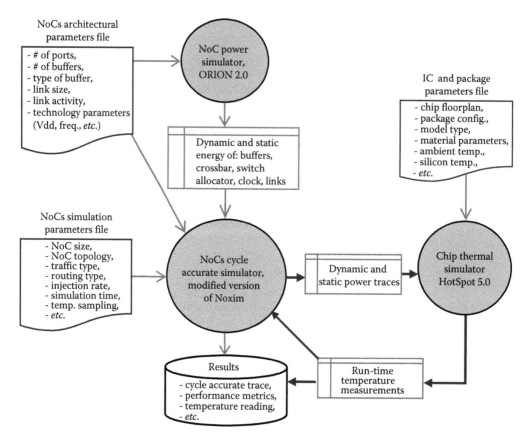

FIGURE 12.8 Automated computational flow of the proposed tool for dynamic thermal optimization for 3D NoC.

and arbitration; (2) flit forwarding energy; (3) flit receiving energy; (4) clock energy, and (5) leakage energy. The flit receiving energy is assumed to be equal to buffer writing energy. The forwarding energy, which dominates the energy consumption of the router, comprises the energies of buffer read, E_{buffer}, crossbar traversal, $E_{crossbar}$, and link traversal, E_{link}:

$$E_{forward} = E_{buffer} + E_{crossbar} + E_{link}. \tag{12.14}$$

The flit forwarding energy along the vertical direction is assumed to be the same as that for the horizontal direction except for the link traversal energy E_{link}. The E_{link} of the vertical link is computed assuming a through-silicon-via (TSV) link length equal to layer thickness.

The energy consumption of any computational unit $U(E_U)$ in the tile is assumed to be modulated by its communication energy ($E_{U\_local\_comm}$). This energy dynamically changes according to local data transfer (data transfer from and to the local router). Thus, the energy of the computational unit E_U is computed as:

$$E_U = \beta \cdot E_{U\_local\_comm} \tag{12.15}$$

where
    β is the ratio of the communication power to the computation power of unit U
    U is any tile unit other than the router

For the TeraFLOPS, these units are: data memory (DMEM), instruction memory (IMEM) and floating-point units (FPMAC0 and FPMAC1). β is estimated based on the results of communication and computation powers for Intel's TeraFLOPS CMP [49].

12.5.3 THERMAL MODEL

A typical modern chip package consists of several layers. Figure 12.9 illustrates these layers for a typical ceramic ball grid array (CBGA) package of a 3D-IC with four vertically stacked silicon dies. This is the packaging scheme adopted in this chapter. The package has several heat conduction layers including heat sink, heat spreader, thermal paste, silicon die(s), C4 pads, ceramic packaging substrate, and solder balls [22]. These layers are designed in such a way as to maximize the heat-flow from the active layer(s), or silicon die(s), to the ambient. This path represents the primary heat-flow in the package. Thus, the heat generated by chip activity could be removed efficiently.

The heat-spreader and heat-sink layers are often made of aluminum, copper, or some other materials of high thermal conductivity. In addition to the primary heat-flow path to the heat sink, there is a secondary heat-flow path from the die(s) to the package, and the PCB is designed in such a way as to minimize the heat-flow in order to protect the board and other installed devices from excessive heat accumulation. To model all of these heat transfer paths, the thermal RC model *HotSpot* [24] is employed. This model is built on top of the dualism between thermal and electrical phenomena, as both are described by the same differential equations. Thermal resistance (R) and thermal capacitance (C) can be computed from the following equations:

$$R = \frac{t}{k \times A} \tag{12.16}$$

$$C = c \times t \times A \tag{12.17}$$

where
t is the thickness of material (in m)
k is the material's thermal conductivity per unit volume (in W/[m K])
A is the cross-sectional area (in m^2)
c is the thermal capacitance per unit volume (in J/[K m^3])

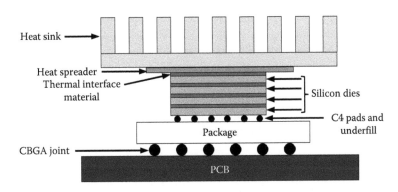

FIGURE 12.9 Illustration of various layers in a typical ceramic ball grid array (CBGA) package of a 3D-IC with four layers. (From Huang, W., Ghosh, S., Velusamy, S., Sankaranarayanan, K., Skadron, K., and Stan, M.R., Hotspot: A compact thermal modeling methodology for early-stage VLSI design, *IEEE Transactions on Very Large Scale Integration (VLSI) Systems*, 14(5), 501–513, © 2006 IEEE.)

Die-level thermal RC modeling can be done at the functional unit level or at finer levels, where the die is divided into regular grid cells in order to gain a more detailed temperature distribution [22]. These cells represent the different architectural blocks in the die. Likewise, the chip can have several dies stacked on top of each other in a 3D-IC, and this can be readily added to the thermal RC model. A more detailed discussion regarding thermal model derivation and calibration, and its validation against a commercial finite element simulator can be found in the original HotSpot papers [22,23,44].

12.6 RESULTS AND DISCUSSION

12.6.1 EXPERIMENTAL SETUP AND TOOLS

To evaluate the proposed 3D NoC thermal optimization, a 3D NoC-based CMP is considered. Tile area and power are computed using the results presented in a previous work [49]. The power and area of the router unit is modified to account for the extension to the 3D mesh router, which consists of seven input/output channels and the overhead introduced by the DP-unit.

For thermal model configuration, layer die thickness is set to 0.15 mm. The thermal resistivity of the heat sink is 0.0025 m K/W. The dies are separated by an interlayer material with a thickness of 0.02 mm. The heat spreader is placed on top of the silicon dies. The thermal interface material (TIM) is used as the filler material in order to separate the heat spreader and the silicon dies. The resistivity of the interlayer material is set to 0.25 m K/W. Ambient temperature is assumed to be 25°C. V_{DD} and frequency are assumed to be 1 V and 3 GHz, respectively.

For synthetic traffic simulation, the traffic patterns considered are *Uniform*, *Transpose*, and *Hotspot*. For random traffic, each tile sends data to all other tiles with equal probability. For the transpose case, *tile(i, j, k)* sends packets to *tile(X − i, Y − j, Z − k)*, where X, Y, and Z are the x, y, and z dimensions of the NoC, respectively. For the Hotspot traffic pattern, the four central tiles of the top layer (layer 3) receive an extra 5% in addition to the uniform (random) traffic. For each of these traffic patterns, the floorplan is arranged as a 3D mesh with a size of 6 × 6 × 4.

In addition to the earlier traffic scenarios, a scenario is included which simulates a layer of shared memory. This is an important application for 3D stacking [4]. To evaluate the proposed RTM in such scenario, the simulator is modified to generate traffic that mimics 3D memory stacking by assuming that top layer is a memory resource shared by the layers of computational cores and which receives 30% of the traffic of these layers. This traffic is called *Memory-Wall* here.

The following schemes are evaluated:

- Odd–even with buffer selection (oe_buff): The original odd–even routing (Rules 1 and 3 are applied for all planes) with buffer level selection strategy and no thermal optimization.
- XYZ with downward routing (dw_xyz): The RTM scheme proposed in [7] for thermal optimization in 3D NoCs, which uses traffic-aware downward routing.
- Odd–even with DP selection (oe_dp): The original 3D odd–even routing with a DP-guided selection strategy for thermal optimization.
- Balanced odd–even with DP selection (boe_dp): Odd–even (Rules 1 and 2) is applied in an even plane, modified odd–even (Rules 4 and 5) is applied in an odd plane, and Rule 3 is used for vertical turns with a DP-guided selection strategy for thermal optimization.

For traffic-aware downward routing (dw_xyz), the authors proposed the use of a downward level, which depends on the packet injection rate and traffic type and, thus, requires calibration. However, in this study, considerable effort is expended in the calibration of the downward level for different traffic patterns and PIRs to ensure fair comparison. This is done according to the method proposed in their paper to achieve similar performance results [6].

12.6.2 TEMPERATURE RESULTS

In the first experiment, simulations are run until the chip temperature is stable after 5 million cycles. The temperature results are illustrated in Figure 12.10 for a PIR of 0.008 *packet/cycle/IP*. For this PIR, all the routing schemes considered achieve the same throughput of 0.0481 *flits/cycle/IP*. This guarantees a fair comparison in terms of the resulting temperature.

The maximum chip temperature and the spatial temperature gradient, which is the difference between the maximum and the minimum temperatures, are shown for the four schemes considered. The results for balanced odd–even with buffer selection (boe_buff) are also included here for

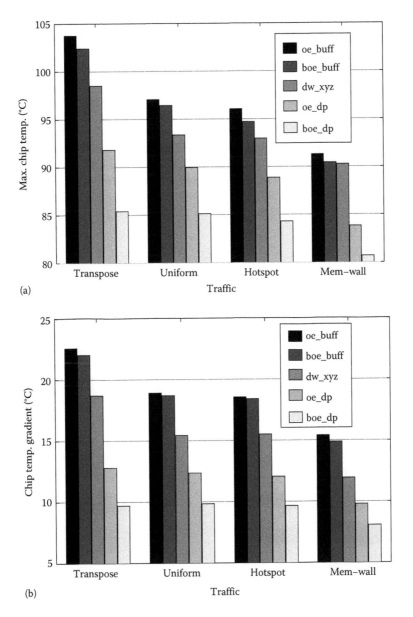

FIGURE 12.10 Comparison of the maximum and spatial gradient (min.–max.) of chip temperature for the considered routing strategies with various traffic scenarios. (a) Max. temperature and (b) temperature gradient.

reference. Taking oe_buffer as a baseline, and given that the four methods have similar throughput, it can be noticed that oe_dp outperforms dw_xyz in its thermal behavior for all the considered traffic patterns. However, boe_dp outperforms both dw_xyz and oe_dp for the four traffic scenarios. For instance, in this experiment, boe_dp achieves more than 18°C cooling compared to the oe_buffer, while dw_xyz could only achieve 5°C cooling for the transpose traffic case (Figure 12.10a). Moreover, it can be noticed that the spatial temperature gradient in the chip for boe_dp is nearly half of that of dw_xyz for all the traffic patterns considered. This implies that higher thermal balancing is achieved by the proposed scheme (see Figure 12.10b).

Figures 12.11 and 12.12 illustrate the spatial temperature and NoC power distributions, respectively, for the four schemes under the transpose traffic case. Figure 12.11 indicates that, besides its higher cooling performance boe_dp, it achieves a more homogeneous spatial thermal distribution compared to dw_xyz. The thermal behavior results can be explained by the power distribution results in Figure 12.12. The cooling performance of an RTM technique is determined by its capability to accommodate to the cooling system of the chip. In this scenario, it can be seen that the proposed boe_dp can migrate power consumption towards the heat sink more efficiently compared to dw_xyz (see Figure 12.12d and c). As a result, better thermal moderation can be achieved by the proposed approach.

To see how each of the considered schemes behave when the PIR changes, Figure 12.13 shows the maximum and gradient of temperature for a PIR range of 0.004–0.016 *packet/cycle/node* with Transpose traffic. In this range of PIR, all of the routing schemes achieve the same throughput. As expected, both the maximum and gradient of temperature significantly increase with PIR for all schemes. However, the trend found in Figure 12.10 can also be seen here. Both DP schemes (boe_dp and oe_dp) outperform dw_xyz in terms of both the maximum and gradient of chip temperature for all the considered PIRs.

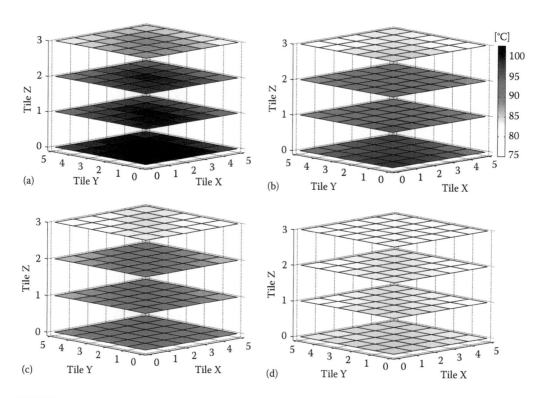

FIGURE 12.11 Spatial thermal distributions (°C) for the four routing strategies. (a) oe_buff, (b) dw_xyz, (c) oe_dp, and (d) boe_dp.

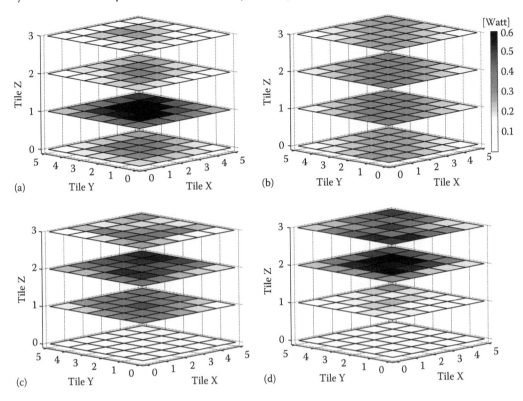

FIGURE 12.12 Spatial power distributions (W) for the four routing strategies. (a) oe_buff, (b) dw_xyz, (c) oe_dp, and (d) boe_dp.

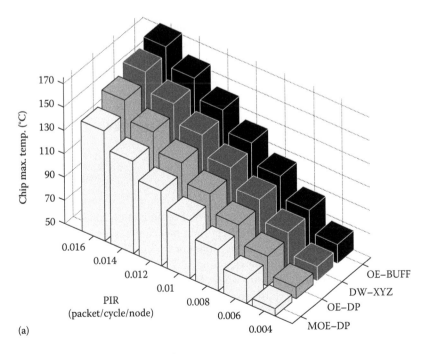

FIGURE 12.13 Maximum and gradient (min.–max.) of chip temperature variation with PIR for the four routing strategies with transpose traffic. (a) Maximum chip temperature. *(Continued)*

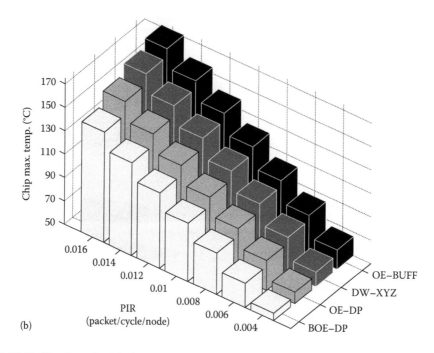

FIGURE 12.13 (*Continued*)   Maximum and gradient (min.–max.) of chip temperature variation with PIR for the four routing strategies with transpose traffic. (b) Temperature gradient (min-max).

These results can be explained by the fact that dw_xyz adapts only to inter-layer thermal variations and not intra-layer thermal variations. Thermal variations within the same layer can be significant, but they are not exploited by dw_xyz. Moreover, dw_xyz is tailored to a cooling system in which the heat sink is placed above the chip. On the other hand, our scheme achieves better results because DP can adapt to inter-layer as well as intra-layer thermal variations due to its adaptive distributed nature.

Moreover, Figures 12.10, 12.11, and 12.13 also indicate that boe_dp considerably outperforms oe_dp in terms of both the maximum and gradient of the temperature. This clearly indicates that the higher balancing of adaptiveness offered by the balanced 3D odd–even routing (boe_dp), compared to the conventional 3D odd–even (oe_dp), is effective in improving the efficiency of the DPN in heat diffusion.

12.6.3   RELIABILITY IMPROVEMENT

To give an insight into the implications of the thermal optimization schemes considered on reliability improvement, the FIT values for different fault mechanisms are evaluated. FIT_{EM}, FIT_{SM}, FIT_{TDDB}, and FIT_{TC} values are computed using Equations 12.1, 12.2, 12.3, and 12.5, respectively. The total FIT (FIT_{TOT}) is computed using Equation 12.7. The material-dependent parameters in these equations are taken from a previous work [45], while the constants (A_{EM}, A_{TDDB}, A_{SM}, and A_{TC}) are taken from another work [46] assuming 65 nm technology. The temperature results presented in Figure 12.10 for $PIR = 0.008$ *packet/cycle/node* are used in FIT computation. The FIT results for the four thermal optimization methods with the four traffic patterns are shown in Table 12.1.

It can be seen that, in general, any reduction in chip temperature results in lower FIT, better reliability, and longer chip lifetime. Taking the oe_buff as a baseline, it can be seen that

TABLE 12.1

Comparison of FIT due to Different Fault Mechanisms for the Four Routing Strategies and Different Traffic Patterns for a 6 × 6 × 4 3D NoC Configuration

Traffic	Routing Strategy	FIT_{EM}	FIT_{SM}	FIT_{TDDB}	FIT_{TC}	FIT_{tot}	Improvement (%)
Transpose	oe_buff	33,268	7360	21,158	3897	65,684	—
	dw_xyz	22,509	5526	17,770	3315	49,120	25.2
	oe_dp	13,426	3744	14,155	2647	33,972	48.3
	boe_dp	8,049	2521	11,342	2088	24,000	63.5
Uniform	oe_buff	20,162	5091	16,924	3163	45,341	—
	dw_xyz	15,154	4106	14,924	2794	36,977	18.5
	oe_dp	11,614	3350	13,288	2479	30,731	32.2
	boe_dp	7,874	2478	11,236	2066	23,654	47.8
Hotspot	oe_buff	18,679	4808	16,362	3061	42,911	—
	dw_xyz	14,700	4012	14,727	2756	36,196	15.7
	oe_dp	10,659	3136	12,803	2383	28,980	32.5
	boe_dp	7,371	2352	10,922	2002	22,647	47.2
Mem. W.	oe_buff	12,889	3629	13,905	2599	33,021	—
	dw_xyz	10,283	3050	12,605	2344	28,281	14.4
	oe_dp	7,075	2278	10,733	1962	22,048	33.2
	boe_dp	5,466	1857	9,616	1726	18,866	42.8

the higher temperature reduction achieved by boe_dp compared to the other methods leads to a significant improvement in reliability. For instance, boe_dp achieved 18.4°C temperature reduction compared to only 5.2°C for dw_xyz in the transpose traffic case. This translates to a 63.46% reduction in FIT_{TOT} for boe_dp compared to only 25% for dw_xyz. This indicates the crucial significance of the thermal optimization achieved by the proposed scheme in increasing reliability and IC lifetime.

12.6.4 PERFORMANCE RESULTS

Figure 12.14 compares the performance of the four schemes in terms of average network delay versus the achievable throughput curves under the four traffic scenarios. It can be noticed that, in general, the performance of dw_xyz is considerably lower than that of the other schemes. Also, it can be seen that the performance of oe_dp and boe_dp is nearly the same, and both slightly outperform the oe_buff for the uniform and transpose traffic cases, while the oe_buff is better than the thermal-aware DP approaches for the hotspot and memory wall traffic patterns. However, for the latter two traffic patterns, both DP approaches are still better than dw_xyz.

Another experiment is conducted to evaluate the performance of the thermal optimization methods considered. The network throughput and the average delay that causes the first violation of a thermal limit are recorded. Table 12.2 summarizes the results of this experiment for the traffic patterns considered and with three thermal limits: 60°C, 70°C, and 80°C. The thermal limits are chosen in the reasonable working temperature range and in line with previous work [43]. This table also shows the boe_dp throughput improvement over oe_buff and dw_xyz. For instance, given a thermal limit of 80°C, the boe_dp routing can achieve 22% and 41% higher throughput compared to dw_xyz and oe_buff, respectively, for transpose traffic. This clearly demonstrates the capability of DP to maneuver packets dynamically at runtime and to exploit the coolest paths. As a result, the thermal violation is delayed to higher PIR and throughput. Similar trend can be seen for other traffic patterns and thermal limits.

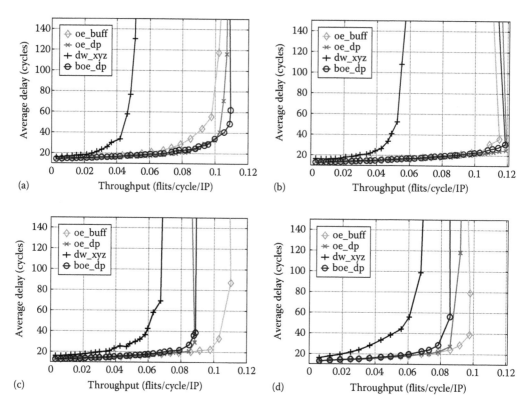

FIGURE 12.14 Performance comparison of the considered routing strategies in terms of delay versus throughput curves for different traffic scenarios. (a) Transpose, (b) uniform, (c) hotspot, and (d) memory wall.

12.6.5 REAL APPLICATION BENCHMARKS

In this section, the proposed DP-based RTM is evaluated with real-application benchmarks. Six real benchmarks with different sizes, topologies, and bandwidth requirements are used. These benchmarks include a generic complex MultiMedia system, which comprises an h263 video encoder and an mp3 audio decoder (MMS) [20], a telecommunication benchmark (TELE), and a video object plane decoder (VOPD) [35]. In addition are three benchmarks, AMI49, AMI25, and MPEG4, which were extracted from the Microelectronics Centre of North Carolina benchmark suite found in [1]. Details of the size and communication bandwidth requirements of these benchmarks are shown in the first three columns of Table 12.3.

Mapping these applications to NoCs is achieved using the algorithm proposed in a previous work [11]. However, the small sizes of these benchmarks mean that they are not suitable to be used in a 3D NoC platform of the appropriate size. Therefore, the sizes of these benchmarks are extended to four layers by mapping each task to a pillar (four IP cores aligned on top of each other) rather than a single IP core. The inter-layer traffic is generated by dividing the original bandwidth between communication pairs to the replicated vertical IP cores in the resulting pillar pairs of the 3D NoC platform.

The proposed DP-based RTM (boe_dp) is evaluated by comparing it to a previous work [6], denoted by dw_xyz. The two routings are compared in terms of maximum, T_{max}, and gradient, $T_{gradient}$, of chip temperature, as well as energy consumption, after draining 5 MB of data for each benchmark. These results, in addition to the percentage reductions in maximum temperature, temperature gradient, and energy consumption achieved by the proposed boe_dp over the dw_xyz, are shown in Table 12.3. The results in Table 12.3 indicate a significant improvement with the boe_dp

TABLE 12.2

Achievable Performance Metrics with Different Routing Algorithms for Different Thermal Limits and Traffic Patterns

		Thermal Limit (°C)					
		60		70		80	
Traffic	Routing Strategy	Delay (Cycle)	Thrpt. (Flit/ Cycle/IP)	Delay (Cycle)	Thrpt. (Flit/ Cycle/IP)	Delay (Cycle)	Thrpt. (Flit/ Cycle/IP)
Transpose	oe_buff	17.8	0.0361	18.4	0.0466	21.7	0.0583
	dw_xyz	57.8	0.046	820.9	0.0554	3887	0.0671
	oe_dp	16.4	0.046	17.0	0.0574	19.3	0.0735
	Boe_dp	17.9	0.052	19.7	0.0665	23.2	0.0817
boe_dp throughput	vs. dw_xyz	13%		21%		22%	
Improvement	vs. oe_buff	44%		43%		41%	
Uniform	oe_buff	15.1	0.0367	16.18	0.0496	17.9	0.0663
	dw_xyz	40.8	0.0488	356.9	0.0630	553.6	0.0643
	oe_dp	15.2	0.0470	16.4	0.0602	17.8	0.0756
	boe_dp	16.2	0.0523	17.5	0.0670	19.6	0.0815
boe_dp throughput	vs. dw_xyz	8%		7%		27%	
Improvement	vs. oe_buff	43%		36%		23%	
Hotspot	oe_buff	15.1	0.0364	16.5	0.0536	18.0	0.0689
	dw_xyz	27.7	0.0489	58.1	0.0629	182.8	0.0683
	oe_dp	15.82	0.0488	17.2	0.0629	19.6	0.0790
	boe_dp	17.0	0.0522	20.2	0.0697	27.0	0.0850
boe_dp throughput	vs. dw_xyz	7%		11%		25%	
Improvement	vs. oe_buff	44%		31%		24%	
Mem. W.	oe_buff	15.80	0.0367	17.33	0.0553	19.04	0.0681
	dw_xyz	38.44	0.0492	98.92	0.0675	554.16	0.0751
	oe_dp	16.96	0.0495	20.20	0.0683	22.33	0.0785
	boe_dp	18.73	0.0555	24.18	0.0728	56.29	0.0856
boe_dp throughput	vs. oe_dp	13%		8%		14%	
Improvement	vs. dw_xyz	51%		40%		26%	

TABLE 12.3

Real Benchmarks Results

			dw_xyz			boe_dp			boe_dp Imprv.		
Bench. (Size)	NoC Size	bw MB/s	T_{max} (°C)	$T_{gradient}$ (°C)	E (mJ)	T_{max} (°C)	$T_{gradient}$ (°C)	E (mJ)	T_{max} (%)	$T_{gradient}$ (%)	E (%)
AMI49	7×7×4	5061	116.4	31.5	8.2	104.4	22.7	7.7	13.0	27.7	6.5
AMI25	5×5×4	3680	91.0	27.9	6.4	84.1	23.6	6.2	10.4	15.5	6.7
MMS	5×5×4	628	69.3	16.8	6.3	65.7	14.9	6.0	8.1	11.3	5.0
TELE	4×4×4	998	79.2	20.1	6.3	72.9	16.3	5.9	11.5	19.0	6.8
VOPD	4×4×4	3731	93.3	25.8	7.8	88.3	22.6	7.5	7.4	12.2	4.0
MPEG4	3×3×4	1951	80.2	20.8	5.7	73.9	16.8	5.5	11.4	19.4	3.7
							Average		10.3	17.5	5.5

Note: Chip maximum, T_{max}, and gradient, $T_{gradient}$, of temperature, NoC energy consumption for dw_xyz [6] and boe_dp in addition to the percentage improvement of boe_dp after draining 5 MB of data.

over the dw_xyz in terms of temperature regulation, where it could achieve up to 13% reduction in maximum chip temperature and up to 27% reduction in temperature gradient. Moreover, it can be seen that, besides better thermal regulation, boe_dp consumes less energy (up to 6.8%) for the same drained data volume compared to dw_xyz even after adding the energy consumption of the DP units. This is due to the fact that dw_xyz is a nonminimal routing and, thus, consumes higher energy compared to boe_dp. Meanwhile, dw_xyz adapts only to vertical paths by employing nonminimal routing to avoid hot layers. On the other hand, the proposed boe_dp adapts to planar paths as well as vertical ones. Paths on the same layer can exhibit significant thermal gradients that are exploited by the proposed technique, due to its global awareness of temperature and distributed control nature, in order to achieve better thermal regulation while adhering to minimal path routing.

12.6.6 HARDWARE IMPLEMENTATION

The DP unit can be implemented using different methods. This section investigates the hardware realization of the DP using synchronous circuits. The aim is to realize a DP hardware that augments the NoC's routers so as to provide an adaptive strategy to diffuse heat throughout the chip geometry. Moreover, the resultant power and area overheads of the DP unit are evaluated.

Figure 12.15 shows the architecture of a router which enables adaptive thermal-aware routing. The architecture supports 3D mesh NoCs. The router circuit is a state-of-the-art design [36] for a 2D mesh with two additional channels for upper and lower layers (labeled as *Up* and *Down*). The design is augmented by an additional block (depicted by a dotted line), which implements the proposed adaptive strategy. The temperature sensor circuit provides the DP computational unit with the local cost.

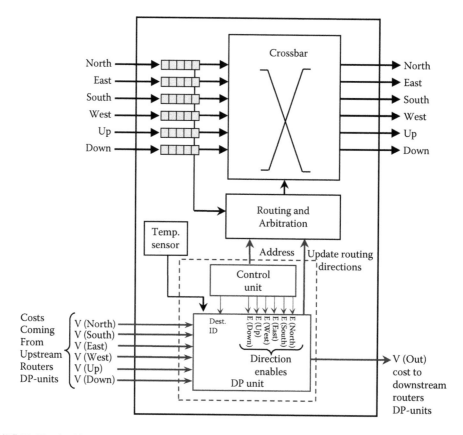

FIGURE 12.15   Architecture of the 3D NoC router including the DP unit to enable dynamic thermal-aware routing.

The local cost and the costs coming from upstream routers are used to compute the *cost-to-go*, which is propagated to all downstream routers. The design consists of combinational circuits. However, the control unit block shown in the figure is a mixture of sequential and combinational circuits. It can be realized using a synchronous counter and a few logic gates. The counter scans the destinations and supplies an address reference the routing table inside the router such that the DP-unit can successively update all destination decisions in the routing table.

The path cost computation is implemented using the DP units, as shown in Figure 12.16. These DP-units are connected as a DP-network. The coolest path computation necessitates a minimum operation to evaluate and compare the costs coming from upstream routers. This can be realized using comparators and data multiplexers (Figure 12.16a). Also, an adder is required to add the local router cost to the optimal cost computed using this circuit. Moreover, another data multiplexer is needed to output the associated action for the minimum expected cost. Therefore, the basic circuit of a DP computational unit for the 3D mesh NoC comprises of one adder, six comparators, and six data multiplexers.

Direction costs that are involved in the optimal (minimum) cost computation are filtered by *direction selection* control circuit, which is shown in Figure 12.16b. The six enable signals (corresponding to six directions) used by these control circuits come from the control unit and they

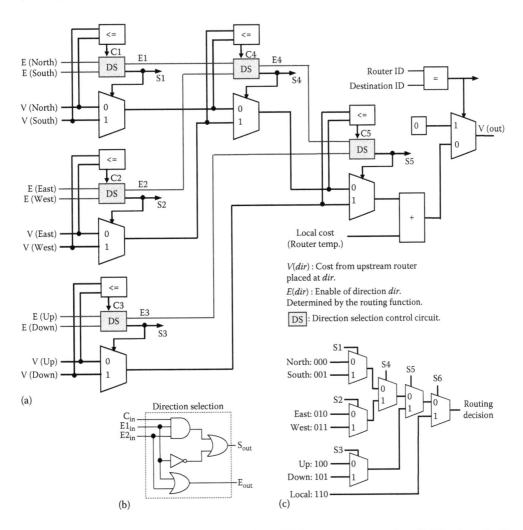

FIGURE 12.16 Hardware realization of the DP unit. (a) Minimum cost computation. (b) Direction selection circuit. (c) Routing decision computation circuit.

TABLE 12.4

Synthesis Results: Router and DP-Unit Power and Area in Addition to DP Unit Relative Overhead

NoC Size	Router Only		Router + DP Unit		DP Unit Overhead	
	Area (mm²)	Power (W)	Area (mm²)	Power (W)	Area (%)	Power (%)
6 × 6 (2D)	0.4751	0.2168	0.4774	0.2171	0.395	0.102
6 × 6 × 4 (3D)	0.7520	0.3290	0.7644	0.3305	1.645	0.456

are determined by the routing algorithm. The direction selection control circuit discards the direction that is not enabled (where the corresponding enable signal is 0). At each comparison stage, it takes two enable signals (one for each input direction), E_{in1} and E_{in2}, in addition to the comparator result, C_{in}, as inputs. It outputs enable, E_{out}, and selection, S_{out}, signals. S_{out} is equal to C_{out} only if both directions are enabled ($E_{in1} = 1$ and $E_{in2} = 1$). If only one direction is enabled, S_{out} must select this direction regardless of its cost. The enable signal E_{out} equals 1 if any of the input directions is enabled. If neither are enabled, E_{out} is 0 and the cost computed at this comparison stage is discarded at the next comparison stage. The direction selection circuits filter directions and, as a result, only allowable directions (allowed by the routing function) are included in the optimal cost computation. This enables the DP-unit to work in compliance to deadlock-free routing algorithm.

The DP-unit circuit also requires the computation of the optimal routing direction for a particular destination (designated *destination ID*). This direction is used to update the routing table of the *Routing and Arbitration* block. The optimal routing direction computation circuit is illustrated in Figure 12.16c. This circuit takes the direction selection signals (S_1–S_6) that result from the optimal cost computation circuit, as multiplexer selects. It requires six 3-bit data multiplexers, assuming seven input directions (six direction plus the local).

Evaluating the hardware power and area overheads for any new proposed solution involved in NoC design is essential. This gives an insight into the trade-offs that exist between the costs paid in terms of power and area and the benefits gained from the proposed technique. To evaluate the area and power overheads of the proposed DP-based routing, the DP computational unit is implemented in Verilog. The implementation is then synthesized using the Synopsys Design Compiler and mapped onto the Faraday UMC 65 nm technology library. Table 12.4 summarizes the results for the area and power estimations of the router with the DP-unit for 2D, 6 × 6, and 3D, 6 × 6 × 4, NoC meshes. The area and power overheads of the DP-unit, as percentages of router total area and power, respectively, are also shown in Table 12.4. The total router area and power have been evaluated using ORION 2.0 [27].

It can be noticed that the overhead for both area and power slightly increases with the increase in the NoC dimensions due to the higher table size needed by table-based routing. However, this overhead is insignificant compared to the total area and power of the router. For instance, the area overhead is 1.64% and the power overhead is 0.45% for the 3D NoC with a size of 144 tiles (6 × 6 × 4).

It is worth mentioning here that although the router frequency is assumed to be 3 GHz, the a DP-unit does not need to operate at this frequency. As described in Section 12.4.5, DP-unit's minimum frequency that guarantees convergence can be estimated from NoC size and the temperature sampling period. Using Equation 12.7 and assuming an NoC size of 6 × 6 × 4 = 144 and an NoC diameter of 16 with a 10 μs temperature sampling time, clocking the DP-unit with 200 MHz guarantees the convergence of all DP units within the sampling period of the temperature.

12.7 SUMMARY AND CONCLUSION

Due to aggressive technology scaling and migration to multilayer 3D VLSI, future 3D NoC-based systems will face serious challenges, the most significant of which is the thermal challenge. In 3D

NoC-based CMPs, communication contributes significantly in heat generation and could modulate the whole chip activity. In the present work, an adaptive distributed thermal optimization strategy for 3D NoCs is proposed. This uses distributed DP units connected via a DPN to manage the routing workload at runtime to achieve global thermal moderation. As a result, the routing adapts such that the heat is diffused from the 3D chip geometry and thermal hotspots are minimized. The DPN is improved such that the cost computation and propagation takes place in compliance with the deadlock-free routing algorithm. Furthermore, 3D adaptive routing algorithms are modified to improve path diversity, balance adaptiveness, and increased DPN performance. The proposed method has been rigorously evaluated and the results show that it outperforms recently proposed RTM methods in terms of adaptation efficiency, thermal regulation, lifetime reliability, and performance. Hardware implementation details are also presented and the overheads introduced by the proposed method are evaluated and reported. These overheads are shown to be insignificant relative to those of the router hardware. This work tackles a major problem in future 3D systems-on-chip and proposes an efficient solution, which could provide better thermal integrity in future many-core systems.

REFERENCES

1. C. Ababei, H.S. Kia, O.P. Yadav, and J. Hu. Energy and reliability oriented mapping for regular networks-on-chip. In *2011 Fifth IEEE/ACM International Symposium on Networks on Chip (NoCS)*, Pittsburgh, PA, pp. 121–128, 2011.
2. R. Al-Dujaily, T. Mak, K. Zhou, K.-P. Lam, Y. Meng, A. Yakovlev, and C.-S. Poon. On-chip dynamic programming networks using 3D-TSV integration. In *2011 International Conference on Embedded Computer Systems (SAMOS)*, Samos, Greece, pp. 318–325, July 2011.
3. L. Benini and G. De Micheli. Networks on chips: A new SoC paradigm. *IEEE Computer*, 35(1):70–78, 2002.
4. S. Borkar. Thousand core chips: A technology perspective. In *Proceedings of the 44th Annual Design Automation Conference, DAC '07*, New York, pp. 746–749, 2007. ACM.
5. D. Brooks, R.P. Dick, R. Joseph, and L. Shang. Power, thermal, and reliability modeling in nanometer-scale microprocessors. *IEEE Micro*, 27:49–62, 2007.
6. C.-H. Chao, K.-Y. Jheng, H.-Y. Wang, J.-C. Wu, and A.-Y. Wu. Traffic- and thermal-aware run-time thermal management scheme for 3D NoC systems. In *2010 Fourth ACM/IEEE International Symposium on Networks-on-Chip (NoCS)*, pp. 223–230, May 2010.
7. K.-C. Chen, C.-H. Chao, S.-Y. Lin, and A.-Y.(Andy) Wu. Traffic- and thermal-aware routing algorithms for 3D network-on-chip (3D NoC) systems. In M. Palesi and M. Daneshtalab (eds.), *Routing Algorithms in Networks-on-Chip*, pp. 307–338. Springer, New York, 2014.
8. G.-M. Chiu. The odd–even turn model for adaptive routing. *IEEE Transactions on Parallel and Distributed Systems*, 11(7):729–738, July 2000.
9. T.H. Cormen, C.E. Leiserson, and R.L. Rivest. *Introduction to Algorithms*. MIT Press and McGraw-Hill Publishers, New York, 2001.
10. A.K. Coskun, T.S. Rosing, and K. Whisnant. Temperature aware task scheduling in MPSoCs. In *Design, Automation Test in Europe Conference Exhibition, 2007. DATE '07*, pp. 1–6, April.
11. N. Dahir, T. Mak, F. Xia, and A. Yakovlev. Minimizing power supply noise through harmonic mappings in networks-on-chip. In *Proceedings of the Eighth IEEE/ACM/IFIP International Conference on Hardware/Software Codesign and System Synthesis, CODES+ISSS '12*, New York, pp. 113–122, 2012. ACM.
12. W.J. Dally and C.L. Seitz. Deadlock-free message routing in multiprocessor interconnection networks. *IEEE Transactions on Computers*, C-36(5):547–553, May 1987.
13. W.J. Dally and B. Towles. Route packets, not wires: On-chip interconnection networks. In *DAC-2001*, pp. 684–689, 2001.
14. J. Donald and M. Martonosi. Techniques for multicore thermal management: Classification and new exploration. In *33rd International Symposium on Computer Architecture, 2006. ISCA '06*, pp. 78–88, 2006.
15. F. Fazzino, M. Palesi, and D. Patti. Noxim: Network-on-chip simulator. http://sourceforge.net/projects/noxim, 2008. Accessed on February 2011.
16. B.S. Feero and P.P. Pande. Networks-on-chip in a three-dimensional environment: A performance evaluation. *IEEE Transactions on Computers*, 58(1):32–45, January 2009.

17. Y. Ge, P. Malani, and Q. Qiu. Distributed task migration for thermal management in many-core systems. In *Proceedings of the 47th Design Automation Conference, DAC '10*, New York, pp. 579–584, 2010. ACM.

18. C.J. Glass and L.M. Ni. The turn model for adaptive routing. In *Proceedings of the 19th Annual International Symposium on Computer Architecture,* Gold Coast, Australia, *1992*, pp. 278–287, 1992.

19. F. Hillier and G. Lieberman. *Introduction to Operations Research.* McGraw-Hill International Editions, 1995.

20. J. Hu and R. Marculescu. Energy-aware mapping for tile-based NoC architectures under performance constraints. In *Proceedings of the ASP-DAC 2003. Asia and South Pacific Design Automation Conference, 2003,* Kitakyushu, Japan, pp. 233–239, 2003.

21. J. Hu and R. Marculescu. Energy-aware communication and task scheduling for network-on-chip architectures under real-time constraints. In *Proceedings of the Design, Automation and Test in Europe Conference and Exhibition, 2004,* Paris, France, Vol. 1, pp. 234–239, 2004.

22. W. Huang, S. Ghosh, S. Velusamy, K. Sankaranarayanan, K. Skadron, and M.R. Stan. Hotspot: A compact thermal modeling methodology for early-stage VLSI design. *IEEE Transactions on Very Large Scale Integration (VLSI) Systems,* 14(5):501–513, May 2006.

23. W. Huang, K. Sankaranarayanan, K. Skadron, R.J. Ribando, and M.R. Stan. Accurate, pre-RTL temperature-aware design using a parameterized, geometric thermal model. *IEEE Transactions on Computers,* 57(9):1277–1288, September 2008.

24. W. Huang, K. Skadron, S. Gurumurthi, R.J. Ribando, and M.R. Stan. Differentiating the roles of IR measurement and simulation for power and temperature-aware design. In *IEEE International Symposium on Performance Analysis of Systems and Software, 2009. ISPASS 2009,* pp. 1–10, April 2009.

25. ITRS. The International Technology Roadmap for Semiconductors (ITRS), Interconnects, http://www. itrs.net/, 2011. Accessed on September 2011.

26. Jedec. Failure mechanisms and models for semiconductor devices. JEDEC Publication JEP122-A, 2002.

27. A.B. Kahng, B. Li, L.S. Peh, and K. Samadi. Orion 2.0: A power-area simulator for interconnection networks. *IEEE Transactions on Very Large Scale Integration (VLSI) Systems,* PP(99):1–5, 2011.

28. A. Lankes, T. Wild, A. Herkersdorf, S. Sonntag, and H. Reinig. Comparison of deadlock recovery and avoidance mechanisms to approach message dependent deadlocks in on-chip networks. In *2010 Fourth ACM/IEEE International Symposium on Networks-on-Chip (NoCS),* pp. 17–24, 2010.

29. S.-Y. Lin, T.-C. Yin, H.-Y. Wang, and A.-Y. Wu. Traffic- and thermal-aware routing for throttled three-dimensional network-on-chip systems. In *2011 International Symposium on VLSI Design, Automation and Test (VLSI-DAT),* pp. 1–4, April 2011.

30. C.-L. Lung, Y.-L. Ho, D.-M. Kwai, and S.-C. Chang. Thermal-aware on-line task allocation for 3D multi-core processor throughput optimization. In *Design, Automation Test in Europe Conference Exhibition (DATE), 2011,* pp. 1–6, March 2011.

31. T. Mak, P.Y.K. Cheung, K.-P. Lam, and W. Luk. Adaptive routing in network-on-chips using a dynamic-programming network. *IEEE Transactions on Industrial Electronics,* 58(8):3701–3716, August 2011.

32. T. Mak, P.Y.K. Cheung, W. Luk, and K.P. Lam. A DP-network for optimal dynamic routing in network-on-chip. In *Proceedings of the 7th IEEE/ACM International Conference on Hardware/Software Codesign and System Synthesis,* Grenoble, France, pp. 119–128. ACM, 2009.

33. R. Marculescu, U.Y. Ogras, L.-S. Peh, N.E. Jerger, and Y. Hoskote. Outstanding research problems in NoC design: System, microarchitecture, and circuit perspectives. *IEEE Transactions on Computer-Aided Design of Integrated Circuits and Systems,* 28(1):3–21, 2009.

34. S. Murali and G. De Micheli. Bandwidth-constrained mapping of cores onto NoC architectures. In *Proceedings of the Conference on Design, Automation and Test in Europe, DATE '04,* Washington, DC, Vol. 2, pp. 896–901, 2004. IEEE Computer Society.

35. P.P. Pande, C. Grecu, A. Ivanov, and R. Saleh. High-throughput switch-based interconnect for future SoCs. In *Proceedings of the Third IEEE International Workshop on System-on-Chip for Real-Time Applications, 2003,* Calgary, Alberta, pp. 304–310, June–2 July 2003.

36. Z. Qian and C.-Y. Tsui. A thermal-aware application specific routing algorithm for network-on-chip design. In *2011 16th Asia and South Pacific Design Automation Conference (ASP-DAC),* Yokohama, Japan, pp. 449–454, 2011.

37. M.S. Rahaman and M.H. Chowdhury. Crosstalk avoidance and error-correction coding for coupled RLC interconnects. In *IEEE International Symposium on Circuits and Systems, 2009. ISCAS 2009,* Taipei, Taiwan, pp. 141–144, 2009.

38. M. Sadri, A. Bartolini, and L. Benini. Single-chip cloud computer thermal model. In *2011 17th International Workshop on Thermal Investigations of ICs and Systems (THERMINIC)*, Paris, France, pp. 1–6, 2011.

39. C. Seiculescu, S. Murali, L. Benini, and G. De Micheli. Sunfloor 3D: A tool for networks on chip topology synthesis for 3-D systems on chips. *IEEE Transactions on Computer-Aided Design of Integrated Circuits and Systems*, 29(12):1987–2000, 2010.

40. O. Semenov, A. Vassighi, and M. Sachdev. Impact of self-heating effect on long-term reliability and performance degradation in CMOS circuits. *IEEE Transactions on Device and Materials Reliability*, 6(1):17–27, March 2006.

41. L. Shang, L.-S. Peh, A. Kumar, and N.K. Jha. Temperature-aware on-chip networks. *IEEE Micro*, 26(1):130–139, 2006.

42. L. Shang, L. Peh, A. Kumar, and N.K. Jha. Thermal modeling, characterization and management of on-chip networks. In *37th International Symposium on Microarchitecture, 2004. MICRO-37 2004*, Portland, Oregon, pp. 67–78, December 2004.

43. K. Skadron, M.R. Stan, W. Huang, S. Velusamy, K. Sankaranarayanan, and D. Tarjan. Temperature-aware microarchitecture. In *Proceedings of the 30th Annual International Symposium on Computer Architecture, 2003*, San Diego, CA, pp. 2–13, June 2003.

44. J. Srinivasan, S.V. Adve, P. Bose, and J.A. Rivers. *The Case for Lifetime Reliability-Aware Microprocessors*. In *Proceedings of the 31st Annual International Symposium on Computer Architecture (ISCA '04)*, München, Germany, pp. 276–287, June 2004.

45. J. Srinivasan, S.V. Adve, P. Bose, and J.A. Rivers. The impact of technology scaling on lifetime reliability. In *2004 International Conference on Dependable Systems and Networks*, Florence, Italy, pp. 177–186, June–1 July 2004.

46. M.B. Taylor, J. Kim, J. Miller, D. Wentzlaff, F. Ghodrat, B. Greenwald, H. Hoffman et al. The raw microprocessor: A computational fabric for software circuits and general-purpose programs. *IEEE Micro*, 22(2):25–35, 2002.

47. S. Vangal, J. Howard, G. Ruhl, S. Dighe, H. Wilson, J. Tschanz, D. Finan, P. Iyer, A. Singh, and T. Jacob. An 80-tile 1.28 TFLOPS network-on-chip in 65 nm CMOS. In *Proceedings of the IEEE International Solid-State Circuits Conference, 2007. ISSCC 2007. Digest of Technical Papers*, Philadelphia, PA, pp. 98–589. IEEE, 2007. IEEE International.

48. S.R. Vangal, J. Howard, G. Ruhl, S. Dighe, H. Wilson, J. Tschanz, D. Finan et al. An 80-tile sub-100-w teraFLOPS processor in 65-nm CMOS. *IEEE Journal of Solid-State Circuits*, 43(1):29–41, 2008.

49. A.Y. Weldezion, M. Grange, D. Pamunuwa, Z. Lu, A. Jantsch, R. Weerasekera, and H. Tenhunen. Scalability of network-on-chip communication architecture for 3-D meshes. In *Proceedings of the 2009 Third ACM/IEEE International Symposium on Networks-on-Chip*, San Diego, CA, pp. 114–123. IEEE Computer Society, 2009.

50. Y. Xie and W. Wolf. Allocation and scheduling of conditional task graph in hardware/software co-synthesis. In *Proceedings of Conference and Exhibition 2001 on Design, Automation and Test in Europe, 2001*, pp. 620–625, 2001.

51. L. Xue, Y. Gao, and J. Fu. A high performance 3D interconnection network for many-core processors. In *2010 Second International Conference on Computer Engineering and Technology (ICCET)*, Vol. 1, pp. V1-383–V1-389, April 2010.

52. N. Dahir, T. Mak, F. Xia, and A. Yakovlev. Minimizing power supply noise through harmonic mappings in networks-on-chip. In *Proceedings of the Eighth IEEE/ACM/IFIP International Conference on Hardware/Software Codesign and System Synthesis, CODES+ISSS '12*, New York, pp. 113–122, 2012. ACM.

13 TSV-to-Device Noise Analysis and Mitigation Techniques

Brad Gaynor, Nauman Khan, and Soha Hassoun

CONTENTS

ABSTRACT

Since 1965, integrated circuit (IC) complexity has doubled approximately every 18 months in accordance with Moore's Law [7,30]. Modern ICs incorporate multiple cores and peripherals comprising billions of transistors (e.g., [28]). Three-dimensional (3D) integration has emerged as a promising technology for IC scaling. With 3D stacking, the length of global wires can be reduced by as much as 50%, wire-limited clock frequency can be increased by 3.9, and wire-limited area can be decreased by 84% [19]. Additionally, power can be reduced by 51% at the 45 nm technology node [3]. 3D-ICs are enabled by vertical front-to-back electrical connections within the die. The vertical interconnect is provided by through-silicon-vias (TSVs) in bulk processes [32] and by through-oxide-vias (TOVs) in silicon-on-insulator (SOI) processes [8]. In both bulk and SOI, a metal interconnect (plug) is fabricated through the die; in bulk processes, an insulator (the liner) is also fabricated to isolate the metal plug from the substrate.

This chapter addresses the analysis and mitigation of TSV-to-device coupling noise within 3D-ICs. Coupling arises due to a signal traveling through the TSV in close proximity to MOS devices (Figure 13.1). The signal propagates through the TSV liner and the substrate, changing the body voltage of nearby devices. While the shift in body voltage resulting from fast (5–50 ps) rise/fall TSV signal-times is small, coupling noise can impact circuit performance, especially for analog circuits requiring a high-degree of noise isolation. Accurate modeling and analysis of TSV-to-device noise is thus needed. Further, noise can be suppressed using noise-mitigation techniques.

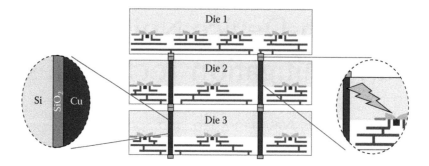

FIGURE 13.1   3D-stacked dies with TSV interconnect. The Cu-TSV is surrounded by an SiO$_2$ liner. TSV-to-device coupling occurs when a signal traveling through the TSV propagates through the liner and the Si substrate, changing the body voltage of neighboring devices.

For 2D-ICs, several techniques, including split power planes, deep-n-well process, and guard ring structures, are employed for noise isolation. While some of these techniques can be utilized in 3D-ICs, high TSV densities and the associated area penalties call for innovative, less costly isolation methods.

This chapter begins by reviewing recent advances in modeling TSV-to-device coupling and providing TSV coupling analysis to MOSFET and FinFET devices. The chapter then describes noise-mitigation techniques appropriate for 3D-ICs. The chapter concludes with a summary and outlines some future challenges. The reviewed material appears in more depth in two theses [16,26] and in related publications [15,23,25].

13.1   EVALUATING TSV-TO-DEVICE NOISE

A simulation methodology to analyze TSV-to-device coupling must capture the physical and electrical characteristics of three critical interfaces (Figure 13.2). The first interface captures the

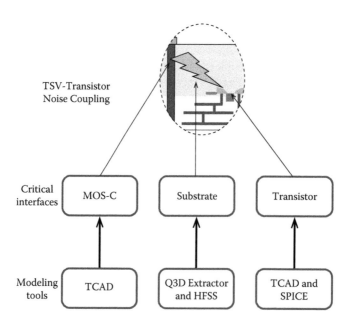

FIGURE 13.2   A simulation methodology must address noise propagation through the MOS-capacitance, the substrate, and the transistor body.

interaction of the TSV-bulk through the oxide liner. The metal-oxide-semiconductor creates a MOS-Capacitance (MOS-C), a combination of the oxide capacitance and the bias-dependent depletion region capacitance. Signals on the TSV generate a MOS-C with time-varying deep-depletion characteristics. Additionally, the trapped charge due to the Si/SiO$_2$ interface must be considered. The first published analytical description of the TSV MOS-C calculates the MOS-C using the full depletion approximation in cylindrical coordinates, identifying that power TSVs within a power distribution network (PDN) follow the high-frequency MOS-C characteristics, while signal TSVs follow deep-depletion MOS-C characteristics [4]. Measurements report duty cycle distortion in TSV-to-substrate noise coupling due to the nonlinear MOS-C [11], highlighting the need to properly model the TSV MOS-C for noise coupling analysis. Later approaches developed analytical models for estimating signal delay through a TSV including considerations for fixed charges in the oxide liner [6,21]. These approaches, however, are not optimal for signal distribution network (SDN) noise coupling analysis because they do not model the MOS-C in deep-depletion. More recently, modeling components to support the time-varying deep-depletion and quasi-static behaviors were developed, but did not support fixed charges in the oxide [9]. This latter model was not verified with device simulation.

The second interface is through the substrate between the TSV and the transistor body. We denote the location corresponding to the transistor body as the observation point (OP). A ground-referenced equivalent circuit model from the TSV to the OP is needed. Using modern CAD tools to extract a substrate equivalent circuit model is complex. Quasi-static field solvers (e.g., Q3D extractor) are designed to extract metal-to-metal coupling; full-wave EM solvers (e.g., HFSS) do not extract lumped impedance values. Because the OP is neither a physical structure in the design nor a conductor, the extraction is not directly supported using either of these physics-based EM solvers.

The third interface is between the OP and the transistor. To capture the impact of substrate noise, detailed transistor models that allow for modeling substrate noise are needed. While such models are available in compact form for planar devices, these detailed models are not available for FinFETs. Detailed TCAD simulations must be used to assess the impact of substrate noise on transistor performance.

Several simulation methodologies have been proposed to evaluate TSV-to-device noise. While used to model TOV-to-device coupling in SOI [33], quasi-static device simulation is not appropriate for modeling signal propagation through the substrate from the TSV to the OP.

Electromagnetic solvers have proven more effective [5,13,29,36]. Brocard et al. [5] performed EM and device co-simulation using DevEM. This approach captures both the EM noise propagation through the substrate and the semiconductor MOS-C effects, although DevEM does not provide support for simulating fixed charges at the Si/SiO$_2$ interface. Their work did not include certain semiconductor effects such as the characterization of substrate noise found at the transistor substrate contact. Le Maitre et al. [29] used DevEM to extract a frequency dependent, compact TSV model of the path between the TSV, and a conductive substrate contact. We adopt the circuit structure of their extracted models and extend their work to extract a model of the coupling between the TSV and the transistor substrate contact.

Xu et al. developed compact models based on an analytical approach for electrical coupling between a TSV and nearby transistors [36]. The models include small-signal TSV MOS-C effects and have been validated with a full-wave EM solver. These models were developed for the analysis of PDN noise and not SDN noise, therefore assuming static boundary conditions derived from a fixed TSV bias voltage. We extend their EM simulation methodology to support analysis of SDN noise, introducing additional modifications to improve accuracy. Their HFSS extraction was performed with lumped ports between a nonphysical OP and a nearby ground. While the introduction of the OP solves the problem of extracting EM coupling to a location in the lossy dielectric, the presence of the OP skews the fields. Also, lumped ports in HFSS artificially constrain the magnitude and orientation of the incident EM fields, introducing additional error. Ideally, the fields only depend on the material properties and geometry of the physical structures in the circuit.

13.1.1 MODELING FRAMEWORK

13.1.1.1 Modeling the TSV MOS-C

We extract the MOS-C C–V curves using the Synopsys TCAD package. We define a 2D TSV model (Figure 13.3) with geometries from [20]: a Cu plug of radius $r_{tsv} = 2.6$ µm, a 120 nm thick SiO_2 liner with radius $r_{liner} = 2.72$ µm, and a substrate larger than the maximum depletion region width under deep-depletion ($r_{sub} = 10$ µm). The substrate is p-type with a boron doping concentration of $1.0 \times 10^{15}/cm^3$. The TSV is driven by a 0.1 µm radial contact from the center of the copper plug and the substrate contact is located at the edge of the substrate region. A mesh is generated with maximum edge length of 25 nm at the Si/SiO_2 interface, 5 µm in the oxide, and 25 µm in the substrate. We define three fixed oxide charge density concentrations at the Si/SiO_2 interface: a baseline value of $7.8 \times 10^{11}/cm^2$ [20]; $-8.43 \times 10^{11}/cm^2$, corresponding to the proposal in [37] for controlling the MOS-C over the intended TSV operating voltage range; and $-0.63 \times 10^{11}/cm^2$, a moderate charge density between the two other values.

We capture the deep-depletion characteristics of the TSV MOS-C by applying a piecewise linear ramp from −10 to +10 V to the TSV contact with a rise time of 1 V/µs, connect the substrate contact to ground, and extract the current flowing between the TSV and substrate. At this ramp rate, substrate parasitics are negligible. Substrate parasitics are extracted separately in the following section. The simulation takes 12,969 CPU-seconds on a Dell PowerEdge R815 with 4, 12-core 2.3 GHz AMD Opteron 6176 processors and 128 GB of memory.

The C–V curves are calculated from the simulation output by applying the formula: $C(V) = I \times (dV/dt)$. The results are shown in Figure 13.4, where the capacitance values include both the fixed liner capacitance and the deep-depletion MOS-C. The extracted data from our model compares favorably with the previously reported measured capacitance data. The baseline MOS-C is in deep-depletion within the operating range of 0–0.7 V, and therefore changes slowly over this range. The controlled fixed oxide charge density MOS-C is accumulated over the operating range, and therefore is roughly constant over this range, although at the greater value of C_{ox}. The moderate fixed oxide charge density MOS-C is in the depletion region in the 0–0.7 V operating range, and therefore experiences the greatest capacitance shift; we will use this value in the following simulations to demonstrate effects of the nonlinear capacitance.

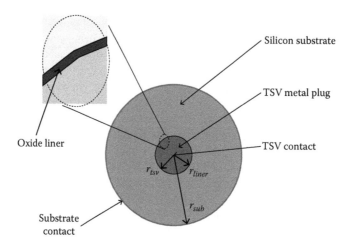

FIGURE 13.3 TCAD TSV model cross section for extracting MOS-C deep-depletion characteristics from a particular TSV geometry.

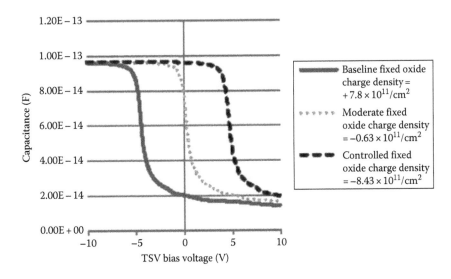

FIGURE 13.4 TSV C–V curves with different fixed charge interface densities.

The extracted TSV MOS-C is compactly modeled in SPICE with a voltage-controlled current source, the HSPICE G-element with the VCCAP parameter. The variable capacitor is specified with a piecewise linear lookup table using the C–V data extracted from the TCAD simulation.

13.1.1.2 Substrate Coupling Extraction

We use a three-step process applied to the CMOS layout from [36] (Figure 13.5). The height of the TSV is 20 µm, the keep-out-zone (KOZ) between the TSV and the center of the nearest NMOS transistor is 2.4 µm, and the V_{ss}/V_{dd} lines alternate every 3 µm, and extend to the left and right of the TSV.

Step 1: Extract the total admittance Y_{TOT} based on the capacitance and conductance between the TSV and the PDN (V_{ss} and V_{dd}) using Q3D Extractor. Because only high-frequencies propagate through the TSV liner and MOS-C, we short V_{ss} and V_{dd} together in the model and perform small-signal RF analysis on the strate. Large-signal effects are limited to the MOS-C extracted in the previous section. We define our substrate materials and geometries to match the model used in [36]. The simulation is driven with a source on the bottom TSV pad and a sink on the top TSV pad. Capacitance and conductance matrices are extracted over frequency from 0.01 to 100.01 GHz in 1 GHz steps. We omit the oxide and TSV MOS-C depletion capacitances from our substrate extraction as these capacitances were

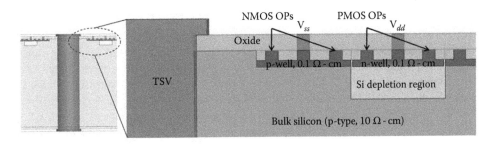

FIGURE 13.5 TSV-substrate model. Left: front view. Right: detail. Transistors are placed in a regular layout with alternating nMOS and pMOS distributed between the PDN traces.

determined previously. They will be composed with the substrate model when performing the overall composite simulation.

Step 2: Extract the potential at the OP due to the coupling between the TSV and the PDN using HFSS. We simulate the same circuit in HFSS, saving the E-field over the same frequency range. This simulation is driven with a 1 V excitation using a wave port at the bottom pad of the TSV (referenced to V_{ss}) and terminated with a wave port at the top pad of the TSV. The OP potential is extracted relative to the nearby V_{ss} using the field calculator to integrate the electric field over a line from V_{ss} to the OP: $V_{OP} = \int_l \vec{E}\,\vec{dl}$.

Step 3: Apportion the total admittance between the TSV/OP and the OP/PDN. For each frequency point, convert the total admittance (Y_{TOT}) to equivalent impedance (Z_{TOT}), divide the total impedance between $Z_{OP} = (1 - V_{OP}) \times Z_{TOT}$, and $Z_{GND} = V_{OP} \times Z_{TOT}$ based on the complex potential at the OP, and convert the two resulting impedances back to admittances Y_{OP} and Y_{GND}. The extracted admittances are compactly modeled in SPICE with S-parameter models (S-elements) with the TYPE set to Y-parameters. The frequency-dependent admittances are captured with small-signal parameter data frequency table models (SP models). The extracted models run much faster (0.21 CPU-seconds) than the initial Q3D extraction (44,393 CPU-seconds) and HFSS extraction (30,149 CPU-seconds).

Our substrate extraction methodology eliminates two sources of error found in previous approaches [24,36]. Whereas we extract the OP voltage by calculating the line integral of the electric field, previous approaches calculated the OP voltage using the Z-parameters extracted from the lumped port: $V = Z_{21}/Z_{11}$. However, lumped ports are only appropriate for use with fields aligned to the port geometry. The lumped port introduces extraction error ranging from a fractional percent at low frequency to over 6% at 100 GHz for the shown example. Previous approaches introduced additional error over the entire frequency range by modeling the OP as a metallic cube. A physical OP shunts the electric field over the OP surface, changing the potential at the extraction point. The combined error ranges from 1.9% at low frequency to 7.9% at 100 GHz for an OP with 0.1 μm per side. Our proposed extraction technique is fully generalizable for impedance extraction to an arbitrary OP in a dielectric so long as the fields are quasi-static for the geometry under evaluation.

Noise propagation through the substrate is a function of distance between the TSV and the transistor body. Our simulations show that the noise attenuation between the TSV and the OP increases as the distance increases, and decreases as the frequency increases. PMOS OPs show significant attenuation at low frequencies due to the depletion region formed between the n-well and the substrate; however, the PMOS depletion region provides only modest attenuation above 5 GHz. The benefit is minimal when we consider the TSV liner capacitance, which provides excellent low-frequency isolation for both NMOS and PMOS.

13.1.1.3 FinFET and Planar Transistor Modeling

Substrate noise at the transistor body impacts transistor performance via the body effect and capacitive coupling through the transistor. These phenomena are well characterized for planar transistors. We use the PTM high-performance (HP) Version 2.1 SPICE models from Arizona State University [38,39] to evaluate the impact of TSV-induced substrate noise on planar transistors in the 22 nm technology node. Our PTM simulations use an NMOS transistor with $L = 34$ nm and $W = 72.0$–78.6 nm to, respectively, match the gate length and effective width of our simulated FinFET.

The impact of substrate noise is not well characterized for FinFETs. The BSIM CMG SPICE models include experimental body effect modeling beginning with version 106.0.0; however, these empirical models have known limitations and there are no open source PDKs available that accurately capture the body effect for FinFET. Therefore, we use a TCAD FinFET model to compare the substrate noise immunity of FinFET to planar technology.

We base our analysis on the 22 nm bulk nFinFET TCAD model adapted from [1]. This model represents the transistor features described in [2]. The key geometries are selected to correspond with Intel's recent bulk FinFET production process [18]. Gate length and fin height are 34 and 35 nm, respectively. The width of the bottom of the active fin is 15 nm while the width of the active fin top is 7 nm. The active fin is undoped with a p-type concentration of $1 \times 10^{15}/cm^3$, and the fin body is doped with a p-type concentration of $1 \times 10^{18}/cm^3$. The corner radius of the rounded fin is set to $(1/2)W_{top}$ to minimize corner effects. We also evaluate the impact of fin shape on noise susceptibility; therefore, we evaluate W_{top} from 1 to 15 nm. All other model parameters take the default value. The simulations include physical models for stress effects, crystal orientation dependent quantum effects, band-to-band-tunneling (BTBT), and drift-diffusion with mobility degradation.

13.1.2 TSV Noise-Coupling Comparison: FinFET vs. Planar Transistors

We combine the extracted models for the TSV and the substrate in SPICE with TCAD and PTM transistor models to evaluate the TSV-induced noise immunity of FinFETs relative to planar. The SPICE netlist for the FinFET models are run within the mixed-mode TCAD environment.

We first simulate the TSV and substrate models to determine the TSV voltage noise observed at the closest OP. A simulation input waveform, V_{tsv}, of a 0.8 V digital signal on the TSV with 50 ps rise (fall) time imparts a transient voltage noise, V_{OP}, of 22.3 mV (−18.9 mV) reported for the OP closest to the TSV. Next we evaluate the impact of a steady-state substrate voltage bias on the transistor output via the body effect. We apply a bias of 20 mV to each transistor configuration, comparable to the worst-case V_{OP}. The FinFET leakage exhibits a significant dependence on fin shape, but the change in leakage due to the body effect is negligible for all fin shapes with the 20 mV substrate bias. The same substrate bias on the planar transistor induces a +16%/−14% change in leakage current. Similarly, while the FinFET saturation current exhibits a significant dependence on fin shape, the change in saturation current due to the body effect is negligible for all fin shapes. The same substrate bias on the planar transistor induces a ±1.9% change in saturation current.

We evaluate the impact of the transient TSV-induced substrate noise on FinFET and planar transistors. Here we only simulate the FinFET with the default $W_{top} = 7$ nm, comparable to modern production devices. We simulate each transistor with the substrate noise from the first 10 ns of our simulation, with the transistor biased off ($V_g = 0$ V) from 0 to 5 ns and on ($V_g = 0.8$ V) from 5 to 10 ns. The TSV-induced substrate noise voltage results in current transients with three orders of magnitude increase from a reference leakage current of approximately 5×10^{-11} A for planar. The FinFET leakage current transients exhibit two orders of magnitude increase relevant to a reference leakage current of approximately 2×10^{-12} A. Therefore, in addition to a reference leakage difference of one order of magnitude less than planar, the FinFET exhibits superior isolation to substrate noise when the transistor is off. The TSV-induced substrate noise voltage results in a 1.8% transient in the saturation current for planar from a reference current of 4.04×10^{-5} A. The TSV-induced substrate noise voltage results in a 0.023% transient in the saturation current for FinFET from a reference current of 2.93×10^{-05} A. The FinFET therefore exhibits an improvement of two orders of magnitude over planar in terms of robustness of the saturation current to TSV noise.

The detailed MOS-C extraction results in a more accurate assessment of the substrate noise relative to a fixed oxide capacitance alone. For the transient simulation, utilizing the fixed capacitance associated with the 0 V TSV bias results in over 7 mV of substrate noise error at the OP with significant ringing around the TSV digital transitions. This 7 mV change increases the peak transient leakage current by approximately 60% and the peak transient saturation current by approximately 0.5%. Modeling noise without our proposed accurate MOS-C extraction may impact simulation results for analog circuits.

13.1.3　KEY OBSERVATIONS

We reviewed a noise coupling analysis methodology and its use in evaluating TSV coupling to planar and FinFET devices. We make the following key observations:

- Noise coupling from TSVs to nearby transistors takes place at interfaces not well characterized by standard design and validation flows. The work reviewed here is effective in modeling noise at each interface and in combining these models to simulate the impact of TSV signal noise on nearby devices.
- Modeling the depletion region was shown previously as critical [4] when electrically modeling TSVs. Modeling the variable MOS-C as a function of the TSV signal voltage is critical in understanding noise coupling. Moreover, it is imperative to capture all relevant semiconductor effects on the MOS-C interface (e.g., fixed oxide charges).
- The parasitic and full-wave EM CAD tools do not provide native support for extracting noise coupling effects at a location within a dielectric. The three-step method reviewed here is general and can be applied in other circumstances to extract coupling to a particular point within the substrate.
- By combining the quasi-static parasitic extraction with the full-wave field analysis, it was possible to obtain more accurate coupling models than using the earlier extraction methodologies. The error ranged from 1.9% to 7.9% at 100 GHz.
- Noise coupling results show that FinFETs exhibit an order of magnitude less leakage current noise transients, and two orders of magnitude less saturation current noise transients, relative to comparable planar technologies.

13.2　TSV-TO-DEVICE NOISE-MITIGATION TECHNIQUES

We review noise-mitigation techniques applicable in cases where noise isolation is of paramount importance when utilizing MOSFETs within 3D circuits. The first technique utilizes a thick dielectric liner, which provides additional shielding that results in reduced coupling between TSV and substrate. The second technique consists of adding a back-side plane connected electrically to circuit GND [22]. This plane provides grounding for the substrate. The third technique consists of surrounding TSVs with guard rings [10]. The guard ring depth is comparable to GND tie depth, and is used to provide noise isolation. The fourth technique is the GND plug, first introduced in [26]. A GND plug is a TSV-like structure that is connected to circuit GND but unlike a TSV, the GND plug may not extend through the complete depth of the substrate. One other noise-mitigation technique consists of using co-axial TSVs [17,22]. However, manufacturing co-axial TSVs is more challenging and costly than comparable conventional TSVs due to structural complexities requiring additional manufacturing steps [35].

13.2.1　EVALUATION FRAMEWORK

We use a three-dimensional evaluation framework composed of a Cu TSV in an Si substrate, ground ties, and voltage OPs. The top view of this setup is shown in Figure 13.6a. We assume a high-R substrate with a resistivity of $10\ \Omega$ cm and relative permittivity of 11.8. This type of substrate is used to fabricate low cost, low-performance devices like memory [12]. We assume a 50 μm × 50 μm substrate. This cross section is sufficiently larger than the TSV and device scale for capturing their interactions without interference from other devices or structures in a die. The boundary condition at the sidewalls of the 50 μm × 50 μm substrate is such that we have zero-current going out of the surfaces.

We assume a cylindrical Cu TSV. Its height is the same as the substrate height and the diameter is fixed to 2 μm. We use an SiO$_2$-based liner, with a resistivity $10^{16}\ \Omega$ cm and relative permittivity 3.9, surrounding the TSV. The default thickness of the dielectric liner is assumed to be 0.1 μm, which is

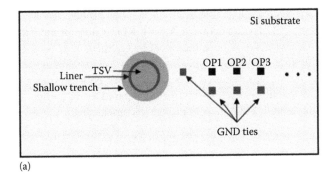

(a)

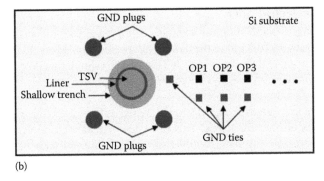

(b)

FIGURE 13.6 TSV-induced noise analysis framework. (a) Top-view. (b) Top-view with four added ground plugs.

consistent with recent design studies [34]. We assume an SiO_2-based shallow trench. Its thickness and depth (into the top surface of substrate) are assumed to be 0.9 and 0.3 μm, respectively. We assume nine equally spaced OPs (OP1, OP2, etc.) located 4–20 μm away from the center of the TSV. These points are modeled as small metallic cubes to enable extracting parasitics between TSV and devices at various distances from the TSV.

Placing GND ties throughout the circuit layout is the conventional approach to control transistor body voltages, and hence, is considered in our setup. We assume a GND tie at a distance of 0.3 μm from the shallow trench edge. Also, we assume that there is at least one GND tie within a 1 μm distance of each OP. This GND tie is not the proposed W-filled GND plug.

Table 13.1 shows the default values of parameters in our TSV-induced noise analysis framework. These values should be assumed when any of the parameters is not being varied for sensitivity

TABLE 13.1

Parameters for Our TSV-Induced Noise Analysis Framework

Parameter	Value
Substrate height	20 μm
Substrate length	50 μm
Substrate width	50 μm
TSV height	20 μm
TSV diameter	2 μm
Liner thickness	0.1 μm
Shallow trench height	0.3 μm
Shallow trench width	0.9 μm
Resistivity of high-R substrate	10 Ω cm

investigation. To extract an equivalent SPICE circuit for our framework, we use a finite element–based 3D extraction tool (Q3D Extractor) from Ansoft. RLC values of the extracted circuit depend upon the extraction frequency. We choose the operating frequency to be 1 GHz for our analysis. A step input, with a rise time of 100 ps and peak voltage of 1 V, is applied at one of the TSV terminals, whereas the other terminal is assumed to be floating. We perform transient analysis and report peak noise at the OPs. Our analysis here is simplified compared to our earlier efforts when analyzing TSV-to-Device coupling as our goal here is to provide a comparative analysis of the various mitigation techniques.

13.2.2 NOISE MITIGATION TECHNIQUES

13.2.2.1 Thicker Dielectric Liner

Increasing liner thickness is the simplest approach to mitigate TSV-induced substrate noise. We vary liner thickness from 0.1 to 3 μm in the default setup shown in Figure 13.6 and extract an RLC circuit for each setup using Q3D Extractor. SPICE is used for simulation. We record peak transient noise at various OPs in the substrate, and make the following observations:

- Peak transient noise ranges from 0.18 to 0.7 V, regardless of distance from TSV, for all examined values of liner thickness. This indicates that standard GND substrate ties are inadequate for creating a reference GND substrate in the presence of a TSV switching signal. Hence, using GND substrate ties alone is not effective in mitigating TSV-induced noise.
- Peak substrate noise decreases with increasing liner thickness. This trend is not uniform and can be divided into three segments. The impact of increasing liner thickness is maximal for liner thickness between 0.1 and 1 μm, reduces for liner thickness between 1 and 2 μm, and saturates for thickness greater than 2 μm.
- Peak substrate noise is ≈18% of VDD for liner thickness of 3 μm. For 2 μm diameter TSV with t_{liner} 0.1 μm and t_{ST} 0.9 μm, increasing t_{liner} by 2.9 μm results in a 6× area penalty (12.6 vs. 75.4 μm²). This large area penalty creates interconnect blockages and reduces the area available for active devices.

13.2.2.2 Backside Ground Plane

During assembly and packaging stages, a 2D die is placed on a grounded metal layer. The same idea can be extended to 3D-ICs; a backside grounded metal, in preferable plate or grid format, is added to the substrate, creating a strong GND reference for substrate. To model this technique, we add a Cu sheet in the default setup shown in Figure 13.6a. The sheet cross section is the same as the substrate cross section. The sheet thickness is assumed to be 2 μm. One side of the sheet connects to the substrate and the other side connects to GND. Substrate thickness, the distance between the devices layer and the backside ground, is the only variable of concern for substrate noise analysis. We vary the substrate thickness from 10 to 40 μm and extract the RLC circuit for each setup. We use SPICE to simulate and record peak transient noise at the OPs. We make the following observations:

- Substrate noise decreases as a function of TSV distance. Using the backside ground plane is more effective than the thicker dielectric in localizing noise.
- Our results show that the capability of the backside ground plane to mitigate substrate noise is a function of substrate thickness. The benefits are minimal for larger values of thickness. Backside ground plane is therefore effective in technology generations where substrates can be aggressively thinned.

13.2.2.3   Guard Rings

A guard ring is a common technique to mitigate substrate noise in 2D-ICs. Guard rings are placed around a sensitive block and are connected to ground pads. We extend the same idea in a substrate with a TSV. We use the default design setup and add a guard ring at 3 μm from the center of the TSV. We vary guard ring depth into the substrate and observe its impact on TSV noise. We record the noise at several OPs. We can conclude the following:

- Although increasing the guard ring height results in smaller TSV-induced substrate noise, the guard ring is not able to localize the TSV-induced noise. The value of peak noise is larger at OPs away from TSV. The main reason for this behavior is that TSVs pass through the whole substrate and shielding only the top part of the substrate does not reduce the noise injected from the lower part of the TSV.
- Using a guard ring of height 1.5 μm and width 1.5 μm results in an area penalty of 6.5× the TSV area, whereas peak substrate noise is still greater than 30% the input voltage.

13.2.2.4   GND Plugs

A GND plug is a TSV-like structure but may not extend through the whole substrate. Multiple GND plugs, fabricated around a TSV, provide better grounding of the substrate and help mitigate the TSV-induced noise. We analyze two types of GND plugs: a "front-side GND plug" and a "back-side GND plug," shown in Figure 13.7. A front-side GND plug is connected to the local interconnect of the same die and occupies substrate area affecting the number of devices. A back-side GND plug, on the other hand, can be connected to the circuit ground from the backside of substrate and does not affect the number of devices in the die. These GND plugs can be fabricated using any of the TSV fabrication techniques using a variety of fill-material including Cu, W, and Poly. Although Cu is a better conductor, we recommend W or Poly as the fill material for two reasons. First, a smaller CTE (Coefficient of Thermal Expansion) mismatch of W/Poly and Si, compared to Cu and Si, will result in less thermal stress in devices. Second, W/Poly does not require any diffusion barrier-like Cu and will provide a direct connection between substrate and circuit ground resulting in better device shielding. W-filled TSVs have been demonstrated for the fabrication of 3D-LSI chips [27].

Figure 13.6b shows the design setup with the added GND plugs. We first investigate the impact of placing different numbers of GND plugs. We evaluate the use of four GND plugs fabricated at 3 μm from the center of TSV. Noise isolation is improved by 2.46× when compared to using only two GND plugs. For the rest of the analysis, we therefore utilize four plugs. Next, we explore the

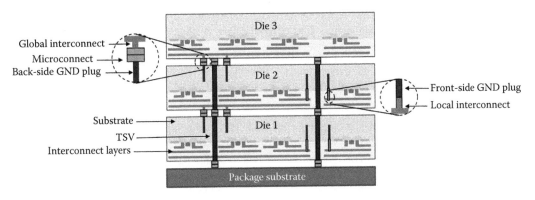

FIGURE 13.7   An illustrative 3D-IC assuming face-to-back metallic bonding with microconnects. A "front-side GND plug" can be connected to a GND net in the same die. A "back-side GND plug" can be connected to a GND net in the neighboring die.

impact of two critical parameters: plug depth and plug diameter. RLC circuits for each setting are extracted using Q3D Extractor and then simulated in SPICE for peak transient voltage at the OPs. We make the following observations:

- A GND plug is effective in reducing the peak noise. A deeper GND plug is more effective than a shallower one in reducing peak noise. Four GND plugs fabricated at 3 μm from the center of the TSV can reduce the peak substrate noise to 9% of the input voltage.
- Analyzing various aspect ratios, increasing the GND plug height is more significant than increasing its diameter. For a 3× increase in plug diameter, from 0.5 to 1.5 μm, noise reduction is only 10%, whereas the same increase in plug height reduces the noise by 25%.

Although front-side GND plugs are effective in reducing the peak substrate noise, they require substrate area. Back-side GND plugs start from the backside of the substrate. We assume four GND plugs fabricated at 3 μm from the center of the TSV and analyze the peak transient noise for various height and size of back-side GND plugs. We make the following observations:

- Back-side GND plugs are also effective in reducing the substrate noise. Increasing the GND plug height or thickness reduces substrate noise. To meet a 10% noise budget, back-side GND plugs of 2 μm diameter and a height of 15 μm are required.
- In contrast to front-side GND plugs, back-side GND plugs are more capable of localizing the noise. In particular, it is only the back-side GND plug that is capable of decreasing noise with increasing distance from the TSV. TSVs pass through the whole substrate. The upper part of the substrate is shielded by devices and grounded by ground ties. The lower part of the TSVs is more responsible for noise injection into devices farther away from the TSV. So, for TSVs, grounding the lower part of substrate is equally important. Devices further away from the TSV are more effectively shielded by the back-side GND plug resulting in better localization of noise as compared to front-side GND plugs.

GND plug diameter and depth are related due to aspect ratio limitations of deep trench formation and filling in Si substrate. Our analysis suggests the need for small diameter yet deep GND plugs. Fabrication of W-filled TSV with an aspect ratio of 50:1 and diameter of 1 μm has been proposed [31], which suggests that the GND plug with a diameter of 0.5 μm and an aspect ratio of 40:1 is achievable. Kikuchi et al. demonstrate W-filled TSV formation using deep-Si-trench etching and Tungsten chemical vapor deposition (CVD) where TSV diameter is maintained till 70% of substrate height for an aspect ratio of approximately 18:1 [27]. GND plugs do not require sidewall isolation nor high uniformity of plug diameter as a function of depth. A higher aspect ratio cone-shaped plug is thus possible, and warrants further investigation.

13.2.3   COMPARISON OF NOISE-MITIGATION TECHNIQUES

In addition to peak noise, area penalty is an important metric to evaluate the effectiveness of any of the noise mitigation approaches. Assuming a baseline TSV setup, we evaluate peak noise and area penalty for each of the approaches. We assume a single layer of high-R substrate of resistivity 10 Ω cm and relative permittivity of 11.8. We utilize the default design parameters described in Table 13.1 and explore the relative merits of each technique.

Table 13.2 reports peak transient noise and substrate area blockages for different noise mitigation techniques normalized to the baseline TSV case where no noise isolation technique is applied. The peak noise is reported at 6, 12, and 18 μm away from TSV center. These results show that the front-side GND plug performs better than all the other techniques and reduces the peak noise by 90% at 6 μm from the center of the TSV. Moreover, its area penalty is less than the area penalty for thicker

TABLE 13.2

Comparison of Peak Transient Noise and Area Blockage for TSV-Induced Noise Mitigation Techniques in High-R Substrate (Relative to the Baseline: TSV Height = 20 μm, Liner Thickness = 0.1 μm)

Technology	Peak Transient Noise (V)			Keep Out Area
	6 μm	12 μm	18 μm	
Thicker liner (liner thickness = 1.5 μm)	0.426	0.402	0.379	2.890
Backside ground plane	0.423	0.149	0.234	1.000
Guard ring (depth = 2 μm, width = 2 μm)	0.241	0.464	0.510	7.958
GND plug (diameter = 0.5 μm, height = 20 μm)	0.102	0.084	0.080	1.950
Back-side GND plug (diameter = 2 μm, height = 15 μm)	0.141	0.071	0.060	1.000

liner and guard ring. The back-side GND plug, on the other hand, does not require any area penalty and is capable of reducing noise by 86% at 6 μm from the center of the TSV.

13.3 CONCLUSION

The chapter reviews a general TSV-to-device noise coupling analysis methodology that extends the state-of-the-art, and uses this methodology to evaluate the impact of TSV noise on nearby devices. The TSV MOS-C model captures complex physical phenomena such as trapped charges in the oxide liner and enables substrate noise analysis due to digital signal transitions on nearby TSVs. We have shown that the typical fixed charge densities found in published TSV structures result in an order of magnitude change in TSV-substrate capacitance over the voltages found in typical digital circuits. The substrate extraction procedure described measures electrical coupling to an OP in the lossy dielectric bulk substrate, with 1.9 (7.9%) more accuracy at low (high) frequencies than prior methods. Additionally, as the substrate extraction technique does not rely on physical OPs or lumped ports, the technique is more general than previous ones, and can be applied in other situations when it is necessary to extract electrical coupling at a particular point within the substrate without altering the field distribution. The noise analysis results for planar and FinFET devices show that FinFETs exhibit an order of magnitude better leakage current noise immunity and two orders of magnitude better saturation current noise immunity relative to planar transistors. The findings regarding the increased noise robustness of FinFETs over planar devices are generalizable to noise sources other than TSVs. While the TSV-induced noise transients presented here are small in magnitude, they may be disruptive to sensitive analog circuits (e.g., amplifiers, phase-locked loop, and oscillator feedback loops).

The chapter also reviews several noise-mitigation techniques including thicker dielectric liner, backside ground plane, guard ring, and front- and back-side GND plugs. Analysis of a 1.5 μm thick dielectric liner shows peak substrate noise of 30% of VDD, thus necessitating further increase in thickness or a significant increase in the KOZ. Furthermore, the resulting area penalty, 3× the size of a 2 μm diameter TSV, creates routing blockages and reduces the area available for active devices. While a backside ground plane or mesh is effective with thinned dies, placing such a metal sheet or mesh between dies in 3D-ICs may not be practical. Guard rings require significant substrate area but do not provide sufficient substrate grounding. We showed that a front-side GND plug with an aspect ratio of 40:1 is effective in reducing noise by an order of magnitude with a smaller area penalty than a thicker liner. Back-side GND plug does not require any substrate area and still results in significant reduction in substrate noise. The proposed GND plug technique thus offers a practical and promising solution to the difficult problem of providing device shielding against TSV-induced substrate noise. The presented results on noise mitigation must be further simulated using the accurate MOS-C model.

While significant progress has been made to analyze and mitigate noise coupling in 3D-ICs, several challenges remain. Simulation results must be tightly integrated with specific designs to accurately predict the impact of noise coupling on a specific IC. The work described used an OP right below the transistor channel as the connecting point between the substrate and transistor simulation models. Planar and FinFET devices, however, have different detailed implementations in this interface between the channel and the substrate, which may have an impact on the extracted substrate coupling model. A more accurate approach is to model the specific details of FinFET and planar IC layouts in separate substrate models within HFSS. Novel and efficient simulation techniques that take into consideration complex layouts and multiple TSVs are still needed. Additionally, it is necessary to identify and analyze other forms of parasitic coupling that may arise due to novel 3D geometries. A recent example is the identification of the parasitic back-gate effect in 3D fully depleted SOI circuits, where interconnect on a backside metal layer acts as a back-gate of transistors on adjacent tiers [14]. Finally, holistic TSV placement and design approaches that consider the interplay between various tradeoffs such as stress, area, performance, and power are needed.

REFERENCES

1. Synopsys application note. Three-dimensional simulation of 20 nm Fin-FETs with round fin corners and tapered fin shape of TCAD sentaurus, 2012.
2. C. Auth, C. Allen, A. Blattner, D. Bergstrom, M. Brazier, M. Bost, M. Buehler et al. A 22 nm high performance and low-power CMOS technology featuring fully-depleted tri-gate transistors, self-aligned contacts and high density MIM capacitors. In *Digest of Technical Papers—Symposium on VLSI Technology*, pp. 131–132, Honolulu, HI, 2012.
3. M. Bamal, S. List, M. Stucchi, A. Verhulst, M. Van Hove, R. Cartuyvels, G. Beyer, and K. Maex. Performance comparison of interconnect technology and architecture options for deep submicron technology nodes. In *2006 International Interconnect Technology Conference, IITC*, Burlingame, CA, pp. 202–204, 2006.
4. T. Bandyopadhyay, R. Chatterjee, D. Chung, M. Swaminathan, and R. Tummala. Electrical modeling of through silicon and package vias. In *2009 IEEE International Conference on 3D System Integration*, San Francisco, CA, 2009.
5. M. Brocard, P. Le Maitre, C. Bermond, P. Bar, R. Anciant, A. Farcy, T. Lacrevaz et al. Characterization and modelling of Si-substrate noise induced by RF signal propagating in TSV of 3D-IC stack. In *Electronic Components and Technology Conference (ECTC)*, pp. 665–672, May 2012.
6. L. Cadix, M. Rousseau, C. Fuchs, P. Leduc, A. Thuaire, R. El Farhane, H. Chaabouni et al. Integration and frequency dependent electrical modeling of through silicon vias (TSV) for high density 3D ICs. In *Interconnect Technology Conference (IITC), 2010 International*, pp. 1–3, June 2010.
7. R. K. Cavin, P. Lugli, and V. V. Zhirnov. Science and engineering beyond Moore's law. *Proceedings of the IEEE*, 100(Special Centennial Issue):1720–1749, May 2012.
8. C. K. Chen, B. Wheeler, D. R. W. Yost, J. M. Knecht, C. L. Chen, and C. L. Keast. SOI-enabled three-dimensional integrated-circuit technology. In *2010 IEEE International SOI Conference*, Piscataway, NJ, 2010.
9. K.-Y. Chen, Y.-A. Sheu, C.-H. Cheng, J.-H. Lin, Y.-P. Chiou, and T.-L. Wu. A novel TSV model considering nonlinear MOS effect for transient analysis. In *2012 IEEE Electrical Design of Advanced Packaging and Systems Symposium*, Taipei, Taiwan, pp. 49–52, 2012.
10. J. Cho, J. Shim, E. Song, J. S. Pak, J. Lee, H. Lee, K. Park, and J. Kim. Active circuit to through silicon via (TSV) noise coupling. In *IEEE 18th Conference on Electrical Performance of Electronic Packaging and Systems*, pp. 97–100, 2009.
11. J. Cho, E. Song, K. Yoon, J. S. Pak, J. Kim, W. Lee, T. Song et al.. Modeling and analysis of through-silicon via (TSV) noise coupling and suppression using a guard ring. *IEEE Transactions on Components, Packaging and Manufacturing Technology*, 1(2):220–233, 2011.
12. F. Clment. Substrate noise coupling analysis in mixed-signal ICs. In *Workshop on Substrate Noise in Mixed-Signal ICs*, September 2001.

13. X. Duan, X. Gu, J. Cho, and J. Kim. A through-silicon-via to active device noise coupling study for CMOS SOI technology. In *Proceedings—Electronic Components and Technology Conference*, Lake Buena Vista, FL, pp. 1791–1795, 2011.

14. B. D. Gaynor and S. Hassoun. Parasitic back-gate effect in 3-D fully depleted silicon on insulator integrated circuits. IEEE Transactions on Components, Packaging and Manufacturing Technology, 4(1):100–108, January 2014.

15. B. D. Gaynor and S. Hassoun. Simulation methodology and evaluation of through silicon via (TSV)-FinFET noise coupling in 3-D integrated circuits. *IEEE Transactions on Very Large Scale Integration (VLSI) Systems*, PP(99):1–1, 2014.

16. B. Gaynor. Simulation of FinFET electrical performance dependence on fin shape and TSV and back-gate noise coupling in 3-D integrated, circuits. PhD thesis, Tufts University, Medford, MA, 2014.

17. S. W. Ho, S. W. Yoon, Q. Zhou, K. Pasad, V. Kripesh, and J. H. Lau. High RF performance TSV silicon carrier for high frequency application. In *58th Electronic Components and Technology Conference*, pp. 1946–1952, 2008.

18. C.-H. Jan, U. Bhattacharya, R. Brain, S.-J. Choi, G. Curello, G. Gupta, W. Hafez et al. A 22 nm SoC platform technology featuring 3-D Tri-gate and high-k/metal gate, optimized for ultra low power, high performance and high density SoC applications. In *Technical Digest—International Electron Devices Meeting, IEDM*, San Francisco, CA, pp. 3.1.1–3.1.4, 2012.

19. J. W. Joyner, R. Venkatesan, P. Zarkesh-Ha, J. A. Davis, and J. D. Meindl. Impact of three-dimensional architectures on interconnects in gigascale integration. *IEEE Transactions on Very Large Scale Integration (VLSI) Systems*, 9(6):922–928, 2001.

20. G. Katti, A. Mercha, M. Stucchi, Z. Tokei, D. Velenis, J. Van Olmen, C. Huyghebaert et al. Temperature dependent electrical characteristics of through-Si-via (TSV) interconnections. In *2010 IEEE International Interconnect Technology Conference*, 2010.

21. G. Katti, M. Stucchi, K. De Meyer, and W. Dehaene. Electrical modeling and characterization of through silicon via for three-dimensional ICs. *IEEE Transactions on Electron Devices*, 57(1):256–262, 2010.

22. N. H. Khan, S. M. Alam, and S. Hassoun. Through-silicon via (TSV)-induced noise characterization and noise mitigation using coaxial TSVs. In *IEEE International Conference on 3D System Integration*, pp. 1–7, 2009.

23. N. H. Khan, S. M. Alam, and S. Hassoun. Mitigating TSV-induced substrate noise in 3-D ICs using GND plugs. In *2011 12th International Symposium on Quality Electronic Design (ISQED)*, pp. 1–6, March 2011.

24. N. H. Khan, S. M. Alam, and S. Hassoun. Mitigating TSV-induced substrate noise in 3-D ICs using GND plugs. In *Proceedings of the 12th International Symposium on Quality Electronic Design*, Santa Clara, CA, pp. 751–756, 2011.

25. N. H. Khan, S. M. Alam, and S. Hassoun. GND plugs: A superior technology to mitigate TSV-induced substrate noise. *IEEE Transactions on Components, Packaging and Manufacturing Technology*, 3(5):849–857, May 2013.

26. N. H. Khan. Through-silicon via analysis for the design of 3-D integrated circuits. PhD thesis, Tufts University, Medford, MA, 2011.

27. H. Kikuchi, Y. Yamada, A. M. Ali, J. Liang, T. Fukushima, T. Tanaka, and M. Koyanagi. Tungsten through-silicon via technology for three-dimensional LSIs. *Japanese Journal of Applied Physics*, 47:2801–2805, April 2008.

28. V. Krishnaswamy, J. L. Shin, S. Turullols, J. M. Hart, G. Konstadinidis, and D. Huang. A 28 nm 3.6 GHz 128 thread SPARC T5 processor and system applications. In *2013 IEEE Asian Solid-State Circuits Conference (A-SSCC)*, Piscataway, NJ, pp. 17–20, 2013.

29. P. Le Maitre, M. Brocard, A. Farcy, and J.-C. Marin. Device and electromagnetic Co-simulation of TSV: Substrate noise study and compact modeling of a TSV in a matrix. In *Proceedings—International Symposium on Quality Electronic Design, ISQED*, Santa Clara, CA, pp. 404–411, 2012.

30. C. A. Mack. Fifty years of Moore's Law. *IEEE Transactions on Semiconductor Manufacturing*, 24(2):202–207, 2011.

31. M. Motoyoshi. Through-silicon via (TSV). *Proceedings of the IEEE*, 97(1):43–48, January 2009.

32. S. Savastiouk, O. Siniaguine, and E. Korczynski. 3-D stacked wafer level packaging. *Advanced Package (USA)*, 9(3):28–34, 2000.

33. A. R. Trivedi and S. Mukhopadhyay. Through-oxide-via-induced back-gate effect in 3-D integrated FDSOI devices. *IEEE Electron Device Letters*, 32(8):1020–1022, 2011.

34. G. Van der Plas, P. Limaye, A. Mercha, H. Oprins, C. Torregiani, S. Thijs, D. Linten et al. Design issues and considerations for low-cost 3D TSV IC technology. In *2010 IEEE International Solid-State Circuits Conference*, pp. 148–149, 2010.

35. R. P. Volant, M. G. Farooq, P. F. Findeis, and K. S. Petrarca. Coaxial through-silicon via. U.S. Patent 8,242,604, August 14, 2012.

36. C. Xu, R. Suaya, and K. Banerjee. Compact modeling and analysis of through-Si-Via-induced electrical noise coupling in three-dimensional ICs. *IEEE Transactions on Electron Devices*, 58(11):4024–4034, 2011.

37. L. Zhang, H. Y. Li, S. Gao, and C. S. Tan. Achieving stable through-silicon via (TSV) capacitance with oxide fixed charge. *IEEE Electron Device Letters*, 32(5):668–670, 2011.

38. W. Zhao and Y. Cao, New generation of predictive technology model for sub-45 nm early design exploration, *IEEE Trans. Electron Devices*, 53(11), pp. 2816–2823, November 2006.

39. Predictive Technology Model (PTM). [Online]. Available: http://ptm.asu.edu/, 2011.

Section IV

CAD Design Tools and Future
Directions for 3D Physical Design

14 Overview of 3D CAD Design Tools

Andy Heinig and Robert Fischbach

CONTENTS

ABSTRACT

In this chapter we will show and explain the current status of commercial available tools, which are needed to bring a 3D system from the idea to a manufactured system. Depending on the design complexity, existing 2D tools can directly be used to design parts of the 3D system. In some cases add-ons for the 2D tools are required, and for complex designs new tools are needed.

Similar to the 2D case, a 3D design flow can be divided into a construction phase and a corresponding verification phase as depicted in Figure 14.1. On the left side of Figure 14.1 the construction or design phase is shown, whereas the verification phase is shown on the left side.

If we look at these two phases, we can divide both into two major sections—a 3D and a 2D section. If the design reaches the 2D section, the 3D system is broken into classical 2D (sub-) systems (chips, interposers) and corresponding system integration elements between them (e.g., balls in Figure 14.2). Thus, the goal of the 3D design is to structure the 3D system into 2D (sub-) systems and some connecting elements. Tools that divide the 3D system into the 2D systems must consider different requirements such as timing, temperature, and fabrication costs.

System-level design exploration should take place at a very early design stage. Most of the required interaction should be done within the 3D design space. On the other hand, to get a detailed evaluation of a chosen design option, the 2D tools can be involved as well. But this kind of interaction should be limited to a minimum as it is often time-consuming.

Most 3D floorplanning approaches fall into the latter category. Examples are 3D-STAF [1] or the floorplanning approach described by Quirring et al. in [2]. Every new floorplan directly affects the layout of each sub-component (here tier). An evaluation requires the calculation of a cost function, which often uses 2D tools for calculation (e.g., routing congestion estimation) increasing the coupling with the 2D design phase. The dependency becomes even clearer when specific architectural decisions need to be made. For example, 3D network-on-chip (NOC) design approaches as described in [3,4] require more detailed layout information as floorplanning itself can offer.

For verification, we separate into 2D and 3D as well. In the 2D section the classical 2D system verification steps take place, such as Design Rule Check (DRC), Layout versus Schematic (LVS), parasitic extraction, timing verification, and optical proximity correction (OPC).

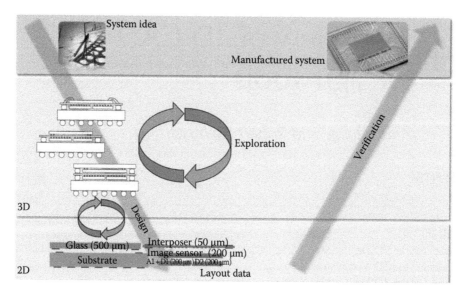

FIGURE 14.1   Design flow for 3D system development.

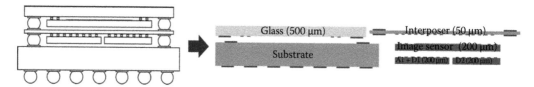

FIGURE 14.2   3D system can be divided into "classical" 2D systems.

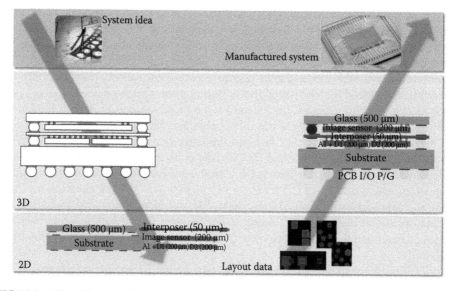

FIGURE 14.3   3D verification (right) requires adopted "classical" verification steps and the consideration of the (sub-) systems interactions (e.g., pad alignment).

These verification steps can be directly integrated into established flows for the 2D chips. Only smaller adoptions to new technologies are required (e.g., rule decks for TSVs or interposers). If we look at 3D, the tools must address the same verification steps as in 2D, but now focused on the interaction between the different (sub-) systems. For example, the 3D DRC must check if the pads between different dies are aligned (see Figure 14.3).

14.1 TOOLS FOR THE 3D DESIGN STEP

As mentioned in the introduction, during the 3D design, the 3D system is broken down into 2D systems. This step can be done manually, using a tool or with a combination of both. Currently, the preferred method is manually, due to the absence of suitable tools for this step.

In parallel, established third party tools are used to consider some second-order effects like thermal, mechanical, or first timing estimations [5].

Common tools are CST Studio [6], COMSOL Multiphysics [7], or ANSYS simulation and workflow tools [8]. CST (Computer Simulation Technology) addresses electromagnetic modeling and simulation. COMSOL enables the consideration of a big variety of different physical effects, such as electromagnetic, mechanical, thermal, fluid mechanical, and chemical. ANSYS combines a multitude of products from geometry preparation over workflow management to sophisticated physical simulation modules.

There are also interesting academic thermal modeling and simulation frameworks such as 3D-ICE [9]. Unfortunately, 3D-ICE considers stacked dies only. Furthermore, 3D integration specific investigations like the studies of Through-silicon-vias (TSV)-specific thermal and power implications described in [10] offer important contributions. All those contributions may be integrated into a higher (system) level approach.

A good example on how electromagnetic simulation can enrich a design flow for electronics is discussed in [11]. The authors in [12] show how to model a thermal flow sensor under the consideration of thermal and electrical effects. Such a model allows valuable predictions of a components' behavior during the design phase.

During the 3D design step, important design decisions are made by the system designer, for example, under the consideration of power hotspots and as a result the thermal hotspots. These decisions have a big influence on the resulting overall performance of the final product.

The usage of different tools to address different problems in different domains is very challenging. Often a single tool only addresses a single problem, resulting in a strongly needed consistent and unique data handling. Unique and consistent data handling means, for example, one data basis for all tools that needs geometrical information to avoid data exchange problems between different modeling formats. In the current approach, different designers use the geometrical engines of their respective programs. Each designer models a system different and often with different knowledge about the system. The different knowledge could result from information gaps between the different design teams.

As an example, different existing finite element method tools (e.g., ANSYS, Comsol) can be used to solve thermal and mechanical problems. All programs provide their own graphical system to describe the geometry, to set the material properties, and to illustrate the results. A thermal result of such a simulation is shown in Figure 14.4.

From the design view, the current system splits into two types—some with only a few connections between the 2D parts (e.g., sensor systems) or with more interconnects but a simple and regular interface (e.g., memory stacks). For both types, a manual design is feasible. Nevertheless, this approach is susceptible for errors. This is why a lot of companies look for suitable supporting tools. The big EDA companies already started to address data handling issues. For example, to design a chip stack, the internal Synopsis database (Milkyway) can handle the design data for each chip. Even the equalization of the separate pad layouts (e.g., mirroring for flip chip) is possible. But still the internal layout design for each chip is a separated process. A more comprehensive approach is required. Currently, universities and research institutes release first tools (prototypes) for 3D unified data handling.

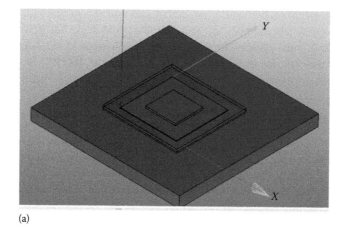

(a)

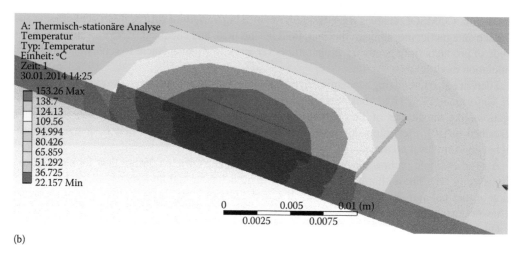

(b)

FIGURE 14.4 Model of the used 3D-stack (PCB, processor, Wide IO memory, and balls) (a). Heat distribution for convective heat removal (b).

14.2 ADVANC3D

Advanc3D is a modular software tool from Fraunhofer IIS/EAS aiming for system integration design support. Figure 14.5 illustrates the different features. The Advanc3D core module basically provides a unified system model to describe advanced system setups, such as 3D stacks, interposer-based solutions, and further System-in-Package (SiP) solutions. The unified system model serves as a central and consistent design database throughout the whole 3D design process. It is the 3D pendant to central 2D design databases (e.g., OpenAccess). The goal is a seamless integration into existing flows. Advanc3D could provide all the necessary 3D design data needed for advanced system integration with existing EDA tools from big vendors, like Cadence, Mentor Graphics, and Synopsys. Within the Advanc3D core module, the combination of geometrical and electrical layout and technological data enables a multitude of valuable functionalities.

One example is the derivation of intermediate netlist modules as shown in Figure 14.6. Once the system description exists within the unified system model, the original top-level netlist of the intended system is mapped against the corresponding elements. Using this information an extended netlist can be derived as follows. For each electrical connection (i.e., signal in top-level netlist), the

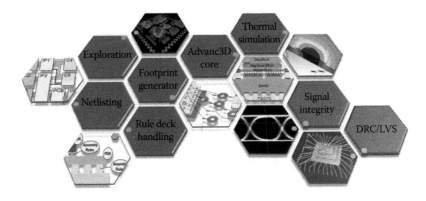

FIGURE 14.5 Modular structure of the Advan3D design support tool.

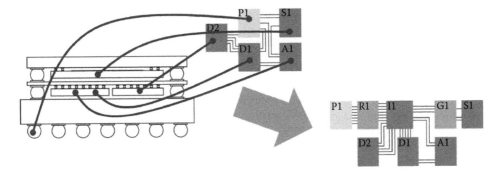

FIGURE 14.6 Netlist derivation based on system description (geometrical, technological, etc.), top-level netlist, and mapping information.

corresponding contact points (typically pads) are determined. The geometric model allows to find valid paths through the system (e.g., over an interposer). Thus, the original signal is directed through a new interposer module within the top-level netlist. Changes within the top-level netlist automatically affect the extended netlist, hereby maintaining consistent design data.

Furthermore there are modules for design exploration, netlist handling, technology management, simulation coupling, etc. Functionalities to support verification are described in the following sections.

In general, Advanc3D promotes a top-down design approach for advanced system integration. The starting point is a basic system description and a top-level netlist, both of which can be detailed later on. The system description format is an XML-based CSG language developed by Fraunhofer EAS [13]. The supported netlist formats are SPICE, Verilog, and VHDL. Thus, interfacing with standard EDA tools is straightforward. Advanc3D integrates into existing flows while simplifying configuration and scripting efforts of the designer. As one example, pad layout creation scripts for various tools can be generated. As another example, tool configurations (e.g., technology setups) for (sub-) system editing can be controlled from Advanc3D, creating a clear path into classical 2D design steps.

14.3 TOOLS FOR THE 2D DESIGN STEP

If we look at the 2D steps after the 3D system is split into its 2D (sub-) systems, established tools can be used. An important step is the placement of the IO pads, which is always coupled with the corresponding counterparts. This step can be done manually or automatically. For the design of the

FIGURE 14.7   Interposer for high-performance processor/memory integration.

classical 2D chips, the well-known tools from Cadence [14] or Synopsys [15] for the analog and digital circuits can be used. For digital/analog circuit design, one could use Cadence Encounter/ Virtuoso or Synopsys IC Compiler. A typical design flow using such tools is described in [16].

If the design is digital, the routing can be done automatically similar to the classical approach. The only challenge is the path from one register in one chip into another register in another chip. The path must be split into two segments, which are routed within the corresponding die. But this approach prevents a co-optimization of segments in different dies. If the design is analog, mostly manual design methods have to be used. The construction of the interposer can be done by standard design tools as well. Often the automatic router of the tools can't be used because they are timing and standard cell driven. Thus, the interposer signals are of different nature, for example, analog, high speed. The better way to design the interposer is to use analog design methods. The TSVs can be considered like normal vias under the consideration of their special geometries.

Another important consideration is the orientation of the GDSII layers of the subsystems. For example, if a chip is attached on the bottom side of an interposer, the chips layers will be reversed in the overall layout (e.g., required for verification).

Figure 14.7 shows a state-of-the-art interposer for a high-performance processor/memory system. Such systems require thousands of pins. In such cases, a manual process is not sufficient any more [17].

14.4   TOOLS FOR THE 2D VERIFICATION STEP

Verification of 2D (sub-) systems can be done using existing tools such as Cadence Assura or Mentor Graphics Calibre. Besides chip level, package and board tools are also used, such as Cadence Allegro or Mentor Graphics Expedition. A challenge for the designer is to correctly choose and configure the right technology setup and verification tool for a particular system component [18]. This is due to the big variety of technological options available as illustrated in Figure 14.8.

Even a simple chip stacked on an interposer can introduce multiple interdependencies. For example, the chosen packaging technology (e.g., bond wires vs. copper pillars) influences the chip's outer metal layers (RDL). Thus, even if the chip technology allows finer features, a part of the chip has to adapt to the environmental conditions (e.g., with a larger pitch). Another problem is the combination of sub-technologies. The verification of the interposer requires an "interposer rule deck", which is typically provided as a static set of rules by the foundry. Sophisticated advanced system integration on the other hand requires a higher flexibility. Therefore, a configurable rule deck is needed, which, for example, allows an easy exchange of the TSV technology.

Advanc3D offers the possibility to generate such rule decks from a given system configuration. Additionally, the designer can control the verification of the (sub-) systems from within Advanc3D,

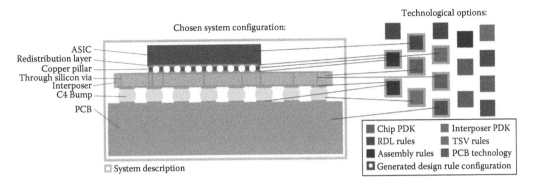

FIGURE 14.8 Technological options for advanced system integration.

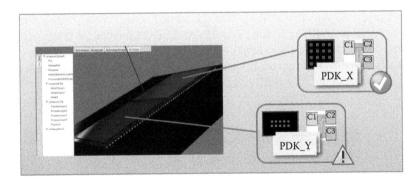

FIGURE 14.9 Technology management using Advanc3D.

properly configuring and externally executing established verification tools, such as Calibre (see Figure 14.9).

On the rule deck side further adaptations are required. TSVs, mainly introduced by 2.5/3D integration technologies, necessitate a particular set of design rules. Their circular shape does not fit well into the chip verification environment with its optimizations for rectangular shapes. For the supported interposer technologies, we developed a rule deck to address this new demand. Figure 14.10 shows typically applied rules (e.g., pad alignment, spacing rules, metal pad covering) to verify a standalone interposer layout with standard verification tools.

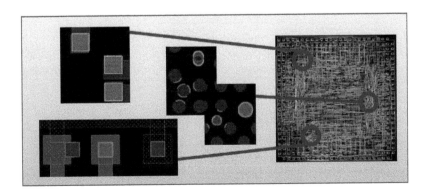

FIGURE 14.10 Interposer/TSV rules required for 2.5/3D integration.

14.5   TOOLS FOR THE 3D VERIFICATION STEP

The tool landscape for 3D verification is not developed as good as for 2D verification [19]. First solutions, such as Mentors Calibre 3DSTACK, emerge and address the increasing demand for package/assembly-level verification. Figure 14.11 illustrates the electrical verification of an interposer-based chip assembly. In the case of 3DSTACK, additional to the layout files, an assembly description and a pin/pad assignment is required. Afterward, density, enclosure, external, internal, off grid, and overlap checks for the combined layout are possible.

The system description within Advanc3D enables the automatic generation of the required 3DSTACK input files, which allows a designer the usage of this tool without the detailed knowledge about the new configuration file format. The same is true for other third party tools.

To allow a design rule check on fabrication level, we currently are working on a 3D DRC engine for Advanc3D. Due to the CSG-based data model, this is possible with a moderate effort.

Figure 14.12 depicts another important part of 3D verification. To compare existing layouts with the corresponding source netlists on different levels, an advanced netlist handling is required [18]. As mentioned earlier, the extended netlist within the unified system model is the central reference. To allow the verification of different (sub-) systems on different integration levels (e.g., chip, package, board), the corresponding (sub-) netlists have to be derived in a consistent way. The derived netlists then serve as input to various tools, such as 2D chip verification or PCB verification. Netlist handling is one module of the Advanc3D platform (compare Figure 14.5).

FIGURE 14.11   Connectivity test of a side-by-side chip assembly.

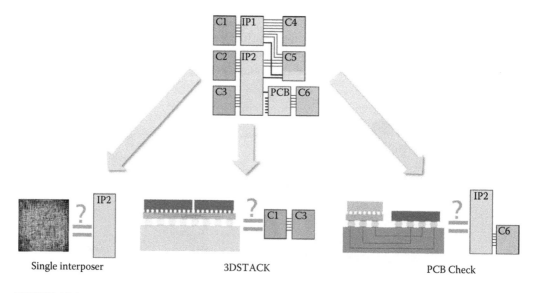

FIGURE 14.12   Advanced netlist handling to enable LVS on different verification levels.

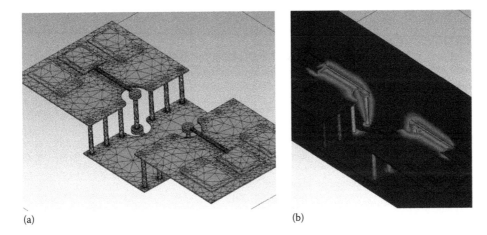

(a)   (b)

FIGURE 14.13   Model (a) and electromagnetic simulation (b) of a TSV test structure.

After all the signals are routed, a final timing analysis can be done. For these analyses, the parasitic information of the 2D chip data must be combined with data from additional elements like balls, bumps, or bond wires. The required (3D) parasitics can be extracted using 3D solvers like CST, Q3D, or HFSS. The choice, of which solver (with different solving methods [finite element method, boundary elements method]) to use, depends on a tradeoff between the necessary accuracy and the allowed runtime for the final results. The different results can be combined into a final parasitic file. This final parasitic file enables an overall timing analysis.

Additionally, a detailed analysis of single critical wires is possible with these tools. For such an analysis the wire can be considered and extracted up to the die. The biggest problem is to provide the right data for the tools. For example, all of the mentioned tools can import design data in the GDSII format, but once a GDSII file is imported, it is not possible to import further 3D data. Thus an external combination of all the required data is necessary.

Different kinds of investigations are possible, such as electromagnetic simulation shown in Figure 14.13 to investigate electrical behavior with regard to new technologies, such as TSVs [20]. Furthermore, detailed thermal, mechanical, or coupled thermo-mechanical simulations are possible in the final verification process. Appropriate tools are provided by Comsol or Ansys, for example.

14.6   CONCLUSION

In this chapter we gave an overview over 3D CAD design tools. We presented a basic design flow and explained important steps on the way from a system idea to the final system. Design and verification can both be split into a 3D and 2D section. For both sections several established tools are available. Nevertheless 3D system design decisions and data handling are still very challenging. At the beginning of a new 3D system design, many interrelated design issues are hard to decide. But step-by-step the design matures.

We presented Advanc3D, a tool to support designers during design and verification of advanced integrated systems. We presented some of its valuable modules, such as netlist derivation and simulation coupling. Finally, we addressed the verification of 3D systems, where technology and rule management is critical.

REFERENCES

1. P. Zhou, Y. Ma, Z. Li, R. Dick, L. Shang, H. Zhou, X. Hong, Q. Zhou, 3D-STAF: Scalable temperature and leakage aware floorplanning for three-dimensional integrated circuits, in *Proceedings of IEEE/ ACM International Conference on Computer-Aided Design ICCAD*, San Jose, CA, 2007, pp. 590–597.

2. A. Quiring, M. Olbrich, E. Barke, Improving 3D-Floorplanning using smart selection operations in meta-heuristic optimization, in *3D Systems Integration Conference (3DIC), 2013 IEEE International*, San Francisco, CA, October 2013, pp. 1, 6, 2–4.

3. F. Miller, T. Wild, A. Herkersdorf, TSV-virtualization for multi-protocol-interconnect in 3D-ICs, in *2012 15th Euromicro Conference on Digital System Design (DSD)*, Cesme, Izmir, Turkey, September 5–8, 2012, pp. 374, 381.

4. I. Loi, F. Angiolini, S. Fujita, S. Mitra, L. Benini, Characterization and implementation of fault-tolerant vertical links for 3-D networks-on-chip, in *IEEE Transactions on Computer-Aided Design of Integrated Circuits and Systems*, San Jose, CA, January 2011, Vol. 30(1), pp. 124, 134.

5. P. Schneider, S. Reitz, J. Stolle, R. Martin, A. Heinig, and A. Wilde, Design methods for 3D IC integration, in *International Wafer-Level Packaging Conference (IWLPC)*, Santa Clara, CA, October 12, 2010.

6. https://www.cst.com. Accessed on March 13, 2015.

7. https://www.comsol.de/. Accessed on March 13, 2015.

8. http://www.ansys.com/. Accessed on March 13, 2015.

9. A. Sridhar, A. Vincenzi, M. Ruggiero, T. Brunschwiler, D. Atienza Alonso, 3D-ICE: Fast compact transient thermal modeling for 3D-ICs with inter-tier liquid cooling, in *International Conference on Computer-Aided Design (ICCAD)*, San Jose, CA, 2010.

10. A. Todri, S. Kundu, P. Girard, A. Bosio, L. Dillilo, A. Virazel, A study of tapered 3D TSVs for power and thermal integrity, in *IEEE Transactions on Very Large Scale Integration (VLSI) Systems*, 2013, Vol. 21(2), pp. 306–319.

11. A. Barchanski, J. Krämer, P. Luzzi, EMC simulation in the design flow of modern electronics, *Microwave Journal*, 12(57):88–95, December 15, 2014.

12. C. Falco, A. De Luca, S. Safaz, F. Udrea, 3-D Multiphysics model of thermal flow sensors, in *COMSOL Conference*, Cambridge, U.K., 2014.

13. S. Wolf, A. Heinig, U. Knoechel, XML-based hierarchical description of 3D systems and SiP, *IEEE Design and Test*, 30(3), 59–69, June 2013.

14. http://www.cadence.com/. Accessed on March 13, 2015.

15. http://www.synopsys.com/. Accessed on March 13, 2015.

16. D. Milojevic, P. Marchal, E.J. Marinissen, G. Van der Plas, D. Verkest, E. Beyne, Design issues in heterogeneous 3D/2.5D integration, in *Design Automation Conference (ASP-DAC)*, Yokohama, Japan, January 22–25, 2013.

17. A. Heinig, R. Fischbach, M. Dittrich, Interposer based Wide IO processor integration, in *International Wafer-Level Packaging Conference (IWLPC)*, San Jose, CA, 2014.

18. R. Fischbach, A. Heinig, P. Schneider, Design rule check and layout versus schematic for 3D integration and advanced packaging, in *3D Integration Circuits Conference (3DIC)*, Cork, Ireland, 2014.

19. T.G. Yip, Y.H. Chuan, V. Iyengar, Challenges in verifying an integrated 3D design, in *Design, Automation and Test in Europe Conference and Exhibition (DATE)*, Dresden, Germany, March 12–16, 2012, pp. 167–168.

20. M. Wojnowski, K. Pressel, G. Beer, A. Heinig, M. Dittrich, J. Wolf, Vertical interconnections using through encapsulant via (TEV) and through silicon via (TSV) for high-frequency system-in-package integration, in *16th Electronics Packaging Technology Conference (EPTC)*, IEEE, Singapore, 2014.

15 Design Challenges and Solutions for Monolithic 3D ICs

Sung Kyu Lim and Yiyu Shi

CONTENTS

ABSTRACT

Monolithic three-dimensional integrated circuit (3D-IC) is a vertical integration technology that builds and stacks two or more tiers of devices sequentially, rather than bonding two independently fabricated dies together using bumps and/or through-silicon vias (TSVs). Compared to other existing 3D integration technologies (wire-bonding, interposer, TSV, etc.), monolithic 3D integration allows ultrafine-grained vertical integration of devices and interconnects, thanks to the extremely small size of intertier vias, typically local-via-sized (50–100 nm in diameter).

In this chapter, we first study the manufacturing process of monolithic 3D-ICs. One of the crucial challenges here is the low-temperature process, as high temperatures used in the fabrication of the second tier may damage the metal layers of the first tier. Next, we review recent work on design and EDA (electronic design automation) tool development for monolithic 3D-IC technology targeting memory and logic applications. For monolithic 3D memory, we explore the static random-access memory (SRAM) design. In the case of digital logic, we explore three design styles. The first is transistor level, where individual logic gate is folded into multiple tiers. Second, we explore gate level, where individual gates occupy a single tier, but they are placed in multiple tiers and connected with intertier vias. The last design is block level, where individual IP blocks occupy a single tier and are connected with others in different tiers using intertier vias. We study the challenges and solutions for tier-to-tier performance variations, power and thermal considerations, and low-power design with respect to these design styles.

15.1 INTRODUCTION

3D-ICs have emerged as a promising solution to extend the 2D scaling trajectory predicted by Moore's law. Current 3D-ICs are enabled by TSVs, where two prefabricated dies are aligned and bonded together. The TSV pitch is limited by the microbump pitch as well as the alignment accuracy.

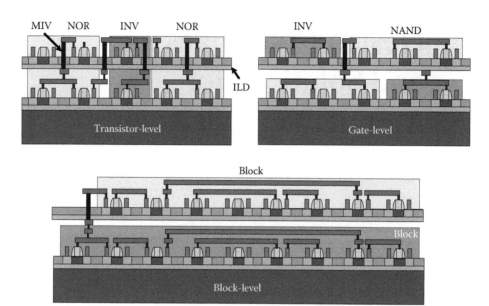

FIGURE 15.1 Various design styles available for monolithic 3D-ICs. (From Panth, S. et al., Design and CAD methodologies for low-power gate-level monolithic 3D-ICs, in: *Proceedings of International Symposium on Low Power Electronics and Design*, La Jolla, CA, 2014, pp. 171–176.)

TSV-based 3D, while ideal for integrating discrete board components onto a single package, does not provide sufficient integration density to solve the on-chip interconnect problem.

Monolithic 3D-IC (M3D) is an emerging technology that enables orders of magnitude of higher integration density than TSV-based 3D due to the extremely small size of the monolithic intertier vias (MIVs). In monolithic 3D integration technology, two or more tiers of devices are fabricated sequentially, one on top of another. This eliminates the need for any die alignment, which enables much smaller via sizes. Each MIV has essentially the same size as a regular local via (<100 nm diameter) [1].

This ultra-high-density enables several design styles, as shown in Figure 15.1. First, with respect to SRAM, the PMOS and NMOS of the bit-cell can be split onto multiple tiers. This gives us the opportunity to tune the PMOS and NMOS processes separately, leading to an optimum process for each device type. Next, a similar separation can be done for standard cells themselves, and this is known as transistor-level M3D. This design style has both intracell and intercell MIVs. Another design style is gate-level M3D, where the standard cells themselves are 2D, but they are placed in a 3D space, and interconnected using MIVs. This design style has only intercell MIVs. Finally, the coarsest level of integration is block-level M3D, where each functional block is 2D, and the 2D blocks are floorplanned onto a 3D space. In this design style, the MIVs are limited to the whitespace between blocks. We now discuss each of these design styles in detail.

15.2 MONOLITHIC 3D-IC FABRICATION

Different from die–stacking-based 3D-ICs, which require large TSVs, monolithic 3D-ICs feature minimal TSVs by directly growing additional device layers on top of the metal layers, which have a much smaller via size [2]. Monolithic 3D-IC provides a very effective method for ultradense through-silicon connections and short wires. The main challenge for monolithic 3D-ICs is that once copper or aluminum is added on the bottom layer, the process temperatures need to be limited to less than 400°C.

There are several techniques to obtain monolithic 3D-ICs: the monolithic 3D-ICs in [17] consist of multiple layers of carbon nanotube field-effect transistors (CNFETs) and carbon nanotube (CNT) interconnects; Germanium (Ge) is introduced into manufacturing in [18] to satisfy the low temperature requirements, which has lower melting and crystallization temperatures as compared with Si; pulsed-laser has been proposed as a very practical integration technology for monolithic 3D-ICs in multiple papers [3,14]; single grain (SG) Si thin-film transistors (TFTs) are proposed for monolithic 3D-ICs in [3], where transistors are fabricated inside a silicon grain, and the location of grain is created by pulsed-laser crystallization-based μ-Czochralski process. As mentioned in [14], most activation or anneal processes needed for CMOS fabrication can be replaced with pulsed-laser annealing process; therefore, it is very practical to fabricate monolithic 3D-IC using the schemes proposed in [14]. Firstly, utilize ion-cut technology to get the thin donor wafer, which will be introduced later, and then the donor wafer will be processed to form transistor with short-pulsed laser exposures providing annealing of process damages and activation of dopants.

The ion-cut technology is proposed for monolithic 3D-ICs based on current silicon process technology [2,13,14,19]; therefore, we will introduce the process flow of ion-cut technology in detail as an example to show an overview of monolithic 3D-ICs manufacture.

Ion-cut (Smart-Cut) utilizes wafer-bonding and hydrogen-implant based cleave, and is the predominant process used in Silicon-On-Insulator (SOI) wafer today. Figure 15.2 shows the process flow of ion-cut, which involves the main steps to build 3D-ICs [2,13,14,19]. Firstly, circuits in bottom wafer (acceptor wafer), which include both transistors and interconnections, are built conventionally. On the other hand, the top wafer (donor wafer) with p − si and n + si is constructed on a new wafer and the dopants are activated under normal high temperature techniques. After that, hydrogen will be implanted into the donor wafer with p − si and n + si regions, which will create a plane for cutting at the desired depth within the wafer. Next, flip the donor wafer and bond it to the acceptor wafer of 3D-ICs, then cleave the donor wafer along the plane created in step 3 using 400°C anneal or sideways mechanical force. Therefore, a thin-doped layer will be left after the cleave, then a recessed channel transistor (RCAT) architecture is utilized to create transistors in this thin layer under low temperature without damaging the connections in acceptor wafer. RCAT is widely used in DRAMS since 90 nm node, and the detailed process flow of step (e) and step (f) in Figure 15.2 for RCAT is shown in Figure 15.3. Finally, standard BEOL (Back-End-Of-Line) resolution and alignment are used to process both intralayer and interlayer interconnections.

15.3 MONOLITHIC 3D DESIGN CHALLENGES AND OPPORTUNITIES

Although TSVs and monolithic inter-layer vias (MIVs) have very different sizes (1–5 μm vs. 0.1 μm), some of the design methodologies for TSV-based and monolithic 3D-ICs can be shared. For instance, 3D floorplanning algorithms developed for TSV-based 3D-ICs can be used to design monolithic 3D-ICs because both of them have exactly the same problem formulation. Similarly, 3D placement algorithms for TSV-based 3D-ICs can also be used for monolithic 3D-ICs. A big difference between them is that TSVs are large, so the TSV count should be carefully controlled in the design of TSV-based 3D-ICs. On the other hand, the size of an MIV is similar to that of a local via, so the MIV count may not need to be carefully controlled. For example, when the die size is 1 mm × 1 mm, utilization is 70%, and the MIV size is 70 nm, the whitespace existing in a layer of the layout can accommodate approximately 15 million MIVs, which is almost infinite. Therefore, 3D placement algorithms for monolithic 3D-ICs can first place cells without considering MIV locations and then insert MIVs for each net during routing. 3D routing and 3D CTS for monolithic 3D-ICs may not need to consider the MIV size similarly to 3D placement, so existing routers and CTS algorithms developed for 2D-ICs can be used for the routing and CTS of 3D-ICs.

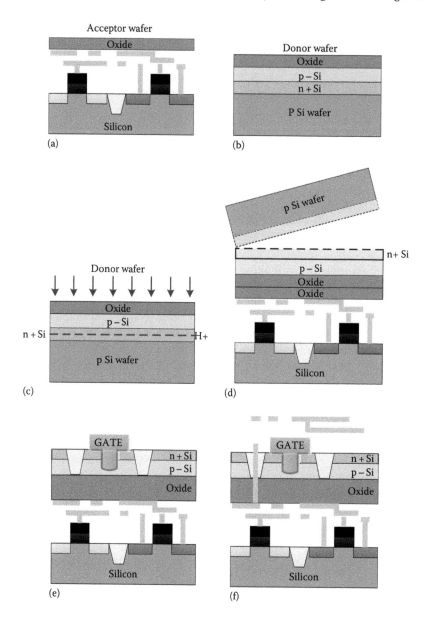

FIGURE 15.2 Process flow of monolithic 3D-ICs using ion-cut and RACT. (a) Build transistors and interconnects for bottom wafer of 3D stack, (b) using new wafer, construct dopant regions and activate at ~1000°C, (c) implant hydrogen for ion-cut, (d) bond and cleave, polish, (e) form RCATs using low-temperature steps, and (f) thin silicon layer enables well-aligned top layer features. (From Geng, H. et al., Monolithic three-dimensional integrated circuits: Process and design implications, in: *ECS Transactions on Electrochemical Society Meeting*, 2014; Sekar, D.C. et al. Monolithic 3D integrated circuits, in: *Future-Fab International*, 2013; Zvi Or-Bach, Paths to Monolithic 3D, http://www.monolithic3d.com/paths-to-monolithic-3d.html, web document.)

15.4 MONOLITHIC 3D SRAM DESIGN

Monolithic 3D offers a unique optimization opportunity for SRAM designs [7]. We can split the PMOS and NMOS into different tiers, which allows us to optimize the process of each type of transistor independently. We pick a state-of-the art 6T SRAM cell as our 2D baseline. This is designed in a 22 nm node, and it has an area of 0.1 μm², as shown in Figure 15.4a. The default 6T

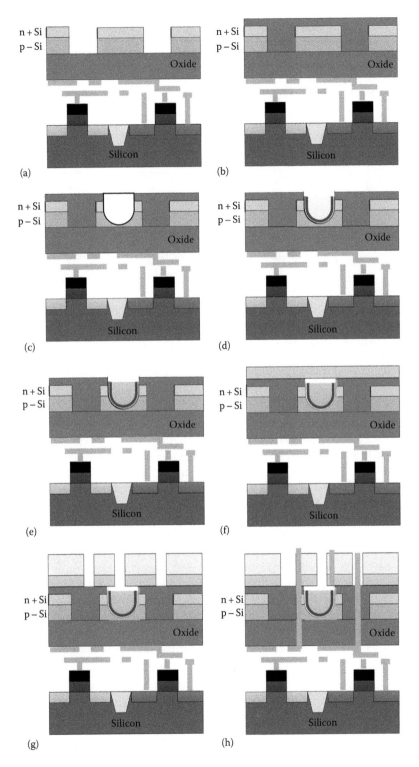

FIGURE 15.3 Process flow of RACT. (a) Etch isolation regions to define RCAT transistor, (b) fill isolation regions with oxide and CMP, (c) etch RCAT gate regions, (d) gate oxide, (e) form gate electrode, (f) add dielectric and CMP, (g) etch through-layer-via and RCAT transistor contacts, and (h) fill in copper. (From Sekar, D.C. et al. Monolithic 3D integrated circuits, in: *Future-Fab International*, 2013.)

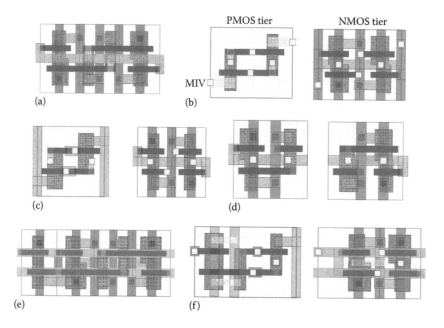

FIGURE 15.4 Layout of different SRAM cell designs. (a) 2D 6T SRAM (=2P4N) cell, (b) 3D 2P4N SRAM without transistor resizing, (c) 3D 2P4N SRAM with 3D-oriented sizing, (d) 3D 3P3N SRAM, (e) 2D 2P6N 8T SRAM, and (f) 3D 8T SRAM with a modified structure. (From Liu, C. and Lim, S.K., Ultra-high-density 3D SRAM cell designs for monolithic 3D integration, in: *Proceedings of IEEE International Interconnect Technology Conference*, San Jose, CA, pp. 1–3, © 2012 IEEE.)

SRAM cell has a 2 PMOS and 4 NMOS (2P4N) configuration. The obvious choice is to blindly split up this bit-cell into two tiers, but we observe that it only gives us a 33% footprint reduction due to the imbalance in the PMOS and NMOS count, as shown in Figure 15.4b. We therefore explore various alternate design options to give us a larger footprint reduction. The first option we explore is the same 2P4N configuration, but with different sizing (Figure 15.4c). We are able to obtain a footprint reduction of 44% with the same static noise margin (SNM) as 2D, but slightly worse write stability. Next, we explore changing the number of PMOS and NMOS in the bit-cell, but keeping the total transistor count the same. We first explore a 3P3N configuration, replacing one pass transistor with an NMOS (Figure 15.4d). The footprint reduction in this case is 45%. Using a single-ended read technique, we are able to achieve a high SNM margin. Lastly, we explore an 8T bit-cell. The conventional 2P6N configuration for 2D is shown in Figure 15.4e. However, since the PMOS and NMOS are on separate tiers, splitting this configuration will lead to area imbalance. Therefore, we move to a 4P4N configuration that can give us a better area balance. This gives us a 40% footprint reduction under the same read margin, write margin, and access time as 2D.

15.5 TRANSISTOR-LEVEL MONOLITHIC 3D-IC DESIGN

Transistor-level M3D is similar to the SRAM case in the sense that the PMOS and NMOS are split into multiple tiers. In this design style, each standard cell is redesigned such that its PMOS and NMOS are on different tiers [4–6]. As in the case of SRAM, the advantage of doing this is that the PMOS and NMOS can be optimized separately.

We begin by constructing a library of 66 monolithic 3D standard cells using a cell-folding technique. An overview of our approach for a simple inverter is shown in Figure 15.5. We draw a cut-line

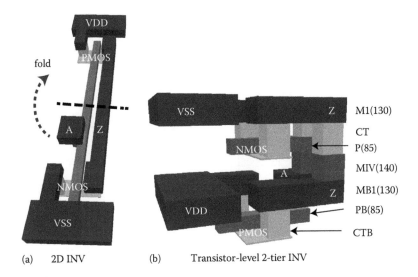

FIGURE 15.5 Layout of an inverter from (a) Nangate 45 nm library and (b) our transistor-level monolithic 3D library. P, M, and CT represent poly, metal, and contact. The suffix "B" means the bottom tier. Top/bottom tier silicon substrate and p/n-wells are not shown for simplicity. Numbers in parentheses mean thickness in nm. (From Lee, Y.-J. et al., Power benefit study for ultra-high-density transistor-level monolithic 3D ICs, in: *Proceedings of ACM Design Automation Conference*, Austin, TX, May 2013, pp. 1–10.)

in the center of the cell, and then fold the cell along this line. All the connections that exist along the cut-line will now become MIVs.

When compared with 2D, we observe a footprint reduction of about 40% because of an imbalance between PMOS and NMOS sizes, and some additional area required for the MIVs themselves. We then perform extraction on each cell, and recharacterize the cells taking into account the new cell internal parasitics. The advantage of this design style is that we can utilize existing 2D P&R tools to perform all the design steps for us. From the tools perspective, the standard cells have pins on different metal layers, and the router is capable of connecting all these pins together, inserting inter-cell MIVs in the process.

Since the total number of pins remain the same and the footprint area is reduced, there is a 1.7×–2× increase in the pin density of the chip, with fewer routing resources than 2D. This therefore causes several routability issues. We explore several interconnect options to mitigate the congestion in transistor-level monolithic 3D-ICs. An overview of the various interconnect options considered are shown in Figure 15.6. We consider three different interconnect options: (1) one metal layer on the bottom tier (1BM), (2) three additional metal layers on the top tier (3TM), and (3) three additional metal layers on the bottom tier (4BM). Upon extraction of the 4BM case, we observe that the additional stacked vias within each cell leads to a significant increase in the cell internal parasitics, which increases the cell delay and power by up to 9.86% and 15.65%, respectively. Among the three options considered, we observe that the 3TM case gives us the best results with up to a 22% reduction in the total power of the chip. To further improve this benefit, we fine tune the 3TM case. For example, the routing congestion is not always on the intermediate metal level. We therefore explore other options like utilizing two intermediate and two global metal layers instead of three intermediate metal layers, and this gives us a further 2.8% power benefit.

Since interconnects play a more dominant role at lower nodes, we also study the benefit of this design style at more advanced and future technology nodes such as 22 and 7 nm. We observe that at the 22 and 7 nm nodes, we get an additional 4% and 23% power benefit, respectively.

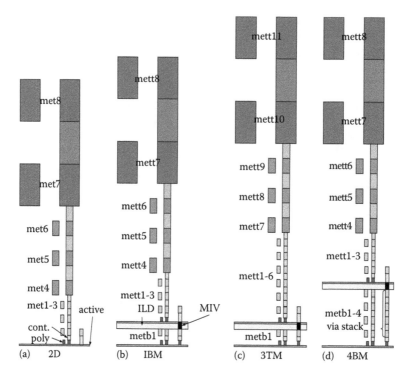

FIGURE 15.6 Metal layer stack options. (a) 2D, (b) baseline MI-T, (c) three local metal layers added to the top tier, and (d) three local metal layers added to the bottom tier. ILD stands for interlayer dielectric between the top and the bottom tier. The bottom tier substrate and ILD for metal layers are not shown for simplicity. Objects are drawn to scale. (From Lee, Y.-J., Morrow, P., and Lim, S.K., Ultra high density logic designs using transistor-level monolithic 3D integration, *Proceedings of IEEE International Conference on Computer-Aided Design*, San Jose, CA, pp. 539–546, © 2012 IEEE.)

15.6   GATE-LEVEL MONOLITHIC 3D-IC DESIGN

In this design style, each standard cell is 2D, and these 2D standard cells are placed onto a 3D space. The advantage of this method is a reuse of existing standard cells, which can avoid the need for library re-characterization. Once the gates are placed in 3D, MIVs are inserted into the existing whitespace available between the cells.

We propose a design flow based on shrunk 2D gate placement [9] (shown in Figure 15.7) that leverages existing commercial 2D placers. This approach first halves the footprint area to represent a monolithic 3D footprint so that there is exactly 0% total silicon area overhead over 2D. Next, the placement capacity is doubled (or the area of standard cells are halved), and the commercial 2D engine is run to give an initial placement. The shrunk 2D placement then needs to be partitioned to give a legal monolithic 3D-IC placement. We define partition bins, and partition the design with local area balance in each placement bin to give us a gate-level M3D design. We demonstrate that this approach can give us up to 30% HPWL savings when compared with 2D-ICs.

To insert MIVs into the layout, the conventional approach is to perform a cell and 3D-via co-placement step. We propose a commercial–router-driven MIV insertion algorithm that improves the routed wirelength (WL) by up to 16.6% and the power delay product (PDP) by up to 6.1%.

To further reduce the routed wirelength, we propose a routability-driven partitioner [10] that utilizes the fine-grained nature of MIVs to reduce routing congestion. By intelligent partitioning, we move nets that are in congested regions in one tier to the other, noncongested tier. Our approach helps give us an additional 4% WL and 4.33% PDP benefit. We also demonstrate that using multiple MIVs per 3D net can help give us a 8.43% WL benefit and 2.25% power benefit.

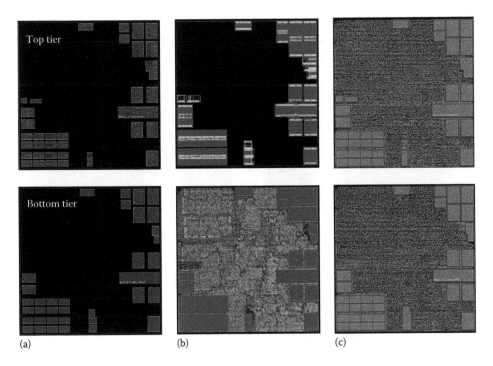

FIGURE 15.7 Shrunk 2D technique for gate-level monolithic 3D-IC placement. (a) Macroplacement, (b) shrunk 2D placement, and (c) partition and repopulate. (From Panth, S. et al., Design and CAD methodologies for low power gate-level monolithic 3D ICs, in: *Proceedings of International Symposium on Low Power Electronics and Design*, La Jolla, CA, 2014, pp. 171–176.)

We also propose techniques for utilizing a commercial tool for timing optimization and clock-tree synthesis (CTS). We demonstrate that keeping the clock backbone on a single tier gives us 29.82% clock power reduction compared to the case where we have one separate clock-tree per tier. Overall, we demonstrate on the OpenSparc T2 design that M3D can give a 15.57% power benefit compared to commercial-quality 2D designs. We also demonstrate that this benefit rises to 16.08% when utilizing dual-Vt libraries.

One of the main challenges facing monolithic 3D-ICs is power-delivery network (PDN) design [16]. In conventional 2D-ICs, the entire top metal layer is available for PDN design. However, in monolithic 3D-ICs, the top metal layer needs to be used for both PDN and MIV landing pads in the top tier [16]. An example of this is shown in Figure 15.8. In Figure 15.8a, we show the MIV landing pad locations without any PDN present in the top tier. We observe that they are spread out all over the footprint of the chip. In Figure 15.8b, we first create a PDN and then perform MIV planning. We observe that the MIV locations are now confined to the space between PDN wires. An isometric view of the MIVs in both tiers is shown in Figure 15.8c. Because of the conflict between PDN and MIV landing pads, adding PDN increases the M3D WL by 20.5%, compared to only a 7.1% WL increase when adding PDN in 2D. This in turn increases the net power and temperature, reducing the benefit of M3D.

We propose a PDN optimization technique that helps reduce the routed wirelength. An overview of this technique is shown in Figure 15.9. In the original case, the VDD and VSS lines are spaced evenly. This leads to fragmented whitespace, which makes MIV planning difficult, increasing the routed wirelength. In our proposed optimization approach, we move the VDD and VSS lines close to each other. This leads to larger continuous whitespace available for MIV insertion. This lowers the routed wirelength by up to 8%. We also perform GDSII-level thermal analysis on the M3D chip with and without the presence of PDN. With the proposed PDN technique, the power consumed by the chip goes down because of the lower routed wirelength. This in turn leads to a 5% reduction in the maximum temperature of the chip, while still meeting the given IR-drop budget.

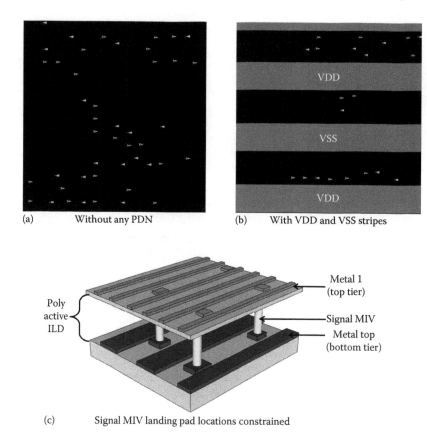

FIGURE 15.8 Impact of PDN on MIV landing pads. (a) MIVs freely distributed without any PDN blockages in top metal, (b) PDN blockages affect MIVs in top metal, and (c) isometric view showing the constraints on signal MIV landing pad locations in top metal and metal of the next tier. (From Samal, S.K., Samadi, K., Kamal, P., Du, Y., and Lim, S.K., Full chip impact study of power delivery network designs in monolithic 3D ICs., *Proceedings of IEEE International Conference on Computer-Aided Design*, San Jose, CA, pp. 565–572, © 2014 IEEE.)

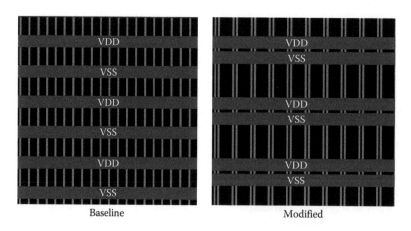

FIGURE 15.9 Baseline PDN versus modified PDN. Note the extra continuous space between the horizontal P/G wires in the modified layout, which enhances MIV insertion and routing. A similar optimization is done on the vertical P/G wires. (From Samal, S.K., Samadi, K., Kamal, P., Du, Y., and Lim, S.K., Full chip impact study of power delivery network designs in monolithic 3D ICs, *Proceedings of IEEE International Conference on Computer-Aided Design*, San Jose, CA, pp. 565–572, © 2014 IEEE.)

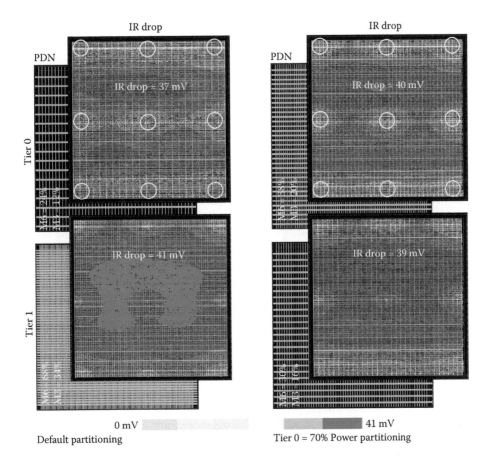

FIGURE 15.10 The impact of power-aware partitioning. Moving the power closer to the package enables a reduction in PDN resources, with <1°C in the maximum temperature of the chip. (From Panth, S. et al., Tier-partitioning for power delivery vs cooling tradeoff in 3D VLSI for mobile applications, in: *Proceedings of ACM Design Automation Conference*, San Francisco, CA, 2015.)

Another solution to this IR-drop problem is moving power hungry cells close to the package [12]. In a conventional package, this causes thermal issues, as the majority of the heat is conducted from the heatsink, which is away from the package. However, we target a mobile system, and since a mobile system has no heatsink, heat is conducted away from both sides of the chip in equal proportions. This means that if high power modules are moved close to the package, there will be negligible thermal impact. We leverage this fact to develop an IR-drop aware partitioning algorithm that improves power-delivery and has a negligible thermal impact. We present a gate-level IR-drop aware partitioning heuristic that reduces IR-drop without hurting design quality. We demonstrate that with the same PDN, our heuristic helps reduce IR-drop by 24.66%. In addition, we demonstrate that when the PDN is optimized to meet IR-drop in M3D, our technique yields 28.57% PDN resource reduction and 4% total wirelength reduction with <1°C temperature increase. Sample PDNs designed to meet the IR-drop targets, along with the IR-drop maps are shown in Figure 15.10.

15.7 BLOCK-LEVEL MONOLITHIC 3D-IC DESIGN

Block-level monolithic 3D-ICs utilize existing functional 2D IP blocks and floorplan them onto a 3D space [8,11]. This design style can be used for SoC-level integration, and it also has the benefit of IP reuse.

We present a simulated annealing framework for M3D floorplanning, which uses a weighted sum of wirelength and area as the cost function. Unlike TSV-based floorplanners, the MIV count is not included in the cost function because MIVs are so small that we need not minimize their number. Once the blocks are floorplanned onto a 3D space, we need to insert MIVs in the whitespace between them. For this purpose, we present a commercial router-driven MIV insertion algorithm. For performance evaluations, we first assume that the monolithic 3D fabrication process is mature, and that both tiers have identical performance. In this case, we demonstrate that we can close the gap to the ideal block-level implementation by up to 50% w.r.t. both power and performance. The ideal block-level implementation is obtained by designing the chip assuming perfect inter-block interconnects, that is, the interblock nets have zero resistance and capacitance. This is the best possible block-level design for a given benchmark, given the same set of blocks.

However, currently we cannot achieve identical tier performance. During the manufacturing process of the top tier, we need to take care not to damage the underlying interconnects and transistors. This can be achieved either by a low-temperature process on the top tier, or by using tungsten on the bottom tier. We model the impact of both these options, and present a variation-aware floorplanning scheme that makes the design tolerant to such manufacturing-induced performance variations. A summary of the power-performance results under these variations are shown in Figure 15.11. These results demonstrate that our variation-tolerant floorplanning scheme improves the chip performance and power by up to 12.6% and 10.6%, respectively. We also demonstrate that tungsten interconnects on the bottom tier are preferable to degraded transistors on the top tier. Finally, we demonstrate that even under such performance variations, we can still close the gap to the ideal block-level implementation by up to 50% w.r.t. performance and 36% w.r.t. power.

The increase in power density associated with 3D-ICs means that thermal-aware design methodologies have become necessary [15]. We first study the thermal properties of monolithic 3D-ICs and observe that it has several unique properties. First, the extremely thin tiers mean that there is negligible lateral thermal coupling. In addition, the absence of any bonding or adhesive layer for underfill implies that heat is not trapped in a given tier, and that the vertical thermal coupling is very

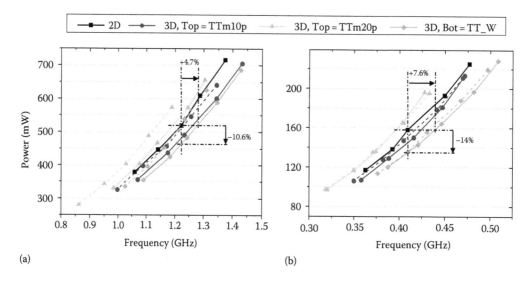

(a)

(b)

FIGURE 15.11 Power-performance trade-off curves for two benchmark designs (a: des3 and b: b19), assuming degraded transistors and interconnects. Dashed lines represent nonvariation-aware floorplanning and solid lines represent variation-aware floorplanning. (TTm10p means 10% worse transistors, TTm20p means 20% worse transistors, and TT_W means tungsten interconnects.) (From Panth, S. et al., Power-performance study of block-level monolithic 3D-ICs considering intertier performance variations, in *Proceedings of ACM Design Automation Conference*, San Francisco, CA, 2014, pp. 1–6.)

high. In addition, the small size of MIVs mean that they do not serve as a conduction path, and their location need not be optimized for thermal reasons.

These properties enable us to develop a multiadaptive regression spline (MARS) model to quickly estimate the temperature of a monolithic 3D-IC. We generate a large sample set considering different block powers, and use it to train our model. We demonstrate that our model has an error of less than 5% when compared to GDSII-level FEA simulations. In addition, our model is extremely fast, and is 10^5 times faster than prior quick thermal approaches. This extremely quick computation means that our model can be used within a simulated annealing floorplanning framework, which makes millions of cost function evaluations to come up with a floorplanning solution. We modify the cost function of our simulated annealing framework to be a weighted sum of area, wirelength, and the maximum temperature of the chip. We first run the nonmodified floorplanner until a certain area and wirelength target are met. Next, the temperature term in the cost function is introduced, and the area and wirelength serve as constraints instead of objectives. This is so that the floorplanner does not increase the chip area to reduce the maximum temperature. Using this approach, we demonstrate up to a 22% reduction in the maximum temperature of the chip, without affecting other design metrics such as wirelength and area. Floorplan screenshots with and without thermal-aware floorplanning are shown in Figure 15.12. We perform two thermal-aware runs, with and without area slack. In the case with area slack, the footprint area constraint is relaxed slightly so that the thermal-aware floorplanner can achieve a better solution. We observe that relaxing the constraint leads to 22% area reduction in addition to temperature reduction.

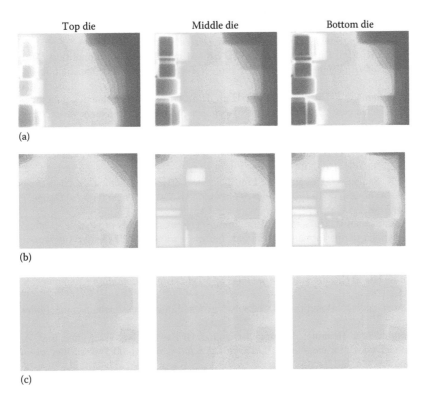

FIGURE 15.12 Three-tier floorplanning layouts with corresponding absolute temperature maps. The thermal-aware floorplans avoid stacking of high-power density blocks and result in 22% temperature reduction in lesser total area. The temperature range is (47°C, 68°C). (a) Nonthermal-aware, (b) thermal-aware w/o area slack, and thermal-aware w/ area slack. (From Samal, S.K. et al., Fast and accurate thermal modeling and optimization for monolithic 3D ICs, in: *Proceedings of ACM Design Automation Conference*, San Francisco, CA, 2014, pp. 1–6.)

15.8 CONCLUSION

We have explored several design styles that are available for monolithic 3D-ICs SRAM, transistor-level, gate-level, and block-level. For each design style, we have presented design flows to obtain GDSII-level signoff-quality power and performance results. We have enumerated various challenges facing M3D, and techniques to overcome them. Overall, ultra-high-density monolithic 3D-ICs offer significant benefit over 2D-ICs.

REFERENCES

1. P. Batude et al. Advances in 3D CMOS sequential integration. In *Proceedings of IEEE International Electron Devices Meeting*, Baltimore, MD, pp. 1–4, 2009.
2. H. Geng, L. Marecsa, B. Cronquist, Z. Or-Bach, and Y. Shi. Monolithic three-dimensional integrated circuits: Process and design implications. In *ECS Transactions on Electrochemical Society Meeting*, Orlando, FL, 2014.
3. R. Ishihara, N. Golshani, J. Derakhshandeh, M.R.T. Mofrad, and C.I.M. Beenakker. Monolithic 3D-ICs with single grain Si thin film transistors. In *Proceedings of International Conference on Ultimate Integration on Silicon*, Cork, Ireland, 2011.
4. Y.-J. Lee, D. Limbrick, and S.K. Lim. Power benefit study for ultra-high density transistor-level monolithic 3D ICs. In *Proceedings of ACM Design Automation Conference*, Austin, TX, pp. 1–10, May 2013.
5. Y.-J. Lee, P. Morrow, and S.K. Lim. Ultra high density logic designs using transistor-level monolithic 3D integration. In *Proceedings of IEEE International Conference on Computer-Aided Design*, San Jose, CA, pp. 539–546, 2012.
6. C. Liu and S.K. Lim. A design tradeoff study with monolithic 3D integration. In *Proceedings of International Symposium on Quality Electronic Design*, Santa Clara, CA, pp. 529–536, 2012.
7. C. Liu and S.K. Lim. Ultra-high density 3D SRAM cell designs for monolithic 3D integration. In *Proceedings of IEEE International Interconnect Technology Conference*, San Jose, CA, pp. 1–3, 2012.
8. S. Panth, K. Samadi, Y. Du, and S.K. Lim. High-density integration of functional modules using monolithic 3D-IC technology. In *Proceedings of Asia and South Pacific Design Automation Conference*, Yokohama, Japan, pp. 681–686, 2013.
9. S. Panth, K. Samadi, Y. Du, and S.K. Lim. Design and CAD methodologies for low power gate-level monolithic 3D ICs. In *Proceedings of International Symposium on Low Power Electronics and Design*, La Jolla, CA, pp. 171–176, 2014.
10. S. Panth, K. Samadi, Y. Du, and S.K. Lim. Placement-driven partitioning for congestion mitigation in monolithic 3D IC designs. In *Proceedings of International Symposium on Physical Design*, Petaluma, CA, pp. 47–54, 2014.
11. S. Panth, K. Samadi, Y. Du, and S.K. Lim. Power-performance study of block-level monolithic 3D-ICs considering inter-tier performance variations. In *Proceedings of ACM Design Automation Conference*, San Francisco, CA, pp. 1–6, 2014.
12. S. Panth, K. Samadi, Y. Du, and S.K. Lim. Tier-partitioning for power delivery vs cooling tradeoff in 3D VLSI for mobile applications. In *Proceedings of ACM Design Automation Conference*, San Francisco, CA, 2015.
13. Zvi Or-Bach, Paths to Monolithic 3D. http://www.monolithic3d.com/paths-to-monolithic-3d.html, web document.
14. B. Rajendran, A.K. Henning, B. Cronquist, and Z. Or-Bach. Pulsed laser annealing: A scalable and practical technology for monolithic 3D IC. In *IEEE International 3D System Integration Conference*, San Francisco, CA, 2013.
15. S.K. Samal, S. Panth, K. Samadi, M. Saedi, Y. Du, and S. Kyu Lim. Fast and accurate thermal modeling and optimization for monolithic 3D ICs. In *Proceedings of ACM Design Automation Conference*, San Francisco, CA, pp. 1–6, 2014.
16. S.K. Samal, K. Samadi, P. Kamal, Y. Du, and S.K. Lim. Full chip impact study of power delivery network designs in monolithic 3D ICs. In *Proceedings of IEEE International Conference on Computer-Aided Design*, San Jose, CA, pp. 565–572, 2014.

17. H. Wei, N. Patil, A. Lin, H.-S.P. Wong, and S. Mitra. Monolithic three-dimensional integrated circuits using carbon nanotube FETs and interconnects. In *Proceedings of IEEE International Electron Devices Meeting*, Baltimore, MD, 2009.

18. S. Wong, A. El-Gamal, P. Griffin, Y. Nishi, F. Pease, and J. Plummer. Monolithic 3D integrated circuits. In *Proceedings of IEEE International Symposium on VLSI Technology, Systems and Applications*, Hsinchu, Taiwan, 2007.

19. D. C. Sekar et al. Monolithic 3D integrated circuits. In *Future-Fab International*, 2013.

16 Design of High-Speed Interconnects for 3D/2.5D ICs without TSVs

Tony Tae-Hyoung Kim and Aung Myat Thu Linn

CONTENTS

ABSTRACT

Transistor scaling has developed the semiconductor industry over the last few decades. However, as the lithography process approaches its limit, research and development cost for a new technology node has increased dramatically. 3D integration has been considered as a promising solution for continuously improving transistor count per area and system performance. Through-silicon via (TSV) technology is the most popular integration technology. However, it has several issues to be addressed such as thermal issues, mechanical stress, high fabrication cost, and placement issues. Several alternative interconnect technologies such as inductive coupling interconnect, capacitive coupling interconnect, and silicon interposer have been introduced.

This chapter will introduce various design techniques utilizing these alternative interconnect methods. First, transceiver design using inductive coupling will be discussed. Several design techniques will be introduced. Second, how to design capacitive coupling interconnect will be explained. Various challenges such as alignment, cross talk, and transceiver design will be covered. Finally, high-speed interconnect utilizing silicon interposer (known as 2.5D) will be briefly explained as an intermediate interconnect technology between conventional PCB and TSV.

16.1   INTRODUCTION

This chapter describes the design of TSV-less high-speed interconnects for 3D/2.5D integration. TSV-less proximity communication for three dimensional integrated circuits (3D-ICs), such as inductive coupling and capacitive coupling interconnects, are presented in this chapter. The key benefit of the proximity communications is that they are CMOS compatible and, thus, wafer processing steps are greatly reduced as compared to TSV-based 3D integration. The inductive coupling interconnect is a current-mode signaling, where the generated magnetic flux can penetrate multiple tiers allowing cross-tier communication. The voltage-mode capacitive coupling interconnect with proximity requirement between two electrodes can only be implemented in face-to-face integration. Several design challenges of these proximity communications in the area of power consumption, alignment, cross talk, and system-level integration are discussed in Sections 16.2 and 16.3. In 2.5D-ICs, a silicon-based interposer is used where the fine-pitch copper wires from back end of line (BEOL) connect two or more chips together through microbumps. A transceiver design and equalization scheme to compensate for channel losses from fine-pitch wire is presented in Section 16.4.

16.2   HIGH-SPEED INTERCONNECTS FOR 3D-ICs UTILIZING INDUCTIVE COUPLING

On-chip near-field communication utilizing inductive coupling has been studied in the literature for more than a decade as an alternative to other well-known 3D integration techniques, such as microbump and TSV. Unlike microbump and TSV, the inductive coupling scheme is more CMOS compatible and thus requires no additional process steps. In the inductive coupling interconnect, the magnetic flux can be sent vertically to neighboring tiers in 3D ICs thanks to the planar structures of the on-chip inductors. Due to the similarity in the permeability of silicon substrate and silicon dioxide (SiO_2), the magnetic flux can penetrate multiple tiers and, therefore, cross-tier communication can be realized without requiring to relay the signal at every tier. The transmission power in this scheme can be controlled according to the communication distance, which is affected more by the current drivability of the transmitter rather than the supply voltage. In other applications, inductive coupling has also been used for delivering power wirelessly to neighboring tiers.

In the following sections, some of the reported transceiver designs based on the inductive coupling interconnect are discussed. Inductor design and other related issues such as cross talk, alignment, and communication distances are also discussed, followed by a few system-level implementation issues.

16.2.1   Transceiver Design and Its Signaling Scheme

One of the most widely used inductive coupling transmitter is the H-bridge type shown in Figure 16.1a. Two inverters are connected to each end of the inductor. The complementary transmit signal to one inverter is skewed by a delay block to limit the bipolar current pulse passing through the inductor coil. The loop of the transmitting inductor is open when it is not transmitting data because the induced current in the idle transmitting inductor can counteract with the change of the magnetic flux, which reduces the induced current in the receiving inductor. The duration of the current pulse can be controlled by adjusting the delay to reduce power consumption; however, this reduces the timing margin and, thus, precise timing and duty cycle control of the sampling clock at the receiver is required [1].

Although the H-bridge transmitter is simple to implement, it draws a substantial amount of power due to the short current flowing through the inductor during magnetic flux generation. One way to reduce the short current is to utilize a current-reuse topology shown in Figure 16.1b. A single-ended transmitter stores the charge during the pull-up in the capacitor and it will be reused when the pull-down is enabled in the transmitter. In this transmitter, the inverter circuits are arranged in four

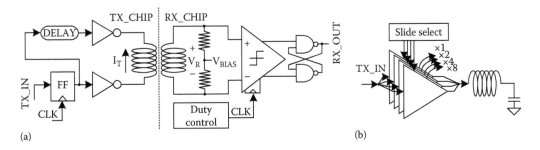

(a) (b)

FIGURE 16.1 (a) H-bridge transmitter and latch-based sense amplifier and (b) single-ended current-reuse transmitter arranged in slides to control the transmit power.

slides, which have different diving strength ratios of 1, 2, 4, and 8. Then, the transmission power is optimized by enabling only a certain number of slide combinations. The same latch-based sense amplifier is used to sample the receiving data [2].

Another work [3] reported that the power consumption of the H-bridge transmitter could be reduced by decreasing the current pulse width (τ) while keeping the slew rate unchanged. This is important since it determines the bit error rate (BER) of the signal. When the pulse width is reduced, the transmitting current (I_p) can be scaled down proportionally. As the energy is defined as $E = V_{DD} \times I_p \times \tau$, three quarters of the energy can be saved by reducing the pulse width by half. The transmitter circuit comprises of a pulse width control circuit using phase interpolators, 4-bit slew rate control with variable capacitors, and a pulse amplitude control by manipulating the driver strength. The combination of an analog high gain receiver with high slew rate clock transmission (not shown) helps the clock frequency recover precisely with only 4.8 ps-rms jitter; this clock is used to sample the data from the receiving inductor using latch-based sense amplifiers. With the precise pulse control, the transceiver power is reduced to 0.14 pJ/b, or by 20 times, compared to the previously reported design. The transceiver architecture is illustrated in Figure 16.2 [3].

In [4], serial data transmission through inductive coupling is proposed. The benefit of the serialization is twofold. An inductor coil is one of the most area-consuming parts in the implementation. Therefore, multiplexing parallel data through one link reduces the overall area usage by inductors. Having a less number of inductors can also reduce the interferences among them and the effective data rate is further improved. The block diagram of the transceiver with the serial data transmission is depicted in Figure 16.3a. An H-bridge transmitter is utilized without a delay block to increase the timing margin. The receiver is implemented by a half-V_{DD} biased inverter with a hysteresis stage composed of cross-coupled PMOS transistors to convert short voltage pulses to full-swing CMOS logics. It achieves a data rate of up to 11 Gbps with a 120 μm diameter inductor over a 15 μm communication distance.

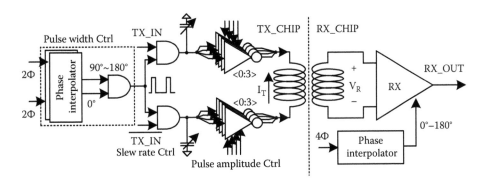

FIGURE 16.2 Transceiver with pulse width control and precise pulse shaping.

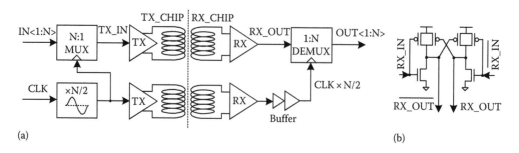

FIGURE 16.3 (a) Block diagram of the transceiver for burst data transmission and (b) receiver front-end circuit.

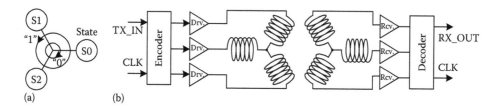

FIGURE 16.4 (a) Explanation of rotary scheme and (b) block diagram of the rotary transceiver.

Inductive coupling interconnect using a rotary coding scheme was introduced in [5]. In the rotary coding scheme, ternary symbols are sent instead of binary symbols. A single bit is represented by a shift in phase and the binary bit is indicated by the rotation direction as shown in Figure 16.4a. The clock can easily be recovered by the rotation period and thus, the power overhead in the clock data recovery (CDR) circuits such as PLL and the symbol overhead in multiple coding systems can be reduced. The block diagram of the rotary transceivers is shown in Figure 16.4b. It comprises of three pairs of inductors to allow ternary symbol transmission. An encoder or a decoder is required to convert binary to ternary symbols or vice versa.

In summary, the H-bridge transmitter is the most widely used transmitter for the inductive coupling interconnect scheme. At relatively low data speeds (~1 Gbps), the single-ended current-reuse topology and pulse shaping method are employed to enhance the power efficiency of the transmitter while employing a latch-based sense amplifier as the receiver. However, in high-speed links (~10 Gbps), analog receivers with high gains are more popular at the receiver front end to improve the signal-to-noise ratio and the jitter performance. Rotary coding can also be integrated into the transceiver design to reduce the complexity of the CDR and thus improve the overall power efficiency of the I/O systems.

16.2.2 INDUCTOR DESIGN AND IMPLEMENTATION

In inductive coupling interconnects, the voltage at the receiving inductor (V_R) is directly proportional to the mutual coupling between the two inductors (M) and the rate of current change at the transmitting inductor. The voltage at the receiving inductor (V_R) can be written as

$$|V_R| = M \frac{dI_T}{dt} = k\sqrt{L_T L_R} \frac{dI_T}{dt} \tag{16.1}$$

where L_T and L_R are the inductances of the transmitting and receiving inductors, respectively. The coupling coefficient k can be defined as

$$k = \left\{ \frac{0.25}{(X/D)^2 + 0.25} \right\}^{1.5} \tag{16.2}$$

where
    X is the distance between two inductors
    D is the diameter of the inductors [6]

In general, D is designed to be larger than X so that the received signal level is significantly larger than ambient noise for easy signal recovery.

Cross-talk issues on the inductive coupling interconnect are discussed in [7]. Two ways to minimize cross-talk issues are discussed. They are space division multiple access (SDMA) and time division multiple access (TDMA). SDMA requires to position the adjacent receiving inductor in a way that makes both outward and inward magnetic flux pass through the inductor to cancel out the induced current (Figure 16.5a). In TDMA, the adjacent transmitting and receiving inductor pairs operate with a 180° phase shift to avoid sampling cross-talk–induced voltage pulses (Figure 16.5b). It is reported that the interference-to-signal ratio (ISR) is −14 dB with an inductor pitch of 120 µm at a communication distance of 60 µm. At the communication distances of 90 and 120 µm, ISR becomes −13 and −5 dB, respectively. Therefore, precise control of the inductor pitch and communication distance is required to achieve good cross-talk cancellation with SDMA. With the combination of SDMA and TDMA, ISR remains between −15 and −16 dB across all three communication distances.

The research work in [2] presents a parallel data transmission through a 3×65 inductive channel array. This implementation applies SDMA and TDMA to reduce cross-talk noise in the inductive array. Transmission power control is built in the transceiver to optimize power consumption according to the communication distances. The total power consumption of the inductor with 50 µm diameter at the communication distances of 15, 30, and 45 µm is at 1.2, 2.2, and 4.1 W, respectively. The work also reports that the cross-talk noise is further reduced with the transmission power control and the aggregate data rate of the channels is improved by 80%.

The inductor structure called XY inductor, which can be embedded in the logic routing, was introduced in [6] (Figure 16.6a). In the conventional inductor design, a dedicated chip area is required for an inductor because the routing wires are not permitted in the inductor area.

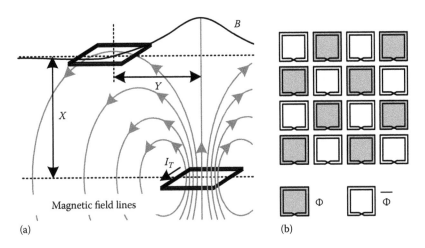

(a)   (b)

FIGURE 16.5 (a) Illustration of cross-talk–cancelled point in space division multiple access (SDMA) and (b) double-edge triggered data transmission in time division multiple access (TDMA).

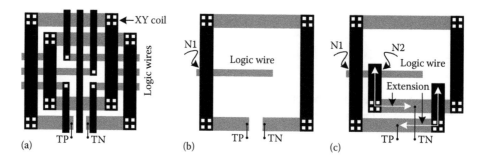

FIGURE 16.6 (a) XY coil structure, (b) XY inductor structure with noise from orthogonal logic wire, and (c) extended XY inductor to cancel cross-talk noise from (b).

The proposed XY inductor structure implemented with two different metal layers can be blended into the logic routing, which crosses the inductor layout parallely or orthogonally. This substantially reduces the area requirement of the inductors at the expense of magnetic flux loss from the eddy current at the logic wires. A power increase of 23% was necessary in [6] to compensate for such a loss. The parallel and orthogonal logic wires also impose differential cross talk to the inductor. A ground shield around the inductor isolates the parallel cross talk. The orthogonal cross talk or the differential noise ($N1$) (Figure 16.6b) is converted into the common-mode noise by supplementing with another noise source ($N2$) at the inductor extension, as illustrated in Figure 16.6c. Due to the propagation delay between the two noise sources, the amount of cancellation will vary. The length of the inductor extension also affects the noise. In [6], the optimal extension is found to be 20% of the $4 \times D$ and the differential noise from the orthogonal wire is suppressed up to 70%. With an inductor diameter of 300 μm, a communication distance of 100 μm was achieved. In this implementation, the inductor pitch is twice its diameter in order to reduce cross-talk interference. The unused inductors are left open to suppress the induced current; otherwise, it will hinder the change of magnetic flux in the receiving inductor.

Data transmission using inductive coupling across 128 stacked NAND flash memory chips is presented in [8]. To reduce the area penalty of large inductors, they are placed on top of the memory cores. By placing square inductor diagonal to bit/word lines, the capacitive and inductive interference to the accessed memory core can be significantly reduced. In this work, each die has three selectable inductor pairs with transceivers to relay the signal to every 8th chip in a top-down manner and all the dies are stacked in a spiral stair manner (Figure 16.7a). Like footsteps in a spiral staircase, the bonding pad area at each step has enough room to perform wire bonding with a package. The controller chip is at the top of the stack controlling the signal relay to the destination chip. Since mutual inductance is degraded by the eddy current in eight memory cores in the relay path, the inductor diameter is increased to 1.1 mm and the distance between every relay inductor is 240 μm. With spiral stair stacking opposed to terraced stacking [9], a higher coupling coefficient is achieved

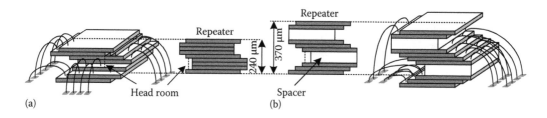

FIGURE 16.7 (a) Spiral stacking of NAND flash memory chips and (b) terraced stacking of NAND flash memory chips.

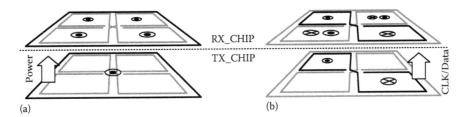

FIGURE 16.8 Illustration of clover coil for interference cancellation in simultaneous (a) power and (b) data transmission.

due to less alignment offset among inductors, and the communication distance is also reduced due to the lack of spacer (Figure 16.7b).

In all the implementations discussed so far, the chip power is supplied through bond wires. If the power supply bond wires are replaced with wireless power transmission, the whole system can be sealed completely without exposing any bonding pads that are prone to erosion. In [10], power-transferring and data-transmitting inductors are allocated side by side, which can incur relatively large interferences between the power inductors and the data inductors. To avoid this, a time-interleaved data and power transmission is adopted. To tackle the interference issue, a nested clover inductor is proposed in [11] to simultaneously transfer data and power without needing time-interleaving between the data and power transmission. Two differential clock and data inductors are rotated by 90° and embedded inside the power inductor as shown in Figure 16.8. The induced current in the clock and data links due to the power inductor is in the form of common-mode interference, which does not interfere with differential-mode signal (Figure 16.8a). Moreover, the magnetic flux interference between the data and clock inductors is cancelled out resulting in better isolation of the two inductors (Figure 16.8b). It is reported that the proposed design is able to achieve a data rate of 6 Gbps and a power transfer of 10 mW simultaneously to a distance of 20 μm. The dimension of the data and clock inductors is 300 μm × 300 μm while that of the power inductor is 700 μm × 700 μm.

16.2.3 SUMMARY

In this section, several inductor design parameters such as alignment tolerance, the effect of X/D ratio to data rate, and transceiver power are statically analyzed.

Due to the magnetic flux spreading from the center of the transmitting inductor, the alignment tolerance of inductive coupling interconnect is relaxed unlike the microbump and TSV technology. The alignment tolerance of up to 50 μm is reported in [12] with a 150 μm diameter inductor coil. Reference 13 reports that alignment tolerance is close to half the inductor diameter. However [12] reports that it is based on the measurement of a single inductor, whereas in [13], inductor pitch is about twice the inductor diameter (based on visual clue) and, thus, cross-talk interference among them is minimal. Although there is no experimental result presented in [2,6,7,14], the alignment tolerance of those closely tied inductor channels is likely to be less than that of [12,13] for the sake of cross-talk interference.

Figure 16.9a is the scattered plot of data rate against X/D ratio. The scattered points are categorized into two groups based on the nature of the implemented channel; they are single link and an array of links. In general, the majority of the single links shown in Figure 16.9a are found to be of a higher data rate compared to the array of links due to the minimum amount of interference in the system. The trend line for the single link also suggests that an exponential improvement in data rate can be achieved by reducing the X/D ratio. However, for the array of links implementation, the relationship between data rate and X/D ratio is linear, and data rate improvement at lower X/D is also insignificant.

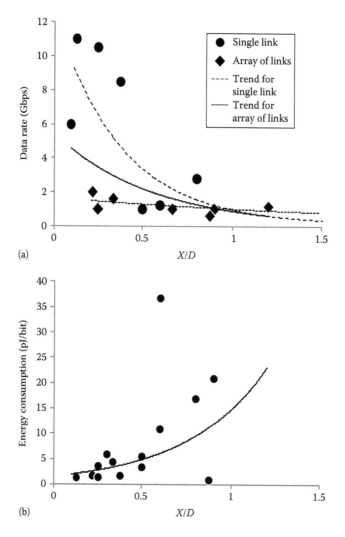

FIGURE 16.9 (a) Scattered plot of *X/D* versus data rate and (b) scattered plot of *X/D* versus energy consumption.

The relationship between *X/D* ratio and transceiver power consumption is illustrated in Figure 16.9b based on the available data. The trend line in the figure supports the theory of more transmission power for longer communication distance regardless of whether the links are single or array type.

According to Figure 16.9a and b, selected *X/D* ratios in many implementation are hugely varied depending on the power budget and the required data rate. The optimal *X/D* ratio for both power and data rate can be found by intersecting two trend lines from two figures. These two trend lines from Figure 16.9a and b are expressed in Equations 16.3 and 16.4, respectively, and Equation 16.4 is normalized according to Equation 16.3:

$$y = 5.41 \times e^{-1.79x} \tag{16.3}$$

$$y = 0.48 \times e^{2.21x} \tag{16.4}$$

The intersection of two trend lines reveals an optimal X/D ratio of 0.6, which can provide both reasonable energy efficiency (6.1 pJ/bit) and data rate (1.8 Gbps).

16.3 HIGH-SPEED INTERCONNECTS FOR 3D-ICs UTILIZING CAPACITIVE COUPLING

Capacitive coupling interconnect is a voltage-driven method, while inductive coupling interconnect is a current-driven one. Due to the proximity requirement, the capacitive coupling interconnect can be realized only in face-to-face die stacking. It has relatively low parasitic capacitance due to its relatively small electrode sizes and the lack of ESD structure. Therefore, the transceivers can operate at low power. In addition, the interconnect density of the capacitive coupling scheme is higher than that of the wire bonding, the microbump, and the inductive coupling scheme [15]. In this section, the modeling of the capacitive coupling interconnects and various capacitive coupling transceivers are discussed.

16.3.1 MODELING OF CAPACITIVE COUPLING INTERCONNECT

Capacitive coupling interconnect is formed when two dies are stacked together in a face-to-face manner [16]. The top metal layer from each die becomes an electrode to realize a coupling capacitor. Signal is transferred from one electrode to the other through capacitive coupling and the amount of coupling is determined depending on the coupling ratio (H) between the coupling capacitance (C_C) and parasitic capacitance (C_P) at the receiving electrode [17]. The electrical model of the capacitive coupling is illustrated in Figure 16.10. The coupling ratio of the interconnect (H) is defined as follows:

$$H = \frac{v_{RX}}{v_{TX}} = \frac{C_C}{C_C + C_P} \tag{16.5}$$

C_P includes both the electrode to substrate capacitance and the gate capacitance of the receiver's input devices. For better signal integrity, a larger coupling ratio is preferred. This can be achieved by increasing C_C and minimizing C_P. Deep submicron CMOS process is beneficial not only because it can switch faster but also because it has small gate capacitance achieving higher H value. C_C can also vary by alignment mismatch in die stacking, which directly affects the coupling ratio. The misalignment issues are discussed in Section 16.3.2.

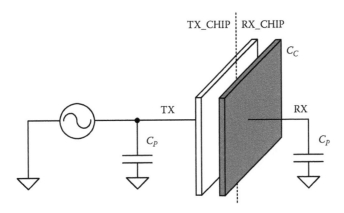

FIGURE 16.10 The parasitic model of the capacitive coupling interconnect.

16.3.2 Alignment

Alignment is critical in capacitive coupling interconnect since it directly affects the coupling ratio and accordingly the signal integrity. In [18], misalignment caused by six degrees of freedom in die stacking is reported. They include three degrees of translation along X, Y, and Z axes and three degrees of rotation at three axes such as yawing (θ), rolling (β), and pitching (α). Among them, the capacitive coupling interconnect, in particular, is most sensitive to Z-axis variation due to thickness variation in dielectric material between electrodes. This directly changes the coupling ratio of the interconnect [18]. Although the impact of the dielectric thickness variation can be mitigated by using overdesigned coupling capacitance values, the variations can be measured by sensing the capacitance of the interconnect using a rectifier circuit as presented in [19] (Figure 16.11). Measuring the capacitance at each corner of a die can reveal the rotation caused by β and α. Several design methods to tackle the misalignment in X, Y, and θ are also proposed in the literature [18,20,21].

The offset in the X, Y, and θ degrades the coupling ratio. When it is large enough, overlapping of electrodes with unintended neighboring electrodes occurs that jeopardizes the signal integrity by the degraded coupling ratio and the cross talk [18]. Moreover, thermal expansion of dies at a high operating temperature is another issue to be tackled when the process technologies of two stacked dies are different [21]. One way to deal with the alignment offset and the thermal expansion is to design the transmitting electrode larger than the receiving electrode so that the overlapping area is maintained under any offset circumstances. For example, if the area of the receiving electrode is $A_{RX} = x^2$ μm², the transmitting electrode area should be greater than or equal to $A_{TX} = [x \times (1 + (\alpha_{TX} - \alpha_{RX}) \times \Delta T) + \sigma]^2$ μm² across the operating temperature range. α_{TX} and α_{RX} are the coefficients of the linear expansion of TX and RX dies, respectively, and σ is the alignment tolerance. Larger receiving electrodes are not preferred since they increase parasitic capacitance and accordingly degrade the coupling ratio.

Figure 16.12a depicts an on-chip Vernier measurement system for confirming chip alignment [18]. Each system consists of 10 transmitting electrode bars with a 12.6 μm pitch and 9 receiving electrode bars (shown in gray color) with a 14 μm pitch. The difference in the pitch of the transmitting bars and that of the receiving bars is the resolution of the system. Although adjacent transmitting bars are alternatively sending complementary data bits, the data pattern at the receivers will be varied depending on the offset in two dies. In perfectly aligned stacking, the central (fifth) receiving bar shows null detection in which no signal transition is detected. By employing two Vernier systems horizontally and vertically, it is possible to detect not only the lateral shift in X- and Y-axes but also the rotation in θ.

An on-chip electronic alignment system to correct the offset after die stacking is proposed in [20] (Figure 16.12b). The system is design to correct X and Y transitions. The main idea is to use multiple microelectrodes instead of one to direct the signal to a receiving electrode. A group of selectable (4 × 4) microelectrodes serves as a transmitting electrode, and suitable microelectrodes are selected through a network of multiplexers to minimize the offset. In the implemented system,

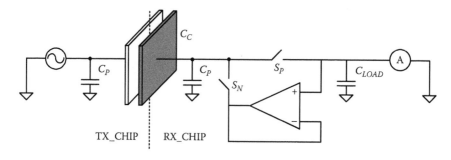

FIGURE 16.11 A rectifier circuit to measure the coupling capacitance C_C.

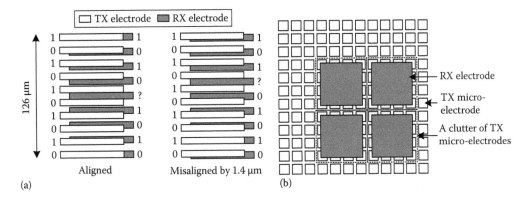

(a)   (b)

FIGURE 16.12 (a) The interconnect layout of the Vernier alignment system and (b) the electronic alignment system with microelectrodes to drive signal to RX electrode.

each microelectrode has a pitch of 12.5 μm, which is a quarter of the receiving electrode and the alignment offset can be corrected up to 6.25 μm.

In summary, the misalignment in X and Y directions can be evaluated and corrected even after die stacking. However, Z, θ, β, and α misalignment can only be measured without correction. Thus, overdesigning is still necessary to provide an acceptable yield from die stacking under misalignment.

16.3.3 Cross Talk

Capacitive coupling is prone to cross talk when electrodes are misaligned and neighboring intercon-nects are in close proximity. Cross talk due to proximity is the focus of this section.

Cross-talk components in the capacitive coupling interconnect is illustrated in Figure 16.13. When two or more interconnects are closely placed to improve the interconnect density, the effect of cross-coupling components such as C_{TR} and C_{RX} become apparent [22] (Figure 16.13). They not only reduce the coupling ratio of the interconnect but also cause more cross talk. The noise from the cross talk modulates the amplitude of the signal and subsequently causes the receiver output to jitter. In a worst-case scenario, signal recovery can fail entirely. In this section, various countermeasures to cross talk are described.

One way to deal with cross talk in a single-ended interconnect array is to block cross-coupling components with grounded wires, as shown in Figure 16.14. In this arrangement, the wires are directly inserted between the RX electrodes to minimize C_{RX}. Using extended TX electrodes over the grounded wires also significantly reduces the fringing capacitance or C_{TR} from neighboring TX electrodes [18].

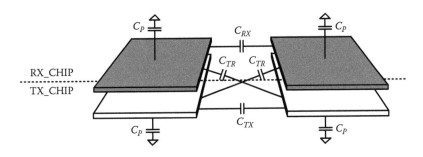

FIGURE 16.13 Cross-talk components between two interconnects.

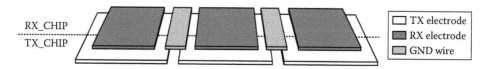

FIGURE 16.14   Shielding of cross-talk components with ground wire between RX electrodes.

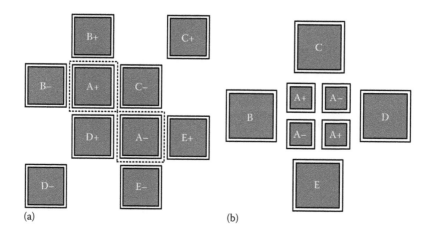

FIGURE 16.15   Cross-talk cancellation in (a) butterfly differential interconnects array and (b) hybrid array.

Another way of mitigating cross talk is to use a differential interconnect array structure. Figure 16.15a describes the butterfly differential array structure [20] illustrating how the cross talk in channel A is cancelled while surrounded by other channels (B, C, D, and E). Among them, channel B and channel E do not contribute any noise to channel A because the coupled noise from B+ and E+ is cancelled out by the respective complementary noise. The remaining channels (C and D) introduce common-mode noise to channel A, which can be rejected by adopting a differential receiver.

Another approach is to use a hybrid interconnect array as reported in [22]. The hybrid interconnect array has both single-ended and differential interconnects interleaved together, as illustrated in Figure 16.15b. The differential interconnect A is in the common-centroid form, which exposes an equal area of differential electrodes to every surrounding single-ended interconnects (B, C, D, and E). In this array arrangement, none of the single-ended interconnects receive any cross-talk noise as the noises they receive are always complementary and are automatically cancelled out. Moreover, the cross talk from the single-ended interconnects to the differential interconnects is common mode in nature, which can also be rejected by a differential receiver.

In summary, the methods of cross-talk countermeasure can be categorized into two groups, that is, passive approach and active approach. In the passive approach, grounded wires are inserted between the RX electrodes to shield the cross-coupled components while using conventional single-ended receivers. In the active approach, some cross-talk noise is cancelled out by their respective complementary noise components and the residual common-mode noise is rejected by the differential receiver.

16.3.4   TRANSCEIVER DESIGNS

The capacitive coupling interconnect is voltage driven and a two-staged static inverter can be employed as a transmitter. This facilitates simple transmitter design with less power consumption compared to H-bridge transmitters from inductive coupling. For differential signal transmission, a

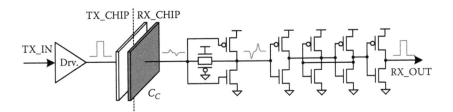

FIGURE 16.16   Illustration of a typical single-ended receiver for capacitive coupling interconnects.

single-ended signal is converted to a differential signal. Three stages of inverters are used to generate a complementary signal, whereas the signal itself goes through two-staged inverters with a delay inserted in between so that two complementary signals at the transmitter outputs have an exact 180° phase shift.

Figure 16.16 explains an inverter-based amplifying receiver with resistive feedback. The gain of the first inverter is very high because its input is biased at its switching threshold through the resistive feedback implemented by a transmission gate. When it is integrated to a capacitive coupling interconnect, it produces voltage pulses at its output whenever there is a data transition at the input. The pulses are further amplified by another inverter before registering it to a latch at the subsequent stage. This design is resilient to inter-symbol-interference (ISI) as long as the duration of the non-return-to-zero (NRZ) pulses from the receiver output is less than the symbol period. It has been demonstrated that its high-speed capability can achieve up to 10 Gbps in 0.18 μm CMOS technology [23]. Another variant of this receiver is presented in [24]. It uses a diode-connected MOS feedback rather than the resistive type. Its operating frequency is less than that in a resistive feedback type because it directly converts the coupled signal close to the full-swing digital logic.

A bistable receiver topology (Figure 16.17) is proposed to reduce the static current consumption at the front-end inverter [16]. This is particularly important in parallel data transmission where the data rate is not very high. In this case, static power overwhelms dynamic power in the aforementioned topology. Instead of using resistive feedback, the input of the front-end inverter is attached to diode-connected PMOS and NMOS to level the received signal at V_{TH} for "0" or $V_{DD} - V_{TH}$ for "1" after data transition. Because the diode-connected MOS has lower threshold compared to that of the transistors at the inverter, either PMOS or NMOS of the inverter is always biased in the subthreshold region. Therefore, the static power at the inverter is dramatically reduced. Moreover, the dual feedback path prevents static current from flowing in the diode-connected devices after data recovery. In this design, the speed of signal recovery is primarily limited by the delay in the dual feedback path.

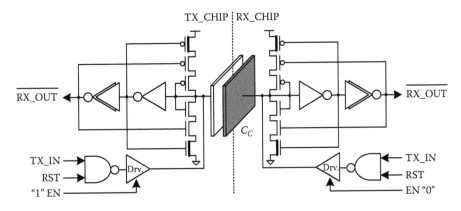

FIGURE 16.17   A bistable topology for single-ended receiver with bidirectional capabilities.

Transceivers can also be configured for bidirectional transmission (Figure 16.17) by employing a transceiver at each side of the interconnect and a control is used to enable or disable an appropriate set of a transmitter and a receiver according to the direction of data transmission. In this approach, the power consumption is higher than that of unidirectional transmission due to larger parasitic capacitance associated at the *TX* and *RX* electrodes. Its maximum data rate is also degraded due to the reduction of coupling ratio from the additional parasitic capacitance.

Recently, a simultaneous bidirectional transmission scheme using three cascaded capacitors was proposed in [25] (Figure 16.18a). In this scheme, four levels of voltages are formed (Figure 16.18b) at the receiving electrodes depending on the transmitting bits (TX1 and TX2) from both sides of the interconnect (namely, "00," "01," "10," and "11"). Two receivers (H and L) are used to track the lower voltage swings (00 and 01) and the upper voltage swings (10 and 11), respectively. When the transmitter transmits logic "1," the upper receiver (H) at the same side is activated; when logic "0" is transmitted, the lower receiver (L) is enabled. Both receiver outputs are wired to 2 × 1 MUX, which is controlled by the transmitting bit to select the correct received data (Figure 16.18c). The voltage clamp circuit is used to maintain the voltage level at the receiving electrode when the data bit does not change for a long period.

Differential receivers are preferred in cross-talk–cancelled interconnects such as the butterfly array [20] and the hybrid array [22] due to the residual common-mode noise mentioned in the previous section. In the butterfly array, a latch-based sense amplifier is implemented for signal recovery; however in the hybrid array, a self-biased fully differential amplifier is employed. The later has two cross-coupled inputs differential amplifiers (Figure 16.19a) to minimize input offset and its current mirror load provides high common mode rejection ratio (CMRR) that is very desirable in the hybrid interconnect array. Figure 16.19b presents the voltage waveforms at the inputs and outputs of the receiver. Inputs are self-biased after the data transition within t_{SB} and the speed of this receiver is limited by t_{SB}.

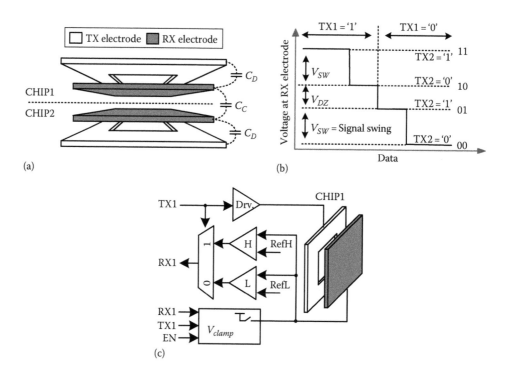

FIGURE 16.18 (a) Three cascaded capacitors interconnect configuration for simultaneous bidirectional signaling, (b) four-level signaling at the interconnect (a), and (c) a transceiver block diagram for simultaneous bidirectional signaling.

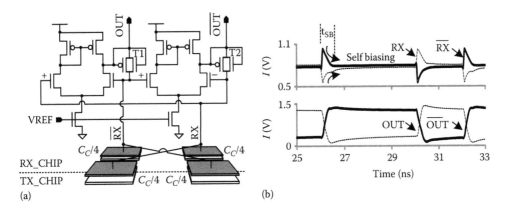

FIGURE 16.19 (a) Self-biased fully differential receiver for the hybrid capacitive coupling array and (b) the signaling voltages at (a).

In general, the capacitive coupling interconnect schemes are more beneficial in parallel data transmission because their electrode sizes are relatively smaller than those of the inductive coupling ones. Therefore, systems requiring multibits data transmission in parallel, such as memory systems, are promising applications. Single-ended interconnects are mainly used where the data rate is not high enough to cause significant signal integrity issue. Differential interconnects are indispensible when cross-talk cancellation is required due to the design constraints in area.

16.3.5 SUMMARY

In face-to-face integration, the capacitive coupling interconnect (CCI) has several advantages over the inductive coupling interconnect (ICI). First of all, the size of CCI is smaller than that of ICI in general. Second, the cross-talk issue in CCI is less prominent than in ICI and, thus, CCI can achieve higher interconnect density, providing more communication parallelism. CCI is also scalable with technology scaling, as shown in Figure 16.20a. The exponential improvement in data rate per square area of the interconnect size is observed when the technology is scaled. Moreover, its transceiver power consumption is scaled down dramatically (Figure 16.20b) as it moves from one technology node to another due to supply voltage scaling and gate capacitance scaling, which results in higher coupling ratio of the interconnect relaxing the voltage gain requirement at the receiver.

16.4 HIGH-SPEED INTERCONNECTS FOR SILICON INTERPOSER-BASED 2.5D-ICs

The 2.5D-ICs have been considered as an intermediate technology before moving toward the true 3D-ICs integration with TSV. With the introduction of silicon-based interposer, its dense back end of line (BEOL) copper wires provide fine-pitch high data rate signal lines for local chip-to-chip communication. The bump size can be reduced down to 50 μm diameter since there is no mismatch in the coefficient of thermal expansion between the interposer and the chips. Figure 16.21 depicts such an example where two flip chips are mounted on a silicon carrier. However, such a fine-pitch wire has inherent channel losses at high data rates leading to significant intersymbol interference (ISI). Therefore, channel equalization is necessary [26].

The channel insertion loss is 17 dB at 5 GHz [26] when the striped line has length = 20 mm, width = 1.2 μm, thickness = 1.2 μm, spacing = 1.2 μm, and distance to ground = 1.6 μm. To increase the high-frequency performance, a low-impedance (10–20 Ω) transmitter together with

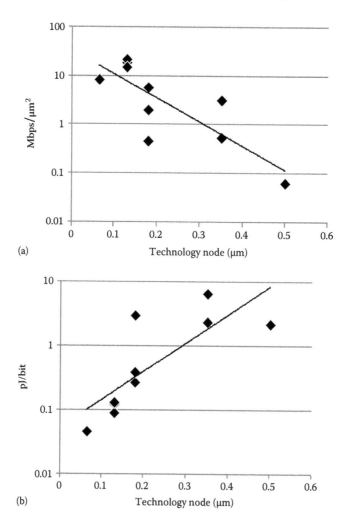

FIGURE 16.20 (a) Area efficiency versus technology node and (b) power efficiency versus technology node.

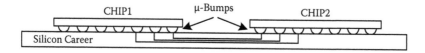

FIGURE 16.21 Cross section of 2.5D integration using silicon-based interposer.

a high-impedance (2 kΩ) receiver termination is used for the I/O transceiver. Although the high-impedance termination allows reducing static power consumption, impedance mismatch creates signal reflection; however, in this lossy channel, such reflections are found to be insignificant when the propagation delay is below 0.5 UI. At the receiver side, a combination of one discrete tap decision feedback equalizer (DFE) and the adjustable continuous-time infinite impulse response (IIR) filter is adopted instead due to the higher taps requirement in conventional DFE architectures. An 8 bit bus-level architecture is presented in Figure 16.22, and the system consumes 4.3 pJ/bit over 20 mm strip line having a 6 μm width, 3 μm thickness, 4 μm spacing, and 3 μm distance to ground [27].

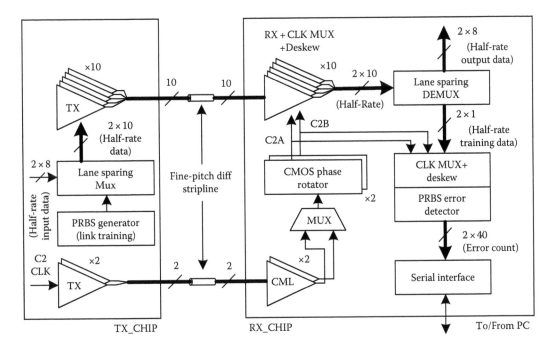

FIGURE 16.22   8-bit bus architecture with two redundant channels.

16.5   SUMMARY

In this chapter, a brief introduction of TSV-less three main interconnect technologies for 3D/2.5D integration is presented. The wireless proximity communication known as capacitive coupling and inductive coupling interconnects is for 3D integration and silicon interposer with fine-pitch BEOL copper wires is for 2.5D integration. The capacitive coupling interconnect is mainly for face-to-face integration of two stacked dies due to proximity requirement between two electrodes; however, the inductive coupling approach can reach larger communication distances allowing communication between more than two tier. Despite the larger communication range of the inductive coupling approach, capacitive coupling schemes are more attractive, especially in face-to-face integration due to their lower power consumption feature. Moreover, the cross talk in capacitive coupling schemes can be canceled easily with passive and active approaches without scarifying data rate, whereas in the inductive coupling approach, data rate degradation is expected even with cross-talk countermeasures (Figure 16.9a). Silicon-based interposer for 2.5D integration provides high I/O counts compared to traditional wire bonding and C4 bumps due to the use of microbump with a diameter as small as 50 μm. However, due to the long wire length and lossy nature of the channel, high-speed I/Os systems require additional processing such as DFE and IIR filter like conventional long-distance data links and, thus, their power consumption is an order of magnitude higher than in wireless approaches in 3D integration.

REFERENCES

1. D. Mizoguchi, Y. B. Yusof, N. Miura, T. Sakura, and T. Kuroda, A 1.2 Gb/s/pin wireless superconnect based on inductive inter-chip signaling (IIS), in *IEEE International Solid-State Circuits Conference Digest of Technical Papers*, San Francisco, CA, Vol. 1, 2004, pp. 142–517.
2. N. Miura, D. Mizoguchi, M. Inoue, H. Tsuji, T. Sakurai, and T. Kuroda, A 195 Gb/s 1.2 W 3D-stacked inductive inter-chip wireless superconnect with transmit power control scheme, in *IEEE International Solid-State Circuits Conference Digest of Technical Papers*, San Francisco, CA, Vol. 1, 2005, pp. 264–597.

3. N. Miura, H. Ishikuro, T. Sakurai, and T. Kuroda, A 0.14 pJ/b inductive-coupling inter-chip data transceiver with digitally-controlled precise pulse shaping, in *IEEE International Solid-State Circuits Conference Digest of Technical Papers*, San Francisco, CA, 2007, pp. 358–608.

4. N. Miura, Y. Kohama, Y. Sugimori, H. Ishikuro, T. Sakurai, and T. Kuroda, A high-speed inductive-coupling link with burst transmission, *IEEE J. Solid-State Circ.*, 44, 947–955, 2009.

5. A. Radecki, N. Miura, H. Ishikuro, and T. Kuroda, Rotary coding for power reduction and S/N improvement in inductive-coupling data communication, in *IEEE Asian Solid State Circuits Conference*, Jeju, Korea, 2011, pp. 205–208.

6. M. Saito, Y. Yoshida, N. Miura, H. Ishikuro, and T. Kuroda, 47% power reduction and 91% area reduction in inductive-coupling programmable bus for NAND flash memory stacking, in *IEEE Transactions on Circuits and Systems I: Regular Papers*, Vol. 57, 2010, pp. 2269–2278.

7. N. Miura, D. Mizoguchi, T. Sakurai, and T. Kuroda, Cross talk countermeasures in inductive inter-chip wireless superconnect, in *Proceedings of IEEE Custom Integrated Circuits Conference*, San Jose, CA, 2004, pp. 99–102.

8. M. Saito, N. Miura, and T. Kuroda, A 2 Gb/s 1.8 pJ/b/chip inductive-coupling through-chip bus for 128-Die NAND-Flash memory stacking, in *IEEE International Solid-State Circuits Conference, Digest of Technical Papers*, San Francisco, CA, 2010, pp. 440–441.

9. Y. Sugimori, Y. Kohama, M. Saito, Y. Yoshida, N. Miura, H. Ishikuro, T. Sakurai, and T. Kuroda, A 2 Gb/s 15 pJ/b/chip inductive-coupling programmable bus for NAND Flash memory stacking, in *IEEE International Solid-State Circuits Conference Digest of Technical Papers*, San Francisco, CA, 2009, pp. 244–245, 245a.

10. Y. Yuan, N. Miura, S. Imai, H. Ochi, and T. Kuroda, Digital rosetta stone: A sealed permanent memory with inductive-coupling power and data link, in *Symposium on VLSI Circuits*, Honolulu, HI, 2009, pp. 26–27.

11. Y. Yuxiang, A. Radecki, N. Miura, I. Aikawa, Y. Take, H. Ishikuro, and T. Kuroda, Simultaneous 6 Gb/s data and 10 mW power transmission using nested clover coils for non-contact memory card, in *IEEE Symposium on VLSI Circuits*, 2010, Kyoto, Japan, pp. 199–200.

12. W. R. Davis, J. Wilson, S. Mick, J. Xu, H. Hua, C. Mineo, A. M. Sule, M. Steer, and P. D. Franzon, Demystifying 3D ICs: The pros and cons of going vertical, *IEEE Design Test Comput.*, 22, 498–510, 2005.

13. A. Iwata, M. Sasaki, T. Kikkawa, S. Kameda, H. Ando, K. Kimoto, D. Arizono, and H. Sunami, A 3D integration scheme utilizing wireless interconnections for implementing hyper brains, in *IEEE International Solid-State Circuits Conference Digest of Technical Papers*, San Francisco, CA, Vol. 1, 2005, pp. 262–597.

14. N. Miura, D. Mizoguchi, M. Inoue, K. Niitsu, Y. Nakagawa, M. Tago, M. Fukaishi, T. Sakurai, and T. Kuroda, A 1 Tb/s 3 W inductive-coupling transceiver for inter-chip clock and data link, in *IEEE International Solid-State Circuits Conference Digest of Technical Papers*, San Francisco, CA, 2006, pp. 1676–1685.

15. K. Kanda, D. D. Antono, K. Ishida, H. Kawaguchi, T. Kuroda, and T. Sakurai, 1.27 Gb/s/pin 3 mW/pin wireless superconnect (WSC) interface scheme, in *IEEE International Solid-State Circuits Conference Digest of Technical Papers*, San Francisco, CA, Vol. 1, 2003, pp. 186–487.

16. A. Fazzi, R. Canegallo, L. Ciccarelli, L. Magagni, F. Natali, E. Jung, P. Rolandi, and R. Guerrieri, 3-D capacitive interconnections with mono- and bi-directional capabilities, *IEEE J. Solid-State Circ.*, 43, 275–284, 2008.

17. A. Fazzi, L. Magagni, M. Mirandola, B. Charlet, L. Di Cioccio, E. Jung, R. Canegallo, and R. Guerrieri, 3-D capacitive interconnections for wafer-level and die-level assembly, *IEEE J. Solid-State Circ.*, 42, 2270–2282, 2007.

18. R. J. Drost, R. D. Hopkins, R. Ho, and I. E. Sutherland, Proximity communication, *IEEE J. Solid-State Circ.*, 39, 1529–1535, 2004.

19. A. Chow, D. Hopkins, R. Ho, and R. Drost, Measuring 6D chip alignment in multi-chip packages, in *Sensors, 2007 IEEE*, Atlanta, GA, 2007, pp. 1307–1310.

20. D. Hopkins, A. Chow, R. Bosnyak, B. Coates, J. Ebergen, S. Fairbanks, J. Gainsley et al., Circuit techniques to enable 430 Gb/s/mm^2 proximity communication, in *IEEE International Solid-State Circuits Conference Digest of Technical Papers*, San Francisco, CA, 2007, pp. 368–609.

21. R. Drost, R. Ho, D. Hopkins, and I. Sutherland, Electronic alignment for proximity communication, in *Solid-State Circuits Conference, 2004. Digest of Technical Papers. ISSCC. 2004 IEEE International*, San Francisco, CA, Vol. 1, 2004, pp. 144–518.

22. A. Myat Thu Linn, E. Lim, T. Yoshikawa, and T. T. Kim, Design of self-biased fully differential receiver and crosstalk cancellation for capacitive coupled vertical interconnects in 3DICs, in *IEEE International Symposium on Circuits and Systems*, Beijing, China, 2013, pp. 966–969.

23. G. Qun, X. Zhiwei, K. Jenwei, and C. Mau-Chung Frank, Two 10 Gb/s/pin Low-Power Interconnect Methods for 3D ICs, in *IEEE International Solid-State Circuits Conference Digest of Technical Papers*, San Francisco, CA, 2007, pp. 448–614.

24. S. A. Kuhn, M. B. Kleiner, R. Thewes, and W. Weber, Vertical signal transmission in three-dimensional integrated circuits by capacitive coupling, in *IEEE International Symposium on Circuits and Systems*, Seattle, WA, Vol. 1, 1995, pp. 37–40.

25. A. Myat Thu Linn, E. Lim, T. Yoshikawa, and T. T. H. Kim, A 3-Gb/s/ch simultaneous bidirectional capacitive coupling transceiver for 3DICs, in *IEEE Tran. Circuits and Systems II: Express Briefs*, 2014, Vol. 61, pp. 706–710.

26. K. Byungsub, L. Yong, T. O. Dickson, J. F. Bulzacchelli, and D. J. Friedman, A 10-Gb/s compact low-power serial I/O with DFE-IIR equalization in 65-nm CMOS, *IEEE J. Solid-State Circ.*, 44, 3526–3538, 2009.

27. T. O. Dickson, L. Yong, S. V. Rylov, B. Dang, C. K. Tsang, P. S. Andry, J. F. Bulzacchelli et al., An 8× 10-Gb/s source-synchronous I/O system based on high-density silicon carrier interconnects, *IEEE J. Solid-State Circ.*, 47, 884–896, 2012.

17 Challenges and Future Directions of 3D Physical Design

Johann Knechtel, Jens Lienig, and Cliff C.N. Sze

CONTENTS

ABSTRACT

The concept of 3D integrated circuits (3D-ICs) provides new opportunities for meeting current and future design criteria, such as performance, functionality, delay, and power consumption. 3D-ICs are thus considered as a promising approach to spur both *More Moore* (i.e., further down-scaling of baseline CMOS device nodes) and *More-than-Moore* (i.e., diversification of functionality; heterogeneous system integration) [1,9] as shown in Figure 17.1. At the same time, 3D-ICs increase complexity for manufacturing and physical design notably.

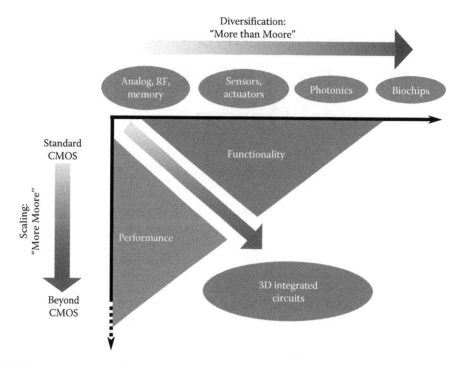

FIGURE 17.1 Besides the well-known trend for downscaling device nodes, slowly but surely reaching its limits for CMOS technology, the need for diversification has been acknowledged [9]. The concept of 3D integrated circuits is considered a promising option to combine both avenues.

In the previous chapters, various 3D design and application approaches have been discussed in detail. In this final chapter of the book, we aim to revise key technological and design challenges, and to point out prospective directions for further adoption of 3D-ICs.

17.1 KEY CHALLENGES FOR 3D CIRCUITS AND APPROACHES REVISED

The challenges for 3D circuits, as for any electrical device in general, can be classified into technological challenges and design challenges. In this section, we highlight both challenges and related approaches to provide an overview of the current state-of-the-art in 3D-IC design.

17.1.1 TECHNOLOGICAL CHALLENGES

Although much progress has been achieved in the recent years, technological challenges are still a hindering factor for mainstream adoption of 3D-ICs.

In the following, we first review different integration approaches for 3D devices, revealing the need for early and proper analysis of suitable technologies for 3D integration. Then, further key challenges, such as power delivery and thermal management, clock delivery and testing are reviewed.

17.1.1.1 Diversity of Integration Approaches

3D integrated devices are typically realized deploying one of the following approaches: package stacking, interposer-based packaging, through-silicon-via (TSV)-based 3D integration or monolithic 3D integration (Figure 17.2). Each of these approaches has its own scope of application, benefits and drawbacks, and requirements for design and manufacturing processes [108,128]. In the following, these approaches are briefly reviewed. Note that package stacking is not reviewed here since its concept is considered a well-known approach.

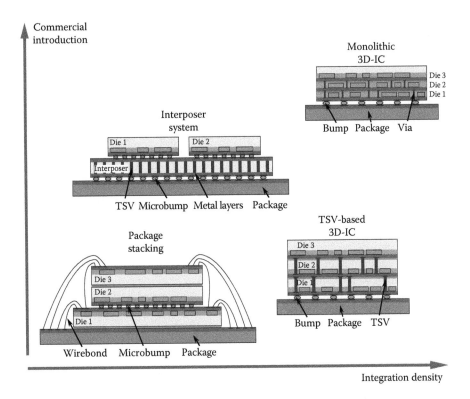

FIGURE 17.2 Evolution of 3D-integration technology. Originating with package stacking, 3D integration has evolved through interposer-based systems (also known as "2.5D integration") toward TSV-based and monolithic 3D-ICs. Both integration density as well as design and manufacturing complexity have increased during this process. Furthermore, heterogeneous integration (mainly memory and logic) is the current key scope of application; homogeneous logic-on-logic integration is not yet foreseeable.

Interposer-based 3D packages are acknowledged as an cost-efficient driver toward 3D integration [67,98,140]. For such 3D packages, usually predesigned dies are stacked in lateral and/or vertical fashion on silicon carriers—the interposers—which comprise metal layers and TSVs for connectivity. Interposers are mainly realized as passive carriers, but can also include embedded components like decoupling capacitors or even glue logic [67]. Interposers support various integration scenarios and applications and are thus widely acknowledged in the industry.

The integration density of interposer-based 3D packages is the lowest among the approaches discussed here. Furthermore, the seemingly straightforward design of such packages is obstructed by the current lack of appropriate design tools [78]. For example, routing an interposer with its few metal layers, typically used to full capacity, falls short of expectations while using current tools. Further, planning and verification of signal integrity across the different domains in interposer-based devices is not sufficiently supported yet.

TSV-based 3D integration has evolved as a "prominent approach" for 3D integration; many researches and industry prototypes are based on TSV technology nowadays, for example, [3,15,29,37, 56,102,124,126,141]. TSVs are metal plugs (mainly made of copper or tungsten) running through silicon dies, which are stacked and bonded, in order to interconnect them. Depending on the type/fabrication of TSVs, different design obstacles are occurring. Via-first and -middle TSVs occupy the active layer, thus resulting in placement obstacles; via-last TSVs and TSVs fabricated after bonding occupy the active layer as well as the metal layers, resulting in placement and routing obstacles [1,57]. There are manifold stacking configurations available, each having its advantages as well as disadvantages [1,128]. The classification mainly comprises wafer-to-wafer, die-to-wafer, and die-to-die stacking.

The concept of TSVs enables chip-level integration, while retaining the benefits of package-level integration [128]. Thus, TSVs are key enablers for 3D integration, as also proven by ample TSV-based prototypes, indicated earlier. Both heterogeneous and homogeneous 3D integration can be realized with TSV-based integration—an important feature to increase acceptance in the industry for such new technologies.

The integration density of TSV-based 3D devices is lower than that of monolithic 3D devices, but higher than that of interposer-based 3D packages. Due to their relatively large size and intrusive character, TSVs cannot be deployed excessively but have to be rather optimized in count and arrangement [63,64]. It is also notable that TSVs do not scale at the same rate as transistors, thus the TSV-to-cell mismatch will likely remain for future nodes and may even increase [101].

Monolithic 3D integration is recently gaining more attention [72,73,104], mainly thanks to advances in process technologies [13]. Active layers are built up sequentially rather than processed in separate and subsequently bonded dies. Due to very small vertical interconnects, monolithic integration enables fine-grain transistor-level integration; it provides the highest integration density among the three mentioned approaches. However, monolithic 3D-ICs also face further challenges, for example, the need for tools and knowledge for a low-temperature manufacturing process [13], or increased delays along with routing congestion [72,81].

Monolithic integration is nevertheless a promising approach, especially for high-density logic-integration [72]. Note that thermal management is even more challenging in such high-density logic-integration scenarios than it already is for "classical" 3D integration. A recent study by Samal et al. [113] has shown, however, that monolithic integration is superior in terms of heat dissipation compared to TSV-based integration.

Choosing the proper 3D-integration approach for a particular design is much more complex than handling the decisions typically required for classical 2D-ICs, for example, the selection of device nodes and packaging concept. As indicated earlier, each integration approach has its scope of application (mainly defined by the integration granularity) and benefits and drawbacks. Furthermore, given an abstract design description in early planning phases, one has to consider the following problems:

- In how many dies should the design be split up, and what technology/device node should be considered for each die?
- How is the system functionality of the design performing when it is split up into several components spread across separate dies? What is the appropriate partitioning strategy?
- How many interconnects are required between the components/subsystems, and what are the requirements for signal transfer? Which interconnect, bonding, and packaging technologies are applicable?
- How can the sub-systems and the overall device be tested?
- What package concept should be applied? Are constraints, such as thermal design, power, and signal integrity, met with the chosen package?
- In what order are the manufacturing steps to be conducted? Which manufacturing party is responsible for what deliverable?

Some of these earlier problems are interacting, and each respective decision does impact the overall performance, reliability and cost of the final 3D chip. It is apparent that addressing these complex problems requires experienced designers and well-defined project structures. Further, design tools that enable (1) a fast, yet accurate, exploration of the technological design space and (2) rapid evaluation of different configurations are crucial. Such tools have only recently become available; further details are discussed in Section 17.2.1.

17.1.1.2 Power Delivery and Thermal Management

The key advantage of 3D integration—high integration density thanks to vertical stacking of active layers or dies—also gives rise to significant challenges for power delivery and thermal management.

Assuming that a 3D-IC has d dies or active layers stacked, its potential power consumption is d times that of a classical 2D IC comprising one die. This is not entirely true, for example, because signal and clock interconnects have a large power consumption* but are much shorter in 3D-ICs than in 2D-ICs, which directly translates into power savings. Still, the *power density* and—along with it—the heat-flow density are notably increased in 3D-ICs. Power delivery is further impacted by the technological implications of stacking multiple dies: power/ground (P/G) TSVs contribute additional, notable resistance and inductance to the power-supply network [38]. Thermal management is similarly complicated by vertical stacking of dies: "thermal barriers" between dies are introduced, which are the layers required for die bonding. These bonding layers are, for example, made of BCB adhesive polymers, which have a thermal resistivity several hundred times that of silicon dies [105].

To account for these major challenges, many studies and approaches for power delivery and thermal management have been proposed. In the following, the most relevant are briefly explained. **Power-delivery networks for 3D-ICs** should be designed considering the following:

- *Proper arrangement of TSVs:* Studies by Healy et al. [38,39] point out that a distributed topology for P/G-TSVs is superior to both single, large TSVs and groups of clustered TSVs. These and other studies, for example, [19,50], also favor irregular TSV placement, in particular such that regions drawing significant current can exhibit a higher TSV density.
- *Optimized power-delivery architectures:* To limit package impedance and external current supply, one can bring DC–DC converters closer to the logic circuitry, as demonstrated in [122] with a dedicated DC–DC die. Another possible architecture is the "multi-story power delivery" [46], where several power domains/supplies (e.g., one per die) reduce the respective load compared to a single, classical power supply. In general, design and optimization of P/G grids needs to account for the overall 3D power-delivery network, not only the part/die it is attached to [42,65].
- *Decoupling capacitor (decap) allocation:* To reduce power-supply noise, classical CMOS decaps and/or metal-insulator-metal (MIM) decaps can be deployed within each die [149] or even in dedicated decap dies [42]. Decap allocation has to be carefully analyzed to comprehend their impact on the complex power-supply noise distribution in 3D-ICs [60].

For thermal management of 3D-ICs, general approaches include

- *Low-power design:* Reducing overall power consumption and, thus, heat dissipation can be achieved by deployment of low-power circuitry.
- *Thermal-aware physical design:* Spreading high-power modules away from each other and arranging them in the dies closer to the heatsink, for example, during thermal-aware floorplanning [45,63,66] and placement [90], are simple but effective design measures.
- *Reducing thermal resistances:* Both internal paths (i.e., paths across and within dies) and external thermal paths (i.e., paths to the heatsink and the package) can be improved by new technologies and/or even simple design techniques. For example, TSVs can be grouped into TSV islands, which are then arranged and aligned such that they serve as effective "heat-pipes" [20]. A notable technology for reducing internal thermal resistance is the deployment of micro-fluidic channels [117].

Besides these separate approaches for power delivery and thermal management, some studies [19,65,71,76] investigated co-optimization of power delivery and thermal management and provided effective techniques. For example, their commonly proposed arrangement of P/G-TSV stacks (i.e., TSVs aligned across the whole 3D-IC) into high-power design regions is self-evident: regions with

---

* For modern 2D ICs, signal interconnects contribute nearly one-third of power consumption [119], and clock networks may consume even up to 50% of total power [150].

high current demand benefit from such P/G-TSV stacks in both terms of increased heat dissipation and reduced power-supply noise.

17.1.1.3 Clock Delivery

For 2D VLSI designs, the clock-network synthesis is one of the most important stages in the design flow because clock-distribution delivers the clock signals to clock sinks (such as latches, flip-flops, and registers) and synchronizes the calculation. Therefore, design automation for clock network is mostly used for block-level designs while designers manually craft the global clock distribution to perfection using SPICE-level simulation. It is not surprising that clock delivery is one major obstacle for general application of 3D-IC technology that, by its nature, targets designs with higher frequency ("More Moore") at a higher cost than 2D designs.

One major challenge of 3D clock distribution is for the clock signals to arrive all clock sinks reliably in different dies. Typically, clock networks in different dies are connected by TSVs and it is shown that 3D clock networks with multiple TSVs yield shorter wirelength [62,99,143,144], which leads to lower power dissipation [147] and shorter clock latency. However, this argument has some potential caveats.

- *TSV reliability.* TSVs are less reliable than other 2D interconnect structures as they are subject to random open defects [86]. Therefore, different TSV-redundancy mechanisms have been proposed [41,55,86], using spare TSVs or TSV groups with reconfigurable routing. One straight-forward implementation of reliable 3D clock distribution is to have double TSVs along the whole clock network, which greatly increase the TSV overhead for clock routing. More research is needed for the industry to understand the tradeoff of TSV utilization between power/latency and fault tolerance.
- *TSV placement.* Other than the fault-tolerant TSV design for clocking, another challenge is TSV placement for clock distribution. Note that a lot of 3D physical-design algorithms extend from 2D algorithms and some assume that TSVs can be placed anywhere except in a set of pre-defined blockages. In reality, TSVs have to be specially designed in order to prevent known reliability problems [96,129] such as mechanical issues [52,96]. A recent work proposes decision-tree-based algorithms to select clock TSVs from a set of TSV arrays [145]. For high-performance 3D designs, it is expected to have specially designed and pre-placed clock TSVs according to a particular clocking style, which will be discussed in details in Section 17.1.2.4.
- *Floorplanning and hierarchical design.* A lot of clocking-automation research in the literature for 2D and 3D designs did not fully consider the fact that a good clock-distribution network has to support other physical-design stages in order to honor all the timing constraints. One example is that clock skew between clock sinks which are constrained by timing checks is much more critical than the global clock skew. This statement obviously also applies to 3D designs where designers usually put each functional partition/module in a single die, instead of separating it into more than one die because of the high connectivity inside the module. In this case, the interdie skew (among different functional blocks) usually is less important than the intradie skew (especially when the clock sinks are related by timing paths). Therefore, replacing the intramodule clock networks with TSVs connecting to different dies in order to reduce clock wirelength, power, and latency is probably not the right thing to do.

  One potential application for 3D design is heterogeneous stacking, for instance, to stack dies from different-technology nodes. Microprocessor designers can design a new core and put it in a different die with new technology node while all other modules remain within the established technology node. For this case, 3D clock design (for both the new core design and TSV planning) has to be able to reuse the existing clocking structure. As a result, 3D clock delivery is usually highly dependent on the choice of the design flow and technology, and discussion on clock-synthesis algorithms without details of the design methodology is far from practical.

- *Testing.* As will be described in the next subsection, testing is vital on the road toward 3D adoption from an industrial perspective. Clocking is highly coupled with testability because prebond and midbond testing are very attractive DfT architectures. For example, a prebond-testable clock tree is presented in [74]. The idea of having a complete clock tree in each die basically limits 3D clock-network-synthesis research trying to replace long global clock wires on each die with TSVs. At the same time, TSVs connecting those clock networks on each die is a feasible clocking structure (Section 17.1.2.4).
- *Power/ground network synthesis.* It is well known that global clock wires have to be shielded to minimize coupling and noise, while one common choice to do so is to use power/ground wires as well as P/G TSVs. In some technologies, metal stacks and wirecodes are pre-defined and characterized. Hence, there is a need for co-optimization between P/G networks and global clock distributions.

In Section 17.1.2.4, the latest design challenges and state-of-the-art clock-distribution algorithms will be explained in detail.

17.1.1.4 Testing

Due to the inherently stacked arrangement of 3D circuits, testing is much more complicated than for 2D circuits and is still considered as a key obstacle for high-volume manufacturing of 3D circuits. Appropriate testing setups need to provide solutions for the following new problems [94]:

- Fault models and tests for wire-based and TSV-based interconnects with, for the latter, consideration of related intradie defects.
- Wafer probing on thinned dies, for dense arrangements of microbumps or TSVs and landing pads, considering stringent mechanical constraints.
- Design-for-Test (DfT) architectures, tailored for testing parts of the stack as well as for testing the whole stack.
- Optimization of the test flow for efficiency and limited cost-time overhead.

State-of-the-art studies which address the earlier problems are outlined next.

Fault models for both interposer-based interconnects [43] and TSV-based interconnects [26,44,85,89] have been proposed. The latter studies focus on specific types of interconnects and/or defects: Loi et al. [85] model and implement fault-tolerant 3D NoCs; Lung et al. [89] address fault-tolerant clock networks; Deutsch et al. [26] propose thermo-mechanical-stress-aware generation of test patterns; and Huang and Li [44] propose built-in self-repair scheme for TSVs.

Wafer probing, that is, early access of the dies' pads for testing purpose, is challenging in the context of 3D integration. More specifically, typical microbumps are too small, too densely arranged, and too fragile to be probed with conventional technologies [92]. Furthermore, the dies thinned for 3D stacking cannot be exposed to large probe weights, which range from 3 to 10 g per probe tip in conventional technologies [69]. New technologies, however, have been successfully developed and are becoming available. In [121], a lithographic-based MEMS, probe card was presented by Cascade Microtech, Inc. and IMEC. This technology is suitable for probing 40 μm- or smaller-pitched arrays, while inducing probe weights of only 1 g per tip. Another option is contact-less probing, as for example demonstrated by ST Microelectronics with capacitively coupled probing [115].

As for DfT architectures, testing facilities have provision for testing separate dies (i.e., *prebond testing*) and for testing the final stack (i.e., *postbond testing*). Further, testing the partial stack (i.e., *midbond testing*) is also relevant [92]. To enable such flexible facilities, modular setups are required. In practice, some wrapper circuitry is to be deployed on each die, which links with test facilities on other dies and, thus, across the whole 3D stack [75,94]. This is also acknowledged for the work-in-progress IEEE standard *P1838* [93]. The related wrapper enables controllability and observability at the die boundaries, which ensures interoperability between different dies (possibly even from different manufacturers). Besides these wrapper components, existing testing facilities should be reused whenever possible. In *P1838* [93], for example, the well-known concepts of test-access ports and scan-chains are applied.

Prebond testability is also associated with another notion, that of integrating only *known-good-dies*. Full and proper testability of separate dies is crucial, and only after these tests succeed, can the 3D stack be safely constructed. In this context, design partitioning also largely affects DfT architectures and testability: the finer the partitioning granularity, the more signals of (partial) modules pass across dies (instead of remaining encapsulated within dies), and the more complex the DfT architectures will become, especially for prebond/known-good-die testing [75].

To optimize the test flow, Noia et al. [103] have studied effective and efficient scheduling of test patterns. Furthermore, they have shown that minor increase in test pins enables great reduction in test time. In another study by Chen et al. [21], it was shown that reducing time and cost for prebond testing is possible despite strictly limiting deployment of additional test pins. Agrawal et al. [8] have proposed a heuristic methodology for test-flow selection, which flexibly adapts for different scenarios of 3D integration.

17.1.2 DESIGN CHALLENGES

Besides technological challenges, design challenges are also still impeding the broad and successful adoption of 3D-ICs. Compared to classical 2D chips, a 3D-IC is a much more complex system; related design algorithms and simulation and verification have to account for complex (and sometimes conflicting) interactions of multiple design criteria and physical domains. Putting the initial (probably too optimistic) expectations for "straightforward 3D solutions" into perspective, researchers and industry experts are now concerned about more pragmatical approaches. This includes to carefully analyze the *scope and applicability* for 3D-ICs, considering the available manufacturing and design approaches and their cost-benefit trade-offs.

Many challenges for physical design of 3D-ICs stem from the simple fact that the solution space is notably increased by adding one dimension compared to 2D-ICs [30]. This naturally escalates complexity for all physical-design steps.

Next, we discuss aspects of design complexity, algorithms, simulation, and verification, following the simplified design flow as shown in Figure 17.3. We also look into state-of-the-art approaches for design challenges.

17.1.2.1 Layout Representations for 3D Circuits

Physical design automation of electronic devices is generally based on abstract models of the corresponding design problems. These models are computationally represented as data structures. The data structures, in combination with accordingly tailored operations, for example, direct access to adjacent blocks, are subsequently referred to as *layout representations*. For 2D floorplanning, it has been shown by Chan et al. [17] that deploying different layout representations induces only minor deviations of final design quality—this is true despite the fact that different mathematical

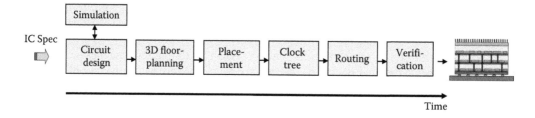

FIGURE 17.3 The major steps in the circuit design flow with a focus on physical design. Please note that this is a simplified view as in reality boundaries are blurred and iterations between these steps are common.

descriptions are at the heart of various representations. Chan et al. observed that a key bottleneck of floorplanning is to evaluate the actual layout (traditionally in terms of packing density and wire-length), and not to handle the abstract representations themselves.

For 3D design, this situation is even more intricate. 3D physical design has to consider a much more sophisticated set of design criteria [45,63,79]: wire-length, fixed outlines, thermal management, packing density, TSV management, power delivery, arrangement of massive interconnects within and across dies, 3D-stack-related noise coupling, etc. Given such complex and often interacting design criteria, it is apparent that 3D layout representations must be tailored carefully. Effective and efficient representations should provide the following features [30]:

- Inherent support for crucial constraints, for example, for spatial constraints given by vertical arrangements of modules interconnected across dies.
- An optimized solution space, that is, minimal redundancy while covering best solutions.
- A versatile set of operations applicable to different "granularities" of physical design. For example, such a set may include global operations, like swapping blocks across dies, and local operations, like shaping soft blocks.
- Fast transformation of abstract solutions into actual layouts and vice versa.
- Straightforward determination of correlations between abstract solutions and prospective design quality, for example, blocks adjacent in the abstract representation will also be adjacent in the actual layout. This is beneficial for speeding up layout evaluation, a key bottleneck for physical design.

In recent years, many effective 3D layout representations have been proposed, for example, layered transitive closure graph [68], T-tree [138], or Corblivar [63,66] (Figure 17.4). While some of these representations are derived from existing 2D representations, others are inherently developed for 3D integration.

Arising from this classification, there are two ways to represent vertical dependencies. The first option is to deploy multiple instances of classical 2D representations, labeled "2.5D" in Figure 17.4. Here, additional mechanisms have to be implemented to consider vertical relations between modules placed among different dies, such as vertical alignment as well as overlapping and nonoverlapping constraints. These representations include a discrete z-direction, such as in the combined bucket and 2D array (CBA) approach in [24]. However, it is obvious that vertical dependencies should be

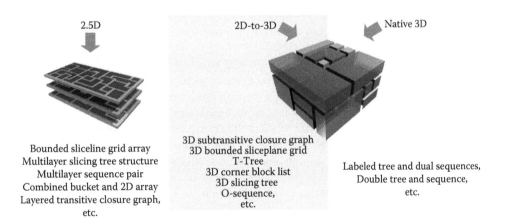

2.5D   2D-to-3D   Native 3D

Bounded sliceline grid array
Multilayer slicing tree structure
Multilayer sequence pair
Combined bucket and 2D array
Layered transitive closure graph,
etc.

3D subtransitive closure graph
3D bounded sliceplane grid
T-Tree
3D corner block list
3D slicing tree
O-sequence,
etc.

Labeled tree and dual sequences,
Double tree and sequence,
etc.

FIGURE 17.4 Categories and examples of layout representations tailored for 3D design [32]. Multiple instances of classical 2D representations are labeled "2.5D," "2D-to-3D" characterizes former 2D representations adjusted to the third dimension and "Native 3D" representations are specifically designed for 3D design.

incorporated directly into the representation. Hence, more recent 3D representations consider multi-layer modules natively in all three dimensions (labeled "2D-to-3D" and "Native 3D" in Figure 17.4). An example for such genuine 3D representation is the 3D Slicing Tree described in [22].

Besides representations themselves, several studies focused on implications of 3D representations and respective design methodologies; some of these studies are outlined next. Wang et al. [131] have shown that consistent correlations between abstract and actual layouts (an important feature for 3D representations, as indicated earlier) are not easily achieved unless $P = NP$. This naturally increases complexity of 3D physical design. Fischbach et al. [30] have developed a methodology to evaluate and compare representations for particular needs. Their methodology is based on Monte-Carlo sampling and analysis of respective solution-space distributions. Quiring et al. [111] have proposed a meta-heuristic methodology, which aims to apply probabilistic optimization techniques (like the well-known simulated annealing) more effectively. The key idea is to track each (past) layout operations' impact on relevant design criteria like thermal management. Then, (future) layout operations are deployed according to most prospective benefits for design quality.

17.1.2.2 Partitioning and Floorplanning

Partitioning divides the design into smaller blocks, each of which can be processed with some degree of independence and parallelism. A *divide-and-conquer strategy* can be implemented by laying out each block individually and reassembling the results as geometric partitions. Historically, this strategy was used for manual partitioning, but became infeasible for large netlists. In contrast, *netlist partitioning* can handle large netlists and redefine a physical hierarchy of an electronic system, ranging from boards to chips and from chips to blocks. Independent of the approach, the resulting partitions, subsequently also called *modules*, range from a small set of electrical components to fully functional ICs.

The initial partitioning step for 3D-ICs divides the circuit into several balanced partitions equal to the number of dies. The goal is, among others, to minimize the connections between dies. This translates into reducing the number of vertical interdie connections and decreasing the area overhead associated with TSVs, as discussed in earlier sections. After dividing the netlist or the circuit into multiple dies, 3D partitioning usually requires a subsequent *intradie partitioning*. Since both inter- and intra-die partitioning steps are extremely technology dependent, they are not further investigated here.

Floorplanning is closely related to partitioning. During floorplanning, the shapes and positions of the partitions/modules (such as digital and analog blocks) are determined. Thus, the floorplanning stage determines the external characteristics—fixed dimensions and external pin locations—of each module. These characteristics are necessary for subsequent placement (see Section 17.1.2.3) and routing steps (see Section 17.1.2.6), both of which determine the internal characteristics of the module.

Conventional floorplanning assumes a single 2D layer on which several modules must be arranged. 3D floorplanning includes new 3D-specific properties that must be represented in the underlying layout representations. For example, modules have vertical dependencies in addition to horizontal ones. As discussed in the previous subsection, layout representations should inherently consider these dependencies to facilitate efficient 3D floorplanning. For example, the 3D Slicing Tree described in [22] provides related features. As illustrated in Figure 17.5, different operations, such as module rotation and swapping, can be carried out efficiently to modify a given tree. A concatenation of these operations allows obtaining any possible slicing tree from any given slicing tree, thus enabling flexible 3D floorplanning. However, solutions from a 3D Slicing Tree are limited to slicing floorplans.

Besides considering vertical dependencies, 3D floorplanning should also account for reducing peak temperatures of 3D designs [63]. In addition to the increased power densities of stacked modules, peak temperatures are closely related to long wires on the chip due to interconnect power consumption [45].

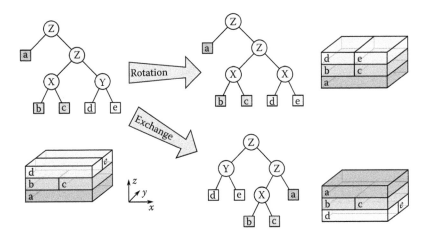

FIGURE 17.5 Illustration of 3D Slicing Tree operations to permute a given 3D floorplan [31]. A rotation alters an inner node (representing a cut through the normal plane) resulting in a physical rotation of modules contained in the subtrees of that node. An exchange swaps two subtrees resulting in a physical exchange of modules contained in these subtrees.

17.1.2.3 Placement

After floorplanning, the design is ready to be placed. That is, the next step in the design flow is to determine the location of each cell within its respective module (partition). The objective of placement is to determine the location and orientation of all cells, given solution constraints (e.g., no overlapping between cells) and optimization goals (e.g., minimizing total wirelength).

Depending on the applied partitioning approach for the 3D design, intra-die (2D) placement is limited to one die (layer), whereas interdie (3D) placement includes optimization between multiple dies. The latter approach requires new placement methodologies because of the "mismatch" between vertical and horizontal granularities: 3D layouts have limited flexibility in the third dimension due to both the relatively small number of dies and the scarce availability of TSVs. According to [36], this favors partitioning approaches (rather than force-directed techniques) at least during global placement. Accordingly, this work initially uses recursive bisectioning to perform global placement, with nets weighted according to the number of TSVs.

Quadratic placement approaches for inter-die placement require the "move force" to be modified such that cell overlap is eliminated in each die separately. More precisely, a move force should not be applied between two cells sharing the same x- and y-coordinate if they are located in different dies.

As mentioned previously, thermal constraints are crucial for reliable 3D designs. Hence, both intra- and interdie placement must spread cells such that a reasonable temperature distribution can be expected. However, due to the increased packing density in 3D-ICs, additional techniques are required to tackle the heat dissipation issue. For example, any vertical metal structure serves as "heat remover"—these structures play an important role in achieving a thermally solid design, and are in this context also called *thermal vias*.

Resulting from thermal constraints, 3D placement must not only place cells and consider regular TSVs but also take thermal vias into account. While intradie placement concerns only one die, the placement of a thermal via is affecting all dies; due to its aligned character, it creates a blockage throughout all dies. As such, thermal vias may represent a severe problem for cell placement and routing. Furthermore, cell placement and thermal-via placement are interacting because the position and size required for a thermal via depend on the thermal energy (i.e., power dissipation) of the cells nearby. Practical solutions, such as work presented by Goplen and Sapatnekar [35], have addressed these problems by designating specific areas within the circuit as potential thermal-via sites (Figure 17.6). Here, thermal conductivity of each region (site) can be considered a design

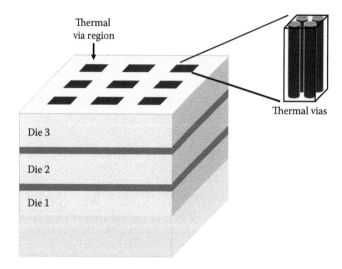

FIGURE 17.6 Regularly arranged thermal-via regions in a 3D-IC. Such regions unify the placement of a specific number of vias. The regions are sized according to the number of vias which are required for meeting the thermal requirements.

variable that is only subsequently translated into a precise number of thermal vias to be placed inside this region. Another advantage is the regularity of these sites (blockages), which can be addressed much easier than spread-out blockages, for example, during subsequent routing.

The aforementioned planning of thermal vias during placement applies also to regular TSVs, which—besides their electrical function—can provide heat-flow paths as well. An approach to grouping them into TSV islands (and thus to reduce their impact on placement, among others) has been presented in [64]. Further studies [12,23,90] consider detailed properties of TSVs in thermal- and wirelength-aware placement algorithms.

17.1.2.4 Clock-Distribution Networks

As mentioned in Section 17.1.1.3, 3D clocking is facing tremendous challenges such as TSV reliability, placement, as well as the dependence on design hierarchy, DfT, and power/ground synthesis. In fact, all the conventional objective and constraints on clock-network synthesis for 2D are getting more complicated, such as power, thermal, skew, slew, clock latency, jitter, glitch, and duty cycle. While Chapter 7 explains problems and algorithms for 3D clock networks, this subsection focuses on the future direction and challenges related to physical-design automation for 3D clock distribution.

Clock-skew minimization is one of the most important objectives for clock-network synthesis and is considered by almost all prior 3D clocking approaches, such as [61,62,89,146]. However, few works use SPICE simulation [109] and undergo Monte-Carlo simulations to obtain the clock-skew distribution [148]. Since 3D technologies are by its nature more fitted for high-performance designs with relatively higher manufacturing cost, it is expected that 3D clock networks have to be tailored for multiple-Gigahertz clock frequency and sub-20 ps of clock skew/jitter. Therefore, accurate timing simulation considering inter- and intra-die variability is a must for skew analysis to be realistic for 3D clock networks. A thorough study of clock variations for different number of TSVs is presented in [148], which is a very good starting point. Along this direction, a statistical clock-skew model considering inter- and intradie variability is presented in [133]. However, future work is urgently needed to understand how TSV usage affects the distribution of timing-impacting clock skew, which is the clock skew between each pair of clock sinks connected with real timing paths. In other words, variability analysis on clock-distribution network has to consider the full timing picture, which is tightly coupled with floorplanning and design hierarchy.

As mentioned previously, it is not likely that a logical partition/module is separated and placed in different dies. Therefore, there are many more timing paths between clock sinks on the same die, and it is more important to keep clock skew to be sub-20 ps for those clock sinks. In this case, we would like to maintain the common clock path of those clock sinks as much as possible and it is not preferable for their upstream clock paths to be on another die. In other words, it is desirable to keep the whole clock network of each module on the same die. An interesting research direction is to study the optimal TSV utilization and placement between the intact clock networks of partitions/modules.

With more variability and tighter skew/slew requirements of 3D designs, it is expected that H-tree and clock grid [109] are more preferable than the classical synthesis algorithms like Method of Mean and Median (MMM) and Deferred-Merge Embedding (DME) [146]. Pavlidis et al. [109] examine 3D clock structures having an H-tree on the middle plane. They found that the mix and match of H-tree, global rings, and local rings yield mixed clocking quality of results with no obvious winning architecture. This reinforces our discussion in Section 17.1.1.3 that each 3D clock-synthesis approach has to be driven and evaluated by a well-defined design methodology. One useful future work is to examine different 3D clock-grid structures where 2D global meshes are linked by TSVs. Since one of the most difficult tasks for 2D clock-grid design is the tuning of mesh wires and buffers driving the mesh, it is essential to achieve 3D clock-grid tuning in reasonable runtime. In summary, we are yet to see how different clock-synthesis algorithms perform in a real 3D hierarchical microprocessor-design flow with industrial timing analysis.

Fault tolerance and testing in 3D clock distribution is much more complex than that for 2D clocks because TSVs are subject to random open defects as mentioned in Section 17.1.1.3. There are different clock-synthesis algorithms considering TSV redundancy [88] and the introduction of fault-tolerant components [89]. While previous works derive fault-diagnosis test sequences to identify single and multiple defective TSVs [112], it is also important to use Monte-Carlo simulation for the timing of clock distribution.

Global TSV planning and codesign are crucial for 3D physical design because clock-network synthesis is highly coupled with almost all other physical-design problems, for example, floorplanning, placement, routing, timing optimization. One example is co-optimization of clocking and power/ground networks, especially when the technology and design rules restrict that clock routing has to be shielded by P/G wires. Another good example is [118] where Shang et al. derive an electrical-thermal model for both signal and thermal TSVs, and use the model to generate thermal-reliable 3D clock trees. In fact, new algorithms are urgently needed to simultaneously plan signal, power/ground, clock and thermal TSVs during 3D physical design.

17.1.2.5 Routability Prediction

One of the last steps for physical design is signal routing, that is, defining the interconnects' geometry (Section 17.1.2.6). As a result of the routing stage, not only the interconnects are deployed but also electrical properties of the circuit are defined. In order to achieve good routing results, all previous design stages have to be optimized with regard to routability. Therefore, *evaluating routability* is an inherent part of most design stages. 3D circuits with complex interconnect topologies require new approaches for routability prediction.

Any routability-prediction method is valuable only if it allows computation in significantly less time compared to actual routing. This can only be achieved by using effective simplifications. These range from fast ("rough") estimations of routing paths to the time-consuming (but more accurate) global-routing procedure.

Global routing for 3D design, that is, global-routing algorithms that find routing paths in several interconnect layers while considering different types of vias, has been solved for some years. Various multilayer global routers are applicable to 3D circuits if vertical routing capacities (i.e., vias) within dies and between dies can be specified independently. This differentiation is necessary in order to respect the different properties of interdie vias (which are often implemented by TSVs) and conventional signal vias.

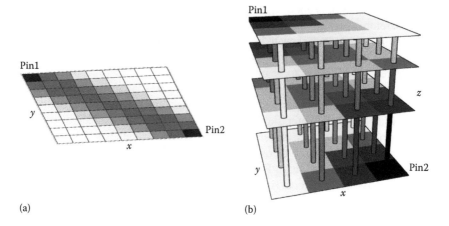

FIGURE 17.7 Routing-density distribution for a two-pin net in one layer (a) and extended 3D distribution in four dies (b). A darker color indicates a higher expected density. All routing paths are assumed to be of the same probability (i.e., no blockages exist).

However, fast 3D-routability prediction without performing global routing is still an open research topic. Such related fast methods are based on simple estimations ("informed guessing") of routing paths. They require to extend routing-density distributions to 3D in order to adapt statistical estimations of routing demand to the requirements of 3D interconnect topologies.

Conventional 2D-routing-density distributions predict the routing demand, overflow, congestion, and thus, routability in a 2D plane (Figure 17.7a). This model must be extended by *vertical* routing capacities and routing-density distributions for each die to render it applicable for 3D-IC design.

A model capable of representing a 3D-routing-density distribution was presented in [31] and is depicted in Figure 17.7b. Depending on the level of abstraction, the layers of the density distribution either correspond to individual routing layers or to the combined layers of one die. Constraints such as blockages and varying densities of interdie vias are considered by means of a varying probability for routing paths [31]. Using this 3D-routing-density-distribution model, it is possible to predict the routing demand for 3D-ICs and to estimate the routing densities in each layer (die) as well as the expected (vertical) interdie-via density.

A more recent model [58] considers the TSVs' impact on estimated routing topologies with particular focus on delay and power consumption.

17.1.2.6 Routing

A net is a set of two or more cell pins or terminals that has the same electrical potential in the final chip design. A circuit netlist includes all of the nets in a design. During the routing stage, all terminals of the nets must be properly connected while respecting constraints (e.g., design rules, routing resource capacities) and optimizing routing objectives (e.g., minimum total wirelength, maximum timing slack).

As already indicated, the main difference between regular (2D) and 3D routing is caused by the multi-die positions of net terminals that lead to net topologies that span more than one die (Figure 17.8). This requires expensive interdie vias (again, often implemented by TSVs) to be used in addition to regular signal vias, which connect metal layers within one die. Furthermore, 3D routing must take additional constraints into account, such as blockages introduced by thermal or interdie vias. These constraints require a much more sophisticated congestions management and blockage avoidance as it is applied for regular 2D routing. Additionally, the limited availability of interdie vias requires a careful allocation of this valuable resource among nets. The increased thermal impact on

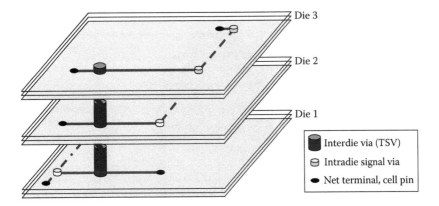

FIGURE 17.8   Example net route spanning multiple dies in a 3D design.

3D designs must also be considered during routing. For example, it is known that the delay of a wire increases with its temperature. Hence, critical nets must avoid hot regions of the chip.

Cong and Zhang presented a thermal-driven 3D router [25] using a multilevel-routing approach composed of recursive coarsening, initial routing, and recursive refinement. Its major feature is a thermal-driven via-planning algorithm. Based on this global view and capabilities for a multilevel scheme, the via-planning step effectively optimizes temperature distribution and wirelength using direct planning of the interdie vias instead of indirect planning through a routing-path search.

Another approach was presented by Zhang et al. [142]. It tackles the temperature-aware 3D-routing problem not only by using thermal vias but also by introducing the concept of *thermal wires*. Thermal wires are dummy objects with the function of spreading thermal energy in the lateral direction. Thermal vias perform the bulk of the conduction toward the heat sink, while thermal wires help distributing the heat paths among multiple thermal vias.

The well-known Steiner routing was also extended for 3D design. In [106], the authors propose a two-step flow: tree construction and tree refinement. The tree-construction step builds a delay-oriented Steiner tree under a given thermal profile. During tree refinement, TSVs are rearranged to further optimize the thermal distribution while preserving the routing topology and considering performance constraints.

17.1.2.7 Multi-Physical Simulation and Verification

Traditionally, physical design is separated from verification which aims to guarantee the intended functionality of a chip [54]. Simulation, on the other hand, is acknowledged as crucial part of physical design, for example, for thermal analysis of a chip. Verification is more detailed and complex than simulation, and typically leverages different simulation and analysis techniques itself. For example, electrical rule checking (ERC) verifies the correctness of power and ground interconnects, capacitive loads, signal transition times, etc. For proper handling of a 3D-IC's complex nature, simulation and verification techniques have to be deployed in a holistic manner in order to ensure design closure. The key reason for that requirement is given by the strong coupling of different physical domains in a 3D chip, and the resulting strong impact on overall design behavior and reliability [116].

The thermal, electrical, and mechanical domains are key subjects for multi-physical 3D-IC simulation and verification—with the domains' coupling being fortified by the high packing density in such chips [78,116] (Figure 17.9). Managing the *thermal domain* is much more challenging for 3D-IC design than for classical 2D design. With large thermal footprints, there is also an increasing impact on the *electrical domain*, that is, the behavior of active components. Since the leakage power of transistors is exponentially dependent on the temperature, a positive feedback mechanism arises, which, in worst-case scenarios, may lead to a thermal runaway and overheating of the 3D-IC.

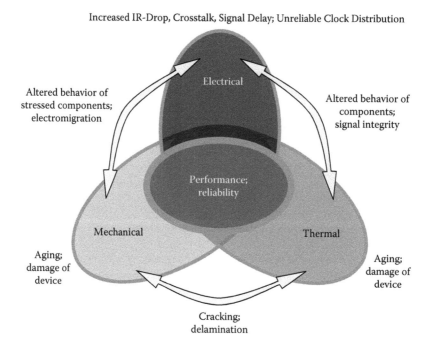

FIGURE 17.9   Coupling of the thermal, mechanical, and electrical domain in 3D-ICs and the related impact on device performance and reliability.

Besides, varying interconnect structures (e.g., metal layers "mixed" with TSVs) also impact the electrical domain: power and signal integrity, coupling, crosstalk and delays are all subject to varying interconnect structures' properties, dominated by large discrepancies in geometry and size. The interfaces between (very large) TSVs and (very small) metal wires are especially prone to electromigration [77,107]. The *mechanical domain* is mainly influenced by the complex composition of 3D-ICs: material properties are varying strongly, for example, due to the "intrusion" of silicon chips by copper TSVs. The coefficient of thermal expansion for copper is approximately six times larger than for silicon, leading to notable thermo-mechanical stress in the surrounding of TSVs [53]. Such stress impacts both the performance and reliability of the chip; it even increases the likeliness for cracks or delamination [18].

These complex multi-physical interactions in 3D-ICs give rise to high demands on simulation and verification approaches. As indicated earlier, simulation and verification should be deployed into 3D physical design as early and holistic as possible. To do so, *hierarchical* modeling and simulation frameworks are a commonly accepted approach [116]. Such hierarchical frameworks include models and respective techniques for different levels of design abstraction or design phases [78]:

   At the lowest level of abstraction, that is, for physical design and verification at transistor level, very detailed models are required. They must capture the composition of all devices with their specific geometries and material properties. Such models are also essential for evaluation and optimization of 3D interconnects technology, for example, for (individual) TSVs with regard to materials and geometries. The models are characterized by high accuracy, accompanied by large computational efforts for simulation. Typically, such models are implemented as fine-grain meshes of the chip's structures, which are then deployed for finite element/finite difference analysis.

   For "medium abstract" design phases, for example, place and route, models are more abstract; they are tailored to represent the system behavior. Therefore, their scope is the (multi-physical) coupling between separate components and the resulting system behavior. For example, an arrangement of multiple TSVs is modeled such that its geometry, the signal crosstalk, the thermomechanical stress, etc., are captured, in order to evaluate the overall reliability and performance of the arrangement. It is often required to derive such behavioral models from low-level simulations where, for example, separate wires, TSVs, and/or active gates are considered. These simulations are independent from the actual design process and can be conducted in advance by experienced engineers, providing their findings into parametric models or design rules. The related models are typically implemented as equivalent networks; this principle can be applied to different physical domains and is sufficiently accurate yet computationally not as demanding as finite element analysis.

The highest level of design abstraction deals with functional or architectural design. Simulation at this abstract level is difficult; the 3D chip and its components can only be modeled as design blocks with abstract properties like design area, number of pins, power dissipation, timing constraints, etc. These properties are furthermore often given as estimations with inherent variations. However, for densely integrated 3D chips, some parameter variations (e.g., in power dissipation) may have a large impact on the final design (e.g., on chip reliability and needs for heat removal). Besides, architectural design requires to analyze many different and diversified 3D-chip compositions in order to determine the appropriate one. This leads inevitably to many analysis iterations, demanding fast computation and simulation. These two opposing requirements—sufficiently high accuracy, also reflecting parameter variations and versatile 3D-chip compositions, and fast computation—make simulation on this design level very challenging. Applied models vary, depending on the required accuracy, available time, and the complexity of the 3D design and chip. In general, models have to be scalable to address these challenges. They are typically implemented as equivalent networks, coarse-grain finite elements, or dedicated models. For the latter, many studies have been proposed in recent years, which also reflects the need for such custom-tailored models. For example, Kim et al. [58] proposed a TSV-aware wirelength distribution model, capable of predicting delays and power consumption.

In summary, simulation and verification for 3D chips is challenging. The need to consider multiphysical coupling as well as the strong impact of technological configurations (like the number, size, and arrangement of separate dies) on each design phases are key issues. In general, hierarchical modeling and simulation frameworks are evolving as method of choice. Therein applied models have to be scalable. Further, the generation of parameterized models from low-level simulations is essential but not supported yet [116]. For verification, it seems practical to adapt available tools and leverage know-how from both classical 2D design and package design. For simulation, especially during high-level design phases, however, new approaches are required. In this context, Lim [79] reviewed key research needs for architectural floorplanning, to evaluate register-transfer-level (RTL)-based designs more accurately in terms of power, performance, and reliability. Lim argued that block-level modeling, TSV management, and chip/package coevaluation are crucial, and should be deployed as early and as effective as possible.

17.2   PROSPECTIVE DIRECTIONS FOR 3D CIRCUITS AND DESIGN

3D integration has been praised as a viable solution to keep up with the constantly increasing demands on electronic systems. The International Technology Roadmap for Semiconductors (ITRS) has prominently featured 3D-ICs for some years now: in the 2009 edition, for example, in the section on Interconnect and the section on Assembly and Packaging [1], or in the "More-than-Moore" whitepaper from 2010 [9].

Throughout these years, researchers and industry experts have been eager to cope with the many challenges arising from complex requirements for manufacturing and design of 3D-ICs, as also discussed earlier. While challenges for mainstream commercialization still remain, they have mainly shifted from manufacturing to design infrastructure, as also confirmed by industry experts. For example, at the GSA 3D-IC Packaging Working Group meeting October 2014, Yazdani [136] argued that path-finding tools (Section 17.2.1) are much-needed to design and evaluate the 3D chip-package-board system.

In the following, we discuss prominent design and manufacturing approaches, which are considered to increase mainstream adaption of 3D-ICs.

17.2.1 PATH-FINDING: SYSTEM-LEVEL DESIGN EXPLORATION AND EVALUATION

Traditionally, high-level models of circuit components have been applied for evaluation of design options. As already discussed, this is much more complex for 3D-ICs than for classical 2D chips. With the ever increasing design complexity and the vast options for manufacturing and integration choices, system-level design of 3D chips cannot be conducted without considering physical-level details. Thus, it is necessary for system-level design tools to handle the complex interactions between performance, power, thermal management, process technology, floorplanning, system architecture, and even dynamic scheduling or workloads. Such extended system-level design exploration and evaluation is known as *path-finding*.

17.2.1.1 Concepts and Approaches for Path-Finding

Early concepts for path-finding in 3D design have already been proposed in 2009, for example, by Milojevic et al. [97]. Their main novelty was to link system-level design exploration with automated synthesis of RTL models and physical-design prototyping. This way, system engineers had been given the opportunity to evaluate their architectures on a much more detailed level, despite not necessarily being equipped with extensive know-how and time for actual physical design.

Research has resurged very recently, and several studies on practical path-finding methodologies have been presented. Martin et al. [95] proposed a methodology for early evaluation of electrical performance. In their study, they deployed building blocks (e.g., of large TSV arrays) using parametrized models. These blocks are then committed to fast electromagnetic solvers for analyzing signal crosstalk. With this methodology, the authors successfully evaluated interposer-based 3D devices and their TSV interconnects. A similar study was conducted by Yazdani and Park [137]; they showcased how to optimize system interconnects in 2.5D integration. More precisely, they conducted and evaluated the placement of buffer cells, arrangement of Cu pillar bumps and package BGA for Wide I/O memory integration on interposers. Thus, their tool enables planning of interconnect structures at early stages and for multiple chips integrated by state-of-the-art memory technology. Priyadarshi et al. [110] proposed transaction-level-based path-finding, complementing known RTL-based approaches. Their tool allows much faster design evaluation, since transaction-based modeling distinguishes computation and communication, thus hiding details not necessarily required for early design simulation. Additionally, they link thermal analysis to transactional modeling and simulation, thereby enabling efficient thermal-aware path-finding.

17.2.1.2 Flows and Tools for Path-Finding

A typical flow for path-finding tools covers three steps (Figure 17.10). Note that feedback loops between these steps are essential; capabilities for passing specifications top-down (e.g., physical constraints or technology details) as well as passing them bottom-up (e.g., simulation results) are needed for flexible and accurate path-finding.

1. *System-level design exploration*: A high-level description (e.g., given in SystemC) is generated. Already at this point, the technology and configuration for 3D integration have to be considered. For example, partitioning modules across separate dies can be modeled in these early phases, to help tackle the vast design space of 3D chips more efficiently.

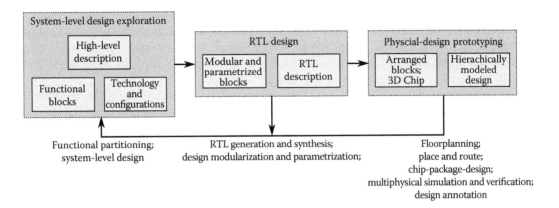

FIGURE 17.10 Path-finding flow, with each step's components labeled in boxes and applied techniques labeled in the following. Note that details of technology (e.g., for a specific 3D integration approach) are to be considered and rendered more specifically for each step.

2. *RTL design*: From high-level descriptions, RTL models are derived. They are serving as "bridge" between system design and physical design. The models should be modularized, in order to represent the high-level design closely and to enable reuse for different arrangements of system components during path-finding. The models can furthermore be annotated during subsequent technology simulation, to provide technology feedback/guidance at this early phase. In short, a parametrized building-block model of the overall system is the objective of this step.

3. *Physical-design prototyping*: The RTL models are then fed to physical-design prototyping. In contrast to actual physical design, more abstract (and thus faster) techniques are applied to obtain estimates of the final design quality. For example, an important step of prototyping is floorplanning. Design blocks are usually annotated, for example, with power consumption and area, to enable more accurate estimates on, for example, thermal distribution and die sizes.

Besides the methodologies outlined in the previous subsection, commercial tools are becoming available. Note that such tools are usually modular and also rely on adapting legacy (2D) tools for simulation and verification.

17.2.2 DESIGN APPROACHES AND STANDARDIZATION

Due to the paradigm shift arising with 3D-ICs—the additional integration in the third dimension—physical-design automation cannot be considered as stand-alone process. In fact, all components of chip design (i.e., technology and manufacturing, system design, and physical-design automation) undergo a notable transition. This wide-ranging shift aggravates the need for reliable and effective design approaches and commonly established standards.

17.2.2.1 Design Approaches for 3D Circuits

Design approaches can be characterized by their *granularity*, that is, the applied partitioning scheme, defining which circuit parts are split and assigned to different dies [84]. On the opposite ends of the granularity scale, the approaches of transistor-level (finest-grain) integration versus core-level (coarsest-grain) integration can be found.

Only recently—mainly due to advances in monolithic manufacturing technologies—transistor-level integration becomes applicable [14,73,81,104]. Here, active layers are built up sequentially rather than processed in separate and subsequently bonded dies. This finest-grain integration style is expected to provide large performance benefits due to shortest-path vertical coupling of transistors. Besides the high demands on very-small-scale vias and other related challenges, this style requires a full redesign, that is, completely prevents design reuse. It also faces further challenges, for example, the need for tools and knowhow for low-temperature manufacturing processes [13] or notably increased delays along with massive routing congestion [72,81].

For the other end of the integration scale, that is, for core-level integration, the efforts are comparable to traditional 2D chip design: only few intercore connects have to be realized by placing and wiring TSVs. Apart from that, the cores can be fully reused. In consequence, gained benefits are low: the properties of such a 3D-IC are still dominated by their stacked 2D chips.

Next, design approaches found in the middle of the granularity scale are reviewed: gate-level and block-level integration (Figure 17.11).

Gate-level integration means to partition cells across multiple dies and use TSVs whenever required for connecting cells across dies. This style promises significant wirelength reduction and great flexibility [84,100,102].

Its adverse effects include, for example, the massive number of necessary TSVs for random logic. Studies by Kim et al. [57] and Mak and Chu [91] reveal that partitioning gates between multiple dies can undermine wirelength reduction unless modules of certain minimal size are preserved and/or TSVs are downscaled. Another study [101] points out that layout effects can largely influence performance for highly regular blocks such as SRAM registers: a mismatch between TSV and cell dimensions introduces wirelength disparities while routing these regular structures to TSVs. Timing-aware placement of partitioned gates is required for design closure [70]; this timing issue is intensified by interdie variation mismatches [33]. Besides, partitioning a design block across multiple dies requires new prebond testing approaches [69,75]. After die stacking, a single failed die renders the whole 3D-IC unusable, thus easily undermining overall yield.

In summary, gate-level integration may be very promising in terms of design flexibility, performance, and wirelength reduction, but it faces many challenges and currently appears—like transistor-level integration—only applicable in a limited scope. Practical scenarios include devices with high demands on efficiency and low power, as demonstrated by, for example, 3D-ICs with complex modules like floating-point units and long-path multipliers [102,124,125].

Block-level integration promises to reduce TSV overhead by assigning only few global interconnects to them. This is possible since blocks typically subsume most of a design's connectivity and are linked by a small number of global interconnects [123].

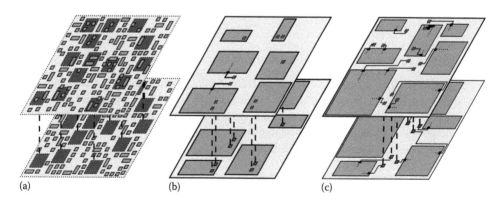

(a)   (b)   (c)

FIGURE 17.11 Design styles for 3D-ICs. TSVs are illustrated as dark-gray boxes and landing pads as dashed, dark-gray boxes. Gates or blocks are represented by light-gray boxes. Face-to-back stacking is considered; TSVs cannot obstruct blocks in lower dies but landing pads may overlap with blocks in upper dies, due to illustration perspective. (a) Gate-level integration, enlarged detail. (b and c) Block-level integration. (b) The *redesigned 2D* style uses predefined TSV sites within for 3D-design adapted blocks. (c) The *legacy 2D* style distributes TSVs between blocks, thus enabling reuse of available design blocks.

Sophisticated 3D systems combining many heterogeneous dies are anticipated in a whitepaper by Cadence [2]. Such devices require distinct manufacturing processes at different-technology nodes for fast and low-power random logic, several memory types, analog and RF circuits, on-chip sensors, microelectromechanical systems, and so on. Block-level integration is imperative for such heterogeneous 3D-ICs where modules cannot be partitioned among different-technology dies.

When assigning entire blocks to separate dies and connecting them with TSVs, we can distinguish two design styles.

- *Redesigned 2D (R2D) style*: 2D blocks designed for 3D integration; TSVs can potentially be embedded within the footprints (Figure 17.11b).
- *Legacy 2D (L2D) style*: 2D blocks not designed for 3D integration; TSVs are to be placed between blocks (Figure 17.11c).

Which style is appropriate also depends on the type of given *intellectual property* (*IP*) blocks. For hard blocks with predefined layout, applying L2D is mandatory. The fixed layout of such blocks cannot include large TSVs simply because the blocks' design was not accounting for TSVs. For soft blocks, that is, blocks given in behavioral description and synthesized during the design flow, the R2D style appears more appropriate but is still challenging in terms of TSV management, as elaborated next.

TSVs introduce design constraints and overheads, mainly due to their (to gates comparably large) dimensions and intrusive character when "injected" into silicon dies. Thus, inserting TSVs into densely packed design blocks is expected to complicate design closure since it (1) introduces placement and routing obstacles [57], (2) induces notable stress for nearby active gates [134], and (3) requires design tools to provide sophisticated TSV-related verification, for example, signal-integrity analysis considering coupling between TSVs [82,83,135].

Grouping TSVs into *TSV islands* is common practice and beneficial for several reasons [44,56,59,63,64,98,127,145]. For example, TSVs introduce stress in the surrounding silicon, which affects nearby transistors [10,47,134], but TSV islands do not need to include active gates. The layout of these islands can be optimized in advance [87,139]. Regular island structures help to limit stress below the yielding strength of copper [51], and limit stress generally to particular design regions [51,52,87]. Placing islands *between* blocks (i.e., applying the L2D style) may thus limit stress on blocks' active gates.

Further benefits of both R2D and L2D styles are described next.

- Design-for-Test (DfT) structures are key components of existing IP blocks and can be used to realize prebond and postbond testing [69].
- Block-level integration can efficiently reduce critical paths, thus simultaneously limiting signal delay, increasing performance and reducing power consumption [11,59,70,84].
- With block-level integration, critical paths are mostly located within 2D blocks—they do not traverse multiple active layers, which limits the impact of process variations on performance [34].
- For yield-optimized matching of "slow dies" and "fast dies," based on accurate delay models with process variations considered [28], block-level integration is mandatory. That is because this matching approach assumes that dies can be delay-tested before stacking, which is only possible when all dies encapsulate self-contained modules.
- Modern chip design mostly relies on predesigned and optimized IP blocks. Existing IP blocks and physical-design automation tools do not account for 3D integration. Even when such tools appear, it will take IP vendors much time and money to upgrade their extensive portfolios for 3D integration. Thus, redesigning existing IP blocks to be spread out on multiple dies (as proposed in gate-level integration) is not practical; in contrast, reusing them as legacy blocks (as proposed in block-level integration) is convenient.

These wide-ranging considerations suggest that block-level integration is a more practical approach for general 3D-IC design; for dedicated applications, other styles may also be considered.

17.2.2.2 Standardization Trends and Examples

Despite the great interest and recent achievements in design and manufacturing of 3D-ICs, this technology is still not yet available in high-volume applications. Besides the aforementioned need for design, verification and test tools, another pressing concern is the lack of standard definitions. However, as for any successful technology in the chip industry, standards will be required for increasing acceptance and establishing supply chains and "ecosystems" [2].

Efforts for standardization initially focus on I/O and interfaces, while later on heterogeneous and/or interposer-based assembling and supply chains need to be addressed. For example for the latter, the JEDEC Multiple Chip Packages Committee is "currently developing mixed technology pad sequence and device package standards to enable SRAM, DRAM, and Flash memory to be combined into a single package that may also contain processor(s) and other devices" [49]. Standards already available and widely acknowledged in the industry cover memory integration and testing, briefly reviewed next.

For memory integration, the JESD229 standard [48], more commonly known as *Wide I/O*, is a prominent example. It defines memory integration with one up to four memory dies stacked on top of a controller die. The standard is considered mature; two versions are available, the first being published December 2011 and the second (*WideIO2*) August 2014. Devices fulfilling the standard provide high-bandwidth memory interfaces, namely four (up to eight for WideIO2) 128-bit-wide memory channels. The standard covers details on functionality, AC and DC characteristics, packages, and micropillar signal assignments. Several studies proposed designs based on Wide I/O and/or tools for related verification [40,53,98,121,137].

With JESD235 [49], better known as *High Bandwidth Memory* (*HBM*), an alternative standard for 3D memory integration is available and currently adopted by industry, for example, by SK Hynix [120].

Yet another memory standard, the *Hybrid Memory Cube* (*HMC*), has recently gained more attention. The related consortium was founded in October 2011 by co-developers Altera, Micron, Open-Silicon, Samsung, and Xilinx. The first specification was released in May 2013 [3], and the second version in December 2014 [7]. This update most notably increases data rates for single channels from 1920 up to 3840 MB/s.

For testing, standards are currently under development. As indicated in Section 17.1.1.4, IEEE standard P1838 [93] evolves as prominent example. The proposed wrapper architecture enables controllability and observability at the die boundaries, which ensures interoperability between different dies, possibly even from different manufacturers. Besides these wrapper components, existing test facilities are proposed for reuse whenever possible: IEEE 1149.x for test access, IEEE 1500 for die test, and IEEE P1687 for internal debugging. Some studies addressed the implementation and review of such standardized DfT structures [94,98].

17.2.3 Demonstrators and Prototyping for 3D Circuits

3D integration has been eagerly discussed and investigated for many years now. There have been many efforts for demonstrators and prototyping of TSV-based 3D-ICs and interposer-based 2.5D systems, both from academic and industrial groups. The following gives a brief overview on demonstrators and outlines prototyping platforms.

17.2.3.1 Academic Efforts

The project *3D-MAPS: A Many-Core 3D Processor with Stacked Memory* [37,56,80] is a prominent example for large-scale demonstrators driven by academia (at Georgia Institute of Technology). In 2010, the first version was developed: a 64-core memory-stacked-on-processor system running at

277 MHz with 64 GB/s memory bandwidth. That bandwidth was confirmed in measurements of taped-out chips; up to 63.8 GB/s was achieved while overall power consumption was approx. only 4 W. The processor was fabricated in 130 nm by GLOBALFOUNDRIES and the TSV technology (1.2 μm size, 2.5 μm pitch, and 6 μm height) was provided by Tezzaron Semiconductors. Note that TSVs had been deployed there only for power delivery and external interconnects. In 2012, a second version was proposed with 128 cores, embedded in two logic dies, and stacked with three DRAM dies.

Another sophisticated academic 3D demonstrator was developed in 2012 at the University of Michigan: *Centip3De*, a large-scale 3D chip with a cluster-based near-threshold computing architecture [29]. The chip comprises a logic die (130 nm) with 64 ARM Cortex-M3 cores, and an SRAM die; both are interconnected via face-to-face bonding. Again, TSVs are deployed from Tezzaron's technology and are used for connecting the chip to the package. A notable result of this demonstrator is a >3× improvement in energy efficiency (measured in DMIPS/W) over traditional chips and operation. Furthermore, the chip can fully operate under a fixed thermal design power of only 250 mW.

Besides these large "flagship" demonstrators, further academic studies have proposed tools, designs, and some also measurements from taped-out 3D chips, for example, [124,141].

A prototyping platform projected for 3D chips was proposed in *FlexTiles: Self Adaptive Heterogeneous many core based on Flexible Tiles* [6,16]. Here, the (so far only conceptional) 3D chip comprises a FPGA die and a many-core die. The FPGA die provides dynamic reconfigurability at runtime, that is, dynamic adaption of system functionality by reloading and/or reconfiguring IP blocks. The many-core die shall comprise general processing cores and dedicated DSPs. The intended scope of the prototyping platform is to evaluate adaptive and heterogeneous designs with state-of-the-art technology. For example, an (so far on two FPGAs instead of the 3D chip) evaluated use case is a "smart" camera, which dynamically adapts for low-power scenarios. Objectives of the project are the definition and development of a heterogeneous many-core with self-adaptation capabilities, its virtualization layer and its tool chain ensuring programming efficiency and low power consumption.

17.2.3.2 Industrial Efforts

Already in 2010, Xilinx has presented the *Virtex-7 FPGA* family [27]. Here, the FPGA is split into four dies (manufactured in 28 nm), which are assembled side-by-side onto a passive silicon interposer (manufactured in 65 nm including TSVs). With this 2.5D approach, cost are limited, while yield and performance are increased compared to previous high-end FPGA systems. For example, the bandwidth-per-watt ratio is over 100× that of standard FPGA interconnects.

Intel presented an energy-efficient, high-performance 80-core system with stacked SRAM in 2011 [15]. The demonstrator provides one tera-FLOPS while consuming less than 100W. At that time, TSVs were not necessarily very reliable, so they had been sparsely (pitch 190 μm) deployed for power delivery and external routing to the package. The SRAM die and the 80-core logic die had been interconnected via face-to-face metal bonding.

IBM demonstrated a 3D version of a processor with up to three dies of eDRAM in 2012 [132]. With deployment of known-good dies, it uses 50 μm pitch of μC4 bumps to join the front side of the processor to the TSV connections on the back side of the thinned memory chips. The demonstrator is based on 45 nm technology and runs at 2 GHz. It achieves a data bandwidth of approx. 56 GB/s.

For 3D memory integration, the industry has passed the prototyping stage, and is recently approaching high-volume manufacturing. For example, since end of 2014, SK Hynix offers HBM modules [120]. With 128 GB/s, these modules provide approx. 4.5× the bandwidth of state-of-the art GDDR5 modules. Alok Gupta, PE at Nvidia, presented at the 3D ASIP 2014 conference [5] a GPU-on-interposer system comprising four HBM modules and achieving a bandwidth of 1 TB/s. Another example are the efforts of Samsung: since 2013, the company is mass-producing so-called vertical-NAND (V-NAND) memory [114]. Here, as the name suggests, the transistors are vertically arranged, that is, the gate and insulator are circularly wrapped around the channel. These transistors are then repeatedly processed onto many stacked layers. With this dedicated design, achieving integration density in the vertical dimension instead of traditionally in the plane, wider bit lines and the deployment of "old" but much more reliable process nodes (e.g., 30 nm) are feasible. These two measures effectively reduce cell-to-cell interferences and small-scale patterning issues, which are major concerns for modern memory technology.

Besides the mentioned Virtex-7 FPGAs from Xilinx, further interposer-based prototypes haven been presented by GLOBALFOUNDRIES. In cooperation with Open-Silicon, a prototype containing two ARM Cortex A-9 chips (28 nm) stacked onto a 65 nm-and-TSV-embedded silicon interposer was presented in 2013 [4]. This demonstrator was mainly intended as proof-of-concept and also to establish EDA flows for design, verification, and test of interposer systems. GLOBALFOUNDRIES' Packaging Director Alapati stated the transition for interposer-based systems as well as for 3D-ICs to high-volume manufacturing in 2015 [130]. Further interposer-based prototypes comprising 20 and 14 nm chips have been presented as well.

17.3 SUMMARY AND CONCLUSION

In comparison to classical 2D chips, 3D-ICs and even interposer-based 2.5D systems are much more complex. As discussed in this chapter (and throughout the book), both design and manufacturing engineers have to cope with quite different and challenging requirements and objectives.

In recent years, much research and development effort, from academia and industry, has been undertaken. Slowly but surely 3D integration is making the transition from a "hyped new technology" toward a viable option for keeping up with constantly increasing demands on performance, functionality, power consumption, and cost of electronic systems. Very recently, few companies (e.g., Samsung and SK Hynix) have introduced 3D-integrated memory products on the market, while other companies (e.g., GLOBALFOUNDRIES) have established tool-chains and design flows to enable high-volume manufacturing of interposer-based systems and even 3D-ICs in very near future.

This implies that technology and manufacturing concerns have been mainly addressed, at least for TSV- and interposer-based systems. It is common consensus that no "show-stoppers" are blocking the adoption of such 3D integration. However, for large-scale heterogeneous integration and especially for logic-on-logic integration, key concerns remain. For example, thermal management, power and clock delivery, testing along with yield and cost are still obstructing such sophisticated 3D systems.

In the recent years, design challenges have largely shifted toward high-level design issues. Besides the fact that 3D-EDA tools are only slowly reaching the market, high-level design features are yet insufficiently supported. Such sought-after features include: multi-physical simulation and verification of the chip-package-board system, including different types of active layers and interconnects; path-finding for efficient design exploration and evaluation; standards-based design of modules and interfaces, for example, NoCs or test structures.

Overall, 3D integration is nevertheless on an "accelerating trajectory," and the next few years will bring this integration approach with its versatile options more and more into mainstream chip development.

REFERENCES

1. International technology roadmap for semiconductor. http://www.itrs.net/Links/2009ITRS/Home2009. htm, 2009. Accessed on December 2014.
2. 3D ICs with TSVs—Design challenges and requirements. http://www.cadence.com/rl/Resources/white_papers/3DIC_wp.pdf, 2010. Accessed on December 2014.
3. Hybrid Memory Cube Specification 1.0. http://hybridmemorycube. org/files/SiteDownloads/HMC_Specification%201_0.pdf, January 2013. Accessed on December 2014.
4. Open-Silicon and GLOBALFOUNDRIES Demonstrate Custom 28 nm SoC Using 2.5D Technology. http://www.open-silicon.com, Nov 2013. Accessed on December 2014.
5. 3D ASIP 2014: All Aboard the 3D IC Train!. http://www .3dincites.com/2014/12/3d-asip-2014-addresses-3d-benefits-challenges-solutions/, 2014. Accessed on December 2014.

6. Flextiles: Self adaptive heterogeneous manycore based on flexible tiles. http://flextiles.eu, 2014. Accessed on December 2014.

7. Hybrid Memory Cube Specification 2.0. http://hybridmemorycube. org/files/SiteDownloads/20141119_HMCC_Spec2.0Release.pdf, February 2014. Accessed on December 2014.

8. M. Agrawal, K. Chakrabarty, and B. Eklow. Test-cost optimization and test-flow selection for 3D-stacked ICs. Technical report, Electrical and Computer Engineering, Duke University, Durham, NC, 2012.

9. W. Arden, M. Brillout, P. Cogez, M. Graef, B. Huizing et al. More-than-Moore white paper. Technical report, ITRS, 2010.

10. K. Athikulwongse, A. Chakraborty, J.-S. Yang, D. Z. Pan, and S. K. Lim. Stress-driven 3D-IC placement with TSV keep-out zone and regularity study. In *Proc. Int. Conf. Comput.-Aided Des.*, San Jose, CA, pp. 669–674, 2010.

11. K. Athikulwongse, D. H. Kim, M. Jung, and S. K. Lim. Block-level designs of die-to-wafer bonded 3D ICs and their design quality tradeoff's. In *Proc Asia South Pacific Des. Autom. Conf.*, Yokohama, Japan, pp. 687–692, 2013.

12. K. Athikulwongse, M. Pathak, and S. K. Lim. Exploiting die-to-die thermal coupling in 3D IC placement. In *Proc. Des. Autom. Conf.*, San Francisco, CA, pp. 741–746, 2012.

13. P. Batude, M. Vinet, B. Previtali, C. Tabone, C. Xu et al. Advances, challenges and opportunities in 3D CMOS sequential integration. In *Proc. Int. Elec. Devices Meeting*, Washington, DC, pp. 7.3.1–7.3.4, 2011.

14. S. Bobba, A. Chakraborty, O. Thomas, P. Batude, T. Ernst et al. Celoncel: Effective design technique for 3-D monolithic integration targeting high performance integrated circuits. In *Proc. Asia South Pacific Des. Autom. Conf.*, Yokohama, Japan, pp. 336–343, 2011.

15. S. Borkar. 3D integration for energy efficient system design. In *Proc. Des. Autom. Conf.*, San Diego, CA, pp. 214–219, 2011.

16. R. Brillu, S. Pillement, A. Abdellah, F. Lemonnier, and P. Millet. Flex-Tiles: A globally homogeneous but locally heterogeneous manycore architecture. In *Proc. Workshop on Rapid Sim. and Perform. Evaluation*, Vienna, Austria, pp. 3:1–3:8, 2014.

17. H. H. Chan, S. N. Adya, and I. L. Markov. Are floorplan representations important in digital design? In *Proc. Int. Symp. Phys. Des.*, San Francisco, CA, pp. 129–136, 2005.

18. Y. S. Chan, H. Y. Li, and X. Zhang. Thermo-mechanical design rules for the fabrication of TSV interposers. *Trans. Compon. Packag. Manuf. Technol.*, 3(4):633–640, 2013.

19. H.-T. Chen, H.-L. Lin, Z.-C. Wang, and T. T. Hwang. A new architecture for power network in 3D IC. In *Proc. Des. Autom. Test Europe*, Grenoble, France, pp. 1–6, 2011.

20. Y. Chen, E. Kursun, D. Motschman, C. Johnson, and Y. Xie. Through silicon via aware design planning for thermally efficient 3-D integrated circuits. *Trans. Comput.-Aided Des. Integr. Circ. Syst.*, 32(9):1335–1346, 2013.

21. Y.-X. Chen, Y.-J. Huang, and J.-F. Li. Test cost optimization technique for the pre-bond test of 3D ICs. In *VLSI Test Symposium*, Maui, HI, pp. 102–107, 2012.

22. L. Cheng, L. Deng, and M. D. F. Wong. Floorplanning for 3-D VLSI design. In *Proc. Asia South Pacific Des. Autom. Conf.*, Shanghai, China, pp. 405–411, 2005.

23. J. Cong, G. Luo, and Y. Shi. Thermal-aware cell and through-silicon-via co-placement for 3D ICs. In *Proc. Des. Autom. Conf.*, pp. 670–675, 2011.

24. J. Cong, J. Wei, and Y. Zhang. A thermal-driven floorplanning algorithm for 3D ICs. In *Proc. Int. Conf. Comput.-Aided Des.*, pp. 306–313, 2004.

25. J. Cong and Y. Zhang. Thermal-driven multilevel routing for 3-D ICs. In *Proc. Asia South Pacific Des. Autom. Conf.*, Shanghai, China, pp. 121–126, 2005.

26. S. Deutsch, K. Chakrabarty, S. Panth, and S. K. Lim. TSV stress-aware ATPG for 3D stacked ICs. In *Proc Asian Test Symp.*, pp. 31–36, 2012.

27. P. Dorsey. Xilinx stacked silicon interconnect technology delivers break-through FPGA capacity, bandwidth, and power efficiency. Technical report, Xilinc, Inc., 2010.

28. C. Ferri, S. Reda, and R. I. Bahar. Strategies for improving the parametric yield and profits of 3D ICs. In *Proc. Int. Conf. Comput.-Aided Des.*, pp. 220–226, 2007.

29. D. Fick, R. G. Dreslinski, B. Giridhar, G. Kim, S. Seo et al. Centip3De: A cluster-based NTC architecture with 64 ARM Cortex-M3 cores in 3D stacked 130 nm CMOS. *J. Solid-State Circ.*, 48(1):104–117, 2013.

30. R. Fischbach, J. Lienig, and J. Knechtel. Investigating modern layout representations for improved 3D design automation. In *Proc. Great Lakes Symp. VLSI*, pp. 337–342, 2011.

31. R. Fischbach, J. Lienig, and T. Meister. From 3D circuit technologies and data structures to interconnect prediction. In *Proc. Int. Workshop Syst.-Level Interconn. Pred.*, pp. 77–84, 2009.

32. R. Fischbach, J. Lienig, and M. Thiele. Solution space investigation and comparison of modern data structures for heterogeneous 3D designs. In *Proc. 3D Syst. Integr. Conf.*, pp. 1–8, 2010.

33. S. Garg and D. Marculescu. 3D-GCP: An analytical model for the impact of process variations on the critical path delay distribution of 3D ICs. In *Proc. Int. Symp. Quality Elec. Des.*, pp. 147–155, 2009.

34. S. Garg and D. Marculescu. Mitigating the impact of process variation on the performance of 3-D integrated circuits. *Trans. VLSI Syst.*, 21(10):1903–1914, 2013.

35. B. Goplen and S. Sapatnekar. Thermal via placement in 3D ICs. In *Proc. Int. Symp. Phys. Des.*, pp. 167–174, 2005.

36. B. Goplen and S. Sapatnekar. Placement of 3D ICs with thermal and interlayer via considerations. In *Proc. Des. Autom. Conf.*, pp. 626–631, 2007.

37. M. B. Healy, K. Athikulwongse, R. Goel, M. M. Hossain, D. H. Kim et al. Design and analysis of 3D-MAPS: A many-core 3D processor with stacked memory. In *Proc. Cust. Integr. Circ. Conf.*, pp. 1–4, 2010.

38. M. B. Healy and S. K. Lim. Power delivery system architecture for many-tier 3D systems. In *Proc. Elec. Compon. Technol. Conf.*, pp. 1682–1688, 2010.

39. M. B. Healy and S. K. Lim. Power-supply-network design in 3D integrated systems. In *Proc. Int. Symp. Quality Elec. Des.*, pp. 223–228, 2011.

40. A. Heinig, R. Fischbach, and M. Dittrich. Thermal analysis and optimization of 2.5D and 3D integrated systems with Wide I/O memory. In *Proc. Therm. Thermomech. Phenom. Electr. Syst. Conf.*, pp. 86–91, 2014.

41. A.-C. Hsieh and T. T. Hwang. TSV redundancy: Architecture and design issues in 3-D IC. *Trans. VLSI Syst.*, 20(4):711–722, 2012.

42. G. Huang, M. Bakir, A. Naeemi, H. Chen, and J. D. Meindl. Power delivery for 3D chip stacks: Physical modeling and design implication. In *Proc. Electri. Perf Elec. Packag. Sys.*, pp. 205–208, 2007.

43. L.-R. Huang, S.-Y. Huang, K.-H. Tsai, and W.-T. Cheng. Parametric fault testing and performance characterization of post-bond interposer wires in 2.5-D ICs. *Trans. Comput.-Aided Des. Integr. Circ. Syst.*, 33(3):476–488, 2014.

44. Y.-J. Huang and J.-F. Li. Built-in self-repair scheme for the TSVs in 3-D ICs. *Trans. Comput.-Aided Des. Integr. Circ. Syst.*, 31(10):1600–1613, 2012.

45. W.-L. Hung, G. M. Link, Y. Xie, N. Vijaykrishnan, and M. J. Irwin. Interconnect and thermal-aware floorplanning for 3D microprocessors. In *Proc. Int. Symp. Quality Elec. Des.*, pp. 98–104, 2006.

46. P. Jain, T.-H. Kim, J. Keane, and C. H. Kim. A multi-story power delivery technique for 3D integrated circuits. In *Proc. Int. Symp. Low Power Elec. Design*, pp. 57–62, 2008.

47. H. Jao, Y. Y. Lin, W. Liao, B. Wu, B. Huang et al. The impact of through silicon via proximity on CMOS device. In *Proc. Microsys. Packag. Assemb. Circ. Technol. Conf.*, pp. 43–45, 2012.

48. JEDEC Solid State Technology Association. JEDEC Standard: JESD229 Wide I/O. http://www.jedec.org/standards-documents/results/jesd229, December 2011. Accessed on December 2014.

49. JEDEC Solid State Technology Association. JEDEC: 3D-ICs. http://www.jedec.org/category/technology-focus-area/3d-ics-0, December 2014. Accessed on December 2014.

50. M. Jung and S. K. Lim. A study of IR-drop noise issues in 3D ICs with through-silicon-vias. In *Proc. 3D Sys. Integr. Conf.*, pp. 1–7, 2010.

51. M. Jung, J. Mitra, D. Z. Pan, and S. K. Lim. TSV stress-aware full-chip mechanical reliability analysis and optimization for 3D IC. In *Proc. Des. Autom. Conf.*, 2011.

52. M. Jung, J. Mitra, D. Z. Pan, and S. K. Lim. TSV stress-aware full-chip mechanical reliability analysis and optimization for 3-D IC. *Trans. Comput.-Aided Des. Integr. Circ. Syst.*, 31(8):1194–1207, 2012.

53. M. Jung, D. Z. Pan, and S. K. Lim. Chip/package mechanical stress impact on 3-D IC reliability and mobility variations. *Trans. Comput.-Aided Des. Integr. Circ. Syst.*, 32(11):1694–1707, 2013.

54. A. B. Kahng, J. Lienig, I. L. Markov, and J. Hu. *VLSI Physical Design: From Graph Partitioning to Timing Closure.* Springer, 2011.

55. U. Kang, H.-J. Chung, S. Heo, S.-H. Ahn, H. Lee et al. 8Gb 3D DDR3 DRAM using through-silicon-via technology. In *Proc. Int. Solid-State Circ. Conf.*, pp. 130–131, 131a, 2009.

56. D. H. Kim, K. Athikulwongse, M. B. Healy, M. M. Hossain, M. Jung et al. 3D-MAPS: 3D massively parallel processor with stacked memory. In *Proc. Int. Solid-State Circ. Conf.*, pp. 188–190, 2012.

57. D. H. Kim, S. Mukhopadhyay, and S. K. Lim. Through-silicon-via aware interconnect prediction and optimization for 3D stacked ICs. In *Proc. Int. Workshop Sys.-Level Interconn. Pred.*, pp. 85–92, 2009.

58. D. H. Kim, S. Mukhopadhyay, and S. K. Lim. TSV-aware interconnect distribution models for prediction of delay and power consumption of 3-D stacked ICs. *Trans. Comput.-Aided Des. Integr. Circ. Syst.*, 33(9):1384–1395, 2014.

59. D. H. Kim, R. O. Topaloglu, and S. K. Lim. Block-level 3D IC design with through-silicon-via planning. In *Proc. Asia South Pacific Des. Autom. Conf.*, Sydney, Australia, pp. 335–340, 2012.

60. K. Kim, J. S. Pak, H. Lee, and J. Kim. Effects of on-chip decoupling capacitors and silicon substrate on power distribution networks in TSV-based 3D-ICs. In *Proc. Elec. Compon. Technol. Conf.*, pp. 690–697, 2012.

61. T.-Y. Kim and T. Kim. Clock tree synthesis for TSV-based 3D IC Designs. *Trans. Des. Autom. Elec. Syst.*, 16(4):48:1-48:21, 2011.

62. T.-Y. Kim and T. Kim. Clock tree embedding for 3D ICS. In *Design Automation Conference (ASP-DAC), 2010 15th Asia and South Pacific*, Taiwan, pp. 486–491, January 2010.

63. J. Knechtel. Interconnect Planning for Physical Design of 3D Integrated, Circuits, volume 445 of Fortschritt-Berichte VDI. VDI-Verlag, Düsseldorf, Germany, 2014.

64. J. Knechtel, I. L. Markov, and J. Lienig. Assembling 2-D blocks into 3-D chips. *Trans. Comput.-Aided Des. Integr. Circ. Syst.*, 31(2):228–241, 2012.

65. J. Knechtel, I. L. Markov, J. Lienig, and M. Thiele. Multiobjective optimization of deadspace, a critical resource for 3-D-IC integration. In *Proc. Int. Conf. Comput.-Aided Des.*, pp. 705–712, 2012.

66. J. Knechtel, E. F. Y. Young, and J. Lienig. Structural planning of 3D-IC interconnects by block alignment. In *Proc. Asia South Pacific Des. Autom. Conf.*, Singapore, pp. 53–60, 2014.

67. J. H. Lau. TSV interposer: The most cost-effective integrator for 3D IC integration. In *SEMATECH Symposium*, Taiwan, 2011.

68. J. H. Law, E. F. Y. Young, and R. L. S. Ching. Block alignment in 3D floorplan using layered TCG. In *Proc. Great Lakes Symp. VLSI*, pp. 376–380, 2006.

69. H.-H. S. Lee and K. Chakrabarty. Test challenges for 3D integrated circuits. *Des. Test Comput.*, 26(5):26–35, 2009.

70. Y.-J. Lee and S. K. Lim. Timing analysis and optimization for 3D stacked multi-core microprocessors. In *Proc. 3D Sys. Integr. Conf.*, pp. 1–7, 2010.

71. Y.-J. Lee and S. K. Lim. Co-optimization and analysis of signal, power, and thermal interconnects in 3-D ICs. *Trans. Comput.-Aided Des. Integr. Circ. Syst.*, 30(11):1635–1648, 2011.

72. Y.-J. Lee and S. K. Lim. Ultrahigh density logic designs using monolithic 3-D integration. *Trans. Comput.-Aided Des. Integr. Circ. Syst.*, 32(12):1892–1905, 2013.

73. Y.-J. Lee, P. M., and S. K. Lim. Ultra high density logic designs using transistor-level monolithic 3D integration. In *Proc. Int. Conf. Comput.-Aided Des.*, pp. 539–546, 2012.

74. D. L. Lewis and H.-H. S. Lee. A scan-island based design enabling prebond testability in die-stacked microprocessors. In *Int. Test Conf.*, pp. 1–8, 2007.

75. D. L. Lewis and H.-H. S. Lee. Test strategies for 3D die stacked integrated circuits. In *Proc. Des. Autom. Test Europe 3D Workshop*, 2009.

76. Z. Li, Y. Ma, Q. Zhou, Y. Cai, Y. Xie et al. Thermal-aware P/G TSV planning for IR drop reduction in 3D ICs. *Integration*, 46(1):1–9, 2013.

77. J. Lienig. Electromigration and its impact on physical design in future technologies. In *Proc. Int. Symp. Phys. Des.*, pp. 33–40, 2013.

78. J. Lienig and M. Dietrich, editors. *Entwurf integrierter 3D-Systeme der Elektronik*. Springer, Berlin and Heidelberg, Germany, 2012.

79. S. K. Lim. Research needs for TSV-based 3D IC architectural floorplanning. *J. Inf. Commun. Converg. Eng.*, 12(1):46–52, 2014.

80. S. K. Lim, H.-H. Lee, and G. Loh. The 3D-MAPS processors. http://www.gtcad.gatech.edu/3d-maps/index.html, 2010. Accessed on December 2014.

81. C. Liu and S. K. Lim. A design tradeoff study with monolithic 3D integration. In *Proc. Int. Symp. Quality Elec. Des.*, 2012.

82. C. Liu, T. Song, J. Cho, J. Kim, J. Kim et al. Full-chip TSV-to-TSV coupling analysis and optimization in 3D IC. In *Proc. Des. Autom. Conf.*, 2011.

83. C. Liu, T. Song, and S. K. Lim. Signal integrity analysis and optimization for 3D ICs. In *Proc. Int. Symp. Quality Elec. Des.*, pp. 42–49, 2011.

84. G. H. Loh, Y. Xie, and B. Black. Processor design in 3D die-stacking technologies. *Micro*, 27:31–48, 2007.

85. I. Loi, F. Angiolini, S. Fujita, S. Mitra, and L. Benini. Characterization and implementation of fault-tolerant vertical links for 3-D networks-on-chip. *Trans. Comput.-Aided Des. Integr. Circ. Syst.*, 30(1):124–134, 2011.

86. I. Loi, S. Mitra, T. H. Lee, S. Fujita, and L. Benini. A low-overhead fault tolerance scheme for TSV-based 3D network on chip links. In *Proc. Int. Conf. Comput.-Aided Des.*, pp. 598–602, 2008.

87. K. H. Lu, X. Zhang, S.-K. Ryu, J. Im, R. Huang et al. Thermo-mechanical reliability of 3-D ICs containing through silicon vias. In *Proc. Elec. Compon. Technol. Conf.*, pp. 630–634, 2009.

88. C.-L. Lung, J.-H. Chien, Y. Shi, and S.-C. Chang. TSV fault-tolerant mechanisms with application to 3D clock networks. In *Int. SoC Des. Conf.*, pp. 127–130, 2011.

89. C.-L. Lung, Y.-S. Su, H.-H. Huang, Y. Shi, and S.-C. Chang. Through-silicon via fault-tolerant clock networks for 3-D ICS. *Trans. Comput.-Aided Des. Integr. Circ. Syst.*, 32(7):1100–1109, 2013.

90. G. Luo, Y. Shi, and J. Cong. An analytical placement framework for 3-D ICs and its extension on thermal awareness. *Trans. Comput.-Aided Des. Integr. Circ. Syst.*, 32(4):510–523, 2013.

91. W.-K. Mak and C. Chu. Rethinking the wirelength benefit of 3-D integration. *Trans. VLSI Syst.*, 20(12):2346–2351, 2012.

92. E. J. Marinissen. Challenges and emerging solutions in testing TSV-based 2 1/2D- and 3D-stacked ICs. In *Proc. Des. Autom. Test Europe*, pp. 1277–1282, 2012.

93. E. J. Marinissen. Status update of IEEE Std P1838. In *Proc. Int. Workshop Testing 3D Stack. Integr. Circ.*, 2014.

94. E. J. Marinissen, C.-C. Chi, J. Verbree, and M. Konijnenburg. 3D DfT architecture for pre-bond and post-bond testing. In *Proc. 3D Sys. Integr. Conf.*, pp. 1–8, 2010.

95. B. Martin, K. Han, and M. Swaminathan. A path finding based SI design methodology for 3D integration. In *Proc. Elec. Compon. Technol. Conf.*, 2014.

96. A. Mercha, G. Van der Plas, V. Moroz, I. De Wolf, P. Asimakopou-los et al. Comprehensive analysis of the impact of single and arrays of through silicon vias induced stress on high-k/metal gate CMOS performance. In *Proc. Int. Elec. Devices Meeting*, pp. 2.2.1–2.2.4, 2010.

97. D. Milojevic, T. E. Carlson, K. Croes, R. Radojcic, D. F. Ragett et al. Automated pathfinding tool chain for 3D-stacked integrated circuits: Practical case study. In *Proc. 3D Sys. Integr. Conf.*, pp. 1–6, 2009.

98. D. Milojevic, P. Marchal, E. J. Marinissen, G. Van der Plas, D. Verkest et al. Design issues in heterogeneous 3D/2.5D integration. In *Proc. Asia South Pacific Des. Autom. Conf.*, Yokohama, Japan, pp. 403–410, 2013.

99. J. Minz, X. Zhao, and S. K. Lim. Buffered clock tree synthesis for 3D ICs under thermal variations. In *Proc. Asia South Pacific Des. Autom. Conf.*, Seoul, Korea, pp. 504–509, 2008.

100. R. K. Nain and M. Chrzanowska-Jeske. Fast placement-aware 3-D floor-planning using vertical constraints on sequence pairs. *Trans. VLSI Syst.*, 19(9):1667–1680, 2011.

101. V. S. Nandakumar and M. Marek-Sadowska. Layout effects in fine-grain 3-D integrated regular microprocessor blocks. In *Proc. Des. Autom. Conf.*, pp. 639–644, 2011.

102. G. Neela and J. Draper. Logic-on-logic partitioning techniques for 3-dimensional integrated circuits. In *Proc. Int. Symp. Circ. Syst.*, pp. 789–792, 2013.

103. B. Noia, K. Chakrabarty, S. K. Goel, E. J. Marinissen, and J. Verbree. Test-architecture optimization and test scheduling for TSV-based 3-D stacked ICs. *Trans. Comput.-Aided Des. Integr. Circ. Syst.*, 30(11):1705–1718, 2011.

104. S. Panth, K. Samadi, Y. Du, and S. K. Lim. High-density integration of functional modules using monolithic 3D-IC technology. In *Proc. Asia South Pacific Des. Autom. Conf.*, Yokohama, Japan, pp. 681–686, 2013.

105. J.-H. Park, A. Shakouri, and S.-M. Kang. Fast thermal analysis of vertically integrated circuits (3-D ICs) using power blurring method. In *Proc. ASME InterPACK*, pp. 701–707, 2009.

106. M. Pathak and S. K. Lim. Performance and thermal-aware steiner routing for 3-D stacked ICs. *Trans. Comput.-Aided Des. Integr. Circ. Syst.*, 28(9):1373–1386, 2009.

107. M. Pathak, J. Pak, D. Z. Pan, and S. K. Lim. Electromigration modeling and full-chip reliability analysis for BEOL interconnect in TSV-based 3D ICs. In *Proc. Int. Conf. Comput.-Aided Des.*, pp. 555–562, 2011.

108. V. F. Pavlidis and E. G. Friedman. *Three-Dimensional Integrated Circuit Design*. Morgan Kaufmann Publishers Inc., Burlington, MA, 2008.

109. V. F. Pavlidis, I. Savidis, and E. G. Friedman. Clock distribution networks in 3-D integrated systems. *Trans. VLSI Syst.*, 19(12):2256–2266, 2011.

110. S. Priyadarshi, W. R. Davis, M. B. Steer, and P. D. Franzon. Thermal pathfinding for 3-D ICs. *Trans. Compon, Packag, Manuf. Technol.*, 4(7):1159–1168, 2014.

111. A. Quiring, M. Olbrich, and E. Barke. Improving 3D-floorplanning using smart selection operations in meta-heuristic optimization. In *Proc. 3D Syst. Integr. Conf.*, pp. 1–6, 2013.

112. J. Rajski and J. Tyszer. Fault diagnosis of TSV-based interconnects in 3-D stacked designs. In *Int. Test Conf.*, pp. 1–9, 2013.

113. S. K. Samal, S. Panth, K. Samadi, M. Saedi, Y. Du et al. Fast and accurate thermal modeling and optimization for monolithic 3D ICs. In *Proc. Des. Autom. Conf.*, 2014.

114. Samsung. 3D vertical-NAND memory. http://www.samsung.com/global/business/semiconductor/html/product/flash-solution/ vnand/overview.html, December 2014. Accessed on December 2014.

115. M. Scandiuzzo, S. Cani, L. Perugini, S. Spolzino, R. Canegallo et al. Input/output pad for direct contact and contactless testing. In *Proc. Europ. Test Symp.*, pp. 135–140, 2011.

116. P. Schneider, A. Heinig, R. Fischbach, J. Lienig, S. Reitz et al. Integration of multi physics modeling of 3D stacks into modern 3D data structures. In *Proc. 3D Syst. Integr. Conf.*, pp. 1–6, 2010.

117. D. Sekar, C. King, B. Dang, T. Spencer, H. Thacker et al. A 3D-IC technology with integrated microchannel cooling. In *Proc. Int. Interconn. Technol. Conf.*, pp. 13–15, 2008.

118. Y. Shang, C. Zhang, H. Yu, C. S. Tan, X. Zhao, and S. K. Lim. Thermal-reliable 3D clock-tree synthesis considering nonlinear electrical-thermal-coupled TSV model. In *Design Automation Conference (ASP-DAC), 2013 18th Asia and South Pacific*, Yokohama, Japan, pp. 693–698, January 2013.

119. R. S. Shelar and M. Patyra. Impact of local interconnects on timing and power in a high performance microprocessor. *Trans. Comput.-Aided Des. Integr. Circ. Syst.*, 32(10):1623–1627, 2013.

120. SK Hynix Inc. SK Hynix HBM Graphics Memory. http://www.skhynix.com/inc/pdfDownload.jsp?path=/datasheet/Databook/Databook_Q4'2014_Graphics.pdf, December 2014. Accessed on December 2014.

121. K. Smith, P. Hanaway, M. Jolley, R. Gleason, E. Strid et al. Evaluation of TSV and micro-bump probing for wide I/O testing. In *Proc. Int. Test Conf.*, pp. 1–10, 2011.

122. J. Sun, J.-Q. Lu, D. Giuliano, T. P. Chow, and R. J. Gutmann. 3D power delivery for microprocessors and high-performance ASICs. In *Proc. Appl. Power Electr. Conf.*, pp. 127–133, 2007.

123. D. Sylvester and K. Keutzer. A global wiring paradigm for deep submicron design. *Trans. Comput.-Aided Des. Integr. Circ. Syst.*, 19(2):242–252, 2000.

124. T. Thorolfsson, S. Lipa, and P. D. Franzon. A 10.35 mw/gflop stacked SAR DSP unit using fine-grain partitioned 3D integration. In *Proc. Cust. Integr. Circ. Conf.*, pp. 1–4, 2012.

125. T. Thorolfsson, G. Luo, J. Cong, and P. D. Franzon. Logic-on-logic 3D integration and placement. In *Proc. 3D Syst. Integr. Conf.*, pp. 1–4, 2010.

126. R. Topaloglu. Applications driving 3-D integration and corresponding manufacturing challenges. In *Proc. Des. Autom. Conf.*, pp. 214–219, 2011.

127. M.-C. Tsai, T.-C. Wang, and T. T. Hwang. Through-silicon via planning in 3-D floorplanning. *Trans. VLSI Syst.*, 19(8):1448–1457, 2011.

128. R. R. Tummala. System on Package: Miniaturization of the Entire System. McGraw-Hill Professional, 2008.

129. G. Van der Plas, P. Limaye, I. Loi, A. Mercha, H. Oprins et al. Design issues and considerations for low-cost 3-D TSV IC technology. *J. Solid- State Circ.*, 46(1):293–307, 2011.

130. F. von Trapp. GLOBALFOUNDRIES has its 3D Ducks in a Row. http://www .3dincites.com/2014/11/globalfoundries-3d-ducks-row/, November 2014. Accessed on December 2014.

131. R. Wang, E. F. Y. Young, and C.-K. Cheng. Complexity of 3-D floor-plans by analysis of graph cuboidal dual hardness. *Trans. Des. Autom. Elec. Syst.*, 15(4):33:1–33:22, 2010.

132. M. Wordeman, J. Silberman, G. Maier, and M. Scheuermann. A 3D system prototype of an eDRAM cache stacked over processor-like logic using through-silicon vias. In *Proc. Int. Solid-State Circ. Conf.*, pp. 186–187, 2012.

133. H. Xu, V. F. Pavlidis, and G. De Micheli. Effect of process variations in 3D global clock distribution networks. *J. Emerg. Tech. in Comp. Sys.*, 8(3):20:1–20:25, 2012.

134. J.-S. Yang, K. Athikulwongse, Y.-J. Lee, S. K. Lim, and D. Z. Pan. TSV stress aware timing analysis with applications to 3D-IC layout optimization. In *Proc. Des. Autom. Conf.*, pp. 803–806, 2010.

135. W. Yao, S. Pan, B. Achkir, J. Fan, and L. He. Modeling and application of multi-port TSV networks in 3-D IC. *Trans. Comput.-Aided Des. Integr. Circ. Syst.*, 32(4):487–496, 2013.

136. F. Yazdani. Readiness of 2.5D/3D IC package design environment. http://www.gsaglobal.org/wp-content/uploads/2012/04/3D-IC-Readiness-BroadPak.pdf, 2014. Accessed on December 2014.

137. F. Yazdani and J. Park. Pathfinding methodology for optimal design and integration of 2.5D/3D interconnects. In *Proc. Elec. Compon. Technol. Conf.*, pp. 1667–1672, 2014.

138. P.-H. Yuh, C.-L. Yang, and Y.-W. Chang. Temporal floorplanning using the T-tree formulation. In *Proc. Int. Conf. Comput.-Aided Des.*, pp. 300–305, 2004.

139. C. Zhang and L. Li. Characterization and design of through-silicon via arrays in three-dimensional ICs based on thermomechanical modeling. *Trans. Electron Devices*, 58(2):279–287, 2011.

140. C. Zhang and G. Sun. Fabrication cost analysis for 2D, 2.5D, and 3D IC designs. In *Proc. 3D Syst. Integr. Conf.*, pp. 1–4, 2012.

141. T. Zhang, K. Wang, Y. Feng, Y. Chen, Q. Li et al. A 3D SoC design for H .264 application with on-chip DRAM stacking. In *Proc. 3D Syst. Integr. Conf.*, pp. 1–6, 2010.

142. T. Zhang, Y. Zhan, and S. S. Sapatnekar. Temperature-aware routing in 3D ICs. In *Proc. Asia South Pacific Des. Autom. Conf.*, Yokohama, Japan, pp. 1–6, 2006.

143. X. Zhao, D. L. Lewis, H.-H.S. Lee, and S. K. Lim. Pre-bond testable low-power clock tree design for 3D stacked ICs. In *Proc. Int. Conf. Comput.-Aided Des.*, pp. 184–190, 2009.

144. X. Zhao and S. K. Lim. Power and slew-aware clock network design for through-silicon-via (TSV) based 3D ICs. In *Proc. Asia South Pacific Des. Autom. Conf.*, Taiwan, pp. 175–180, 2010.

145. X. Zhao and S. K. Lim. TSV array utilization in low-power 3D clock network design. In *Proc. Int. Symp. Low Power Elec. Design*, pp. 21–26, 2012.

146. X. Zhao, J. Minz, and S. K. Lim. Low-power and reliable clock network design for through-silicon via (TSV) based 3D ICs. *Trans. Compon. Packag. Manuf. Technol.*, 1(2):247–259, 2011.

147. X. Zhao, J. R. Tolbert, C. Liu, S. Mukhopadhyay, and S. K. Lim. Variation-aware clock network design methodology for ultra-low voltage (ULV) circuits. In *Proc. Int. Symp. Low Power Elec. Design*, pp. 9–14, 2011.

148. X. Zhao, S. Mukhopadhyay, and S. K. Lim. Variation-tolerant and low-power clock network design for 3D ICS. In *Electronic Components and Technology Conference (ECTC), 2011 IEEE 61st*, pp. 2007–2014, May 2011.

149. P. Zhou, K. Sridharan, and S. S. Sapatnekar. Congestion-aware power grid optimization for 3D circuits using MIM and CMOS decoupling capacitors. In *Proc. Asia South Pacific Des. Autom. Conf.*, pp. 179–184, Yokohama, Japan, 2009.

150. Q. K. Zhu. *High-Speed Clock Network Design*. Kluwer Academic Publishers, Boston, CA, 2003.

Index